Lecture Notes in Computer Science   10197

Commenced Publication in 1973
Founding and Former Series Editors:
Gerhard Goos, Juris Hartmanis, and Jan van Leeuwen

Bin Hu · Manuel López-Ibáñez (Eds.)

Evolutionary Computation in Combinatorial Optimization

17th European Conference, EvoCOP 2017
Amsterdam, The Netherlands, April 19–21, 2017
Proceedings

Springer

Editors
Bin Hu
Austrian Institute of Technology
Vienna
Austria

Manuel López-Ibáñez
University of Manchester
Manchester
UK

ISSN 0302-9743   ISSN 1611-3349 (electronic)
Lecture Notes in Computer Science
ISBN 978-3-319-55452-5   ISBN 978-3-319-55453-2 (eBook)
DOI 10.1007/978-3-319-55453-2

Library of Congress Control Number: 2017933870

LNCS Sublibrary: SL1 – Theoretical Computer Science and General Issues

Printed on acid-free paper

This Springer imprint is published by Springer Nature
The registered company is Springer International Publishing AG
The registered company address is: Gewerbestrasse 11, 6330 Cham, Switzerland

Preface

Combinatorial optimization is concerned with finding the optimal solution of problems with discrete variables. The field originates from applied mathematics and computer science, but it has seen contributions from operational research, decision-making, artificial intelligence, and machine learning. It is key to tackling diverse problems in science, industry, and business applications. These problems usually cannot be solved by exact methods within a reasonable time limit, and instead require the use of heuristic methods to provide high-quality or low-cost solutions in as short a time as possible. Heuristic methods include not only problem-specific heuristics, but most prominently metaheuristics, which are general-purpose methods that are relatively simple to apply to new problems. Among the earliest and most successful metaheuristics are evolutionary algorithms, originally inspired by the evolution of species by natural selection, together with various other stochastic local search methods, such as simulated annealing. More recent methods include ant colony optimization, inspired by the foraging behavior of some species of ants, and hybrid methods, such as matheuristics that combine exact and heuristic methods. The successful application of these methods to real-world combinatorial optimization problems is one of the main topics of these proceedings.

This volume contains the proceedings of EvoCOP 2017, the 17th European Conference on Evolutionary Computation in Combinatorial Optimization, that was held in Amsterdam, The Netherlands, during April 19–21, 2017. EvoCOP was held in 2001 as the first workshop specifically devoted to evolutionary computation in combinatorial optimization. It became an annual conference in 2004. EvoCOP is one of the four events of Evostar 2017. The other three are EuroGP (20th European Conference on Genetic Programming), EvoMUSART (6th International Conference on Evolutionary and Biologically Inspired Music, Sound, Art and Design), and EvoApplications (20th European Conference on the Applications of Evolutionary Computation, formerly known as EvoWorkshops).

Previous EvoCOP proceedings were published by Springer in the series *Lecture Notes in Computer Science* (LNCS volumes 2037, 2279, 2611, 3004, 3448, 3906, 4446, 4972, 5482, 6022, 6622, 7245, 7832, 8600, 9026, 9595). The table on the next page reports the statistics for each conference.

This year, 16 out of 39 papers were accepted after a rigorous double-blind process, resulting in a 41% acceptance rate. We would like to thank the quality and timeliness of our Program Committee members' work, especially since the reviewing period coincided with the Christmas holidays. Decisions considered both the reviewers' report and the evaluation of the program chairs. The 16 papers accepted cover both empirical and theoretical studies on a wide range of academic and real-world applications. The methods include evolutionary and memetic algorithms, large neighborhood search, estimation of distribution algorithms, beam search, ant colony optimization, hyper-heuristics, and matheuristics. Applications include both traditional domains, such

EvoCOP	LNCS vol.	Submitted	Accepted	Acceptance (%)
2017	10197	39	16	41.0
2016	9595	44	17	38.6
2015	9026	46	19	41.3
2014	8600	42	20	47.6
2013	7832	50	23	46.0
2012	7245	48	22	45.8
2011	6622	42	22	52.4
2010	6022	69	24	34.8
2009	5482	53	21	39.6
2008	4972	69	24	34.8
2007	4446	81	21	25.9
2006	3906	77	24	31.2
2005	3448	66	24	36.4
2004	3004	86	23	26.7
2003	2611	39	19	48.7
2002	2279	32	18	56.3
2001	2037	31	23	74.2

as the knapsack problem, vehicle routing, scheduling problems and SAT; and newer domains such as the traveling thief problem, location planning for car-sharing systems, and spacecraft trajectory optimization. Papers also study important concepts such as pseudo-backbones, phase transitions in local optima networks, and the analysis of operators. This wide range of topics makes the EvoCOP proceedings an important source for current research trends in combinatorial optimization.

We would like to express our appreciation to the various persons and institutions making this a successful event. First, we thank the local organization team led by Evert Haasdijk and Jacqueline Heinerman from the Vrije University Amsterdam. We thank Marc Schoenauer from Inria Saclay for his continued assistance in providing the MyReview conference management system and Pablo García Sánchez from the University of Cádiz for EvoStar publicity and website. Thanks are also due to SPECIES (Society for the Promotion of Evolutionary Computation in Europe and its Surroundings); in particular, Marc Schoenauer (President), Anna I Esparcia-Alcázar (Secretary and Vice-President), Wolfgang Banzhaf (Treasurer), and Jennifer Willies (EvoStar coordinator). Finally, we wish to thank the keynote speakers, Kenneth De Jong and Arthur Kordon.

Special thanks also to Christian Blum, Francisco Chicano, Carlos Cotta, Peter Cowling, Jens Gottlieb, Jin-Kao Hao, Jano van Hemert, Peter Merz, Martin Middendorf, Gabriela Ochoa, and Günther R. Raidl for their hard work and dedication at past editions of EvoCOP, making this one of the reference international events in evolutionary computation and metaheuristics.

February 2017

Bin Hu
Manuel López-Ibáñez

Organization

EvoCOP 2017 was organized jointly with EuroGP 2017, EvoMUSART 2017, and EvoApplications 2017.

Organizing Committee

Program Chairs

Bin Hu	AIT Austrian Institute of Technology, Austria
Manuel López-Ibáñez	University of Manchester, UK

Local Organization

Evert Haasdijk	Vrije University Amsterdam, The Netherlands
Jacqueline Heinerman	Vrije University Amsterdam, The Netherlands

Publicity Chair

Pablo García-Sánchez	University of Cádiz, Spain

EvoCOP Steering Committee

Christian Blum	IKERBASQUE and University of the Basque Country, Spain
Francisco Chicano	University of Málaga, Spain
Carlos Cotta	University of Málaga, Spain
Peter Cowling	University of York, UK
Jens Gottlieb	SAP AG, Germany
Jin-Kao Hao	University of Angers, France
Jano van Hemert	Optos, UK
Peter Merz	Hannover University of Applied Sciences and Arts, Germany
Martin Middendorf	University of Leipzig, Germany
Gabriela Ochoa	University of Stirling, UK
Günther Raidl	Vienna University of Technology, Austria

Society for the Promotion of Evolutionary Computation in Europe and its Surroundings (SPECIES)

Marc Schoenauer	President
Anna I Esparcia-Alcázar	Secretary and Vice-President
Wolfgang Banzhaf	Treasurer
Jennifer Willies	EvoStar coordinator

Program Committee

Adnan Acan	Eastern Mediterranean University, Turkey
Enrique Alba	University of Málaga, Spain
Richard Allmendinger	University of Manchester, UK
Thomas Bartz-Beielstein	Cologne University of Applied Sciences, Germany
Matthieu Basseur	University of Angers, France
Hans-Georg Beyer	Vorarlberg University of Applied Sciences, Germany
Benjamin Biesinger	Austrian Institute of Technology, Austria
Christian Blum	IKERBASQUE - Basque Foundation for Science, Spain
Sandy Brownlee	University of Stirling, UK
Pedro Castillo	Universidad de Granada, Spain
Francisco Chicano	University of Málaga, Spain
Carlos Coello Coello	CINVESTAV-IPN, Mexico
Peter Cowling	University of Bradford, UK
Luca Di Gaspero	University of Udine, Italy
Karl Doerner	Johannes Kepler University Linz, Austria
Benjamin Doerr	LIX-Ecole Polytechnique, France
Carola Doerr	Max Planck Institute for Informatics, Germany
Paola Festa	Universitá di Napoli Federico II, Italy
Bernd Freisleben	University of Marburg, Germany
Carlos García-Martínez	University of Córdoba, Spain
Adrien Goeffon	University of Angers, France
Jens Gottlieb	SAP, Germany
Walter Gutjahr	University of Vienna, Austria
Said Hanafi	University of Valenciennes, France
Jin-Kao Hao	University of Angers, France
Emma Hart	Edinburgh Napier University, UK
Geir Hasle	SINTEF Applied Mathematics, Norway
Andrzej Jaszkiewicz	Poznan University of Technology, Poland
István Juhos	University of Szeged, Hungary
Graham Kendall	University of Nottingham, UK
Ahmed Kheiri	Cardiff University, UK
Mario Köppen	Kyushu Institute of Technology, Japan
Frédéric Lardeux	University of Angers, France
Rhyd Lewis	Cardiff University, UK
Arnaud Liefooghe	Lille 1 University, France
José Antonio Lozano	University of the Basque Country, Spain
Gabriel Luque	University of Malaga, Spain
David Meignan	University of Osnabrück, Germany
Juan Julian Merelo	University of Granada, Spain
Krzysztof Michalak	University of Economics, Wroclaw, Poland
Martin Middendorf	University of Leipzig, Germany
Christine L. Mumford	Cardiff University, UK
Nysret Musliu	Vienna University of Technology, Austria
Gabriela Ochoa	University of Stirling, UK

Beatrice Ombuki-Berman   Brock University, Canada
Luis Paquete   University of Coimbra, Portugal
Mario Pavone   University of Catania, Italy
Paola Pellegrini   French Institute of Science and Technology
                            for Transport, France

Francisco J.B. Pereira   Universidade de Coimbra, Portugal
Matthias Prandtstetter   Austrian Institute of Technology, Austria
Jakob Puchinger   SystemX-Centrale Supélec, France
Rong Qu   University of Nottingham, UK
Günther Raidl   Vienna University of Technology, Austria
Maria Cristina Riff   Universidad Técnica Federico Santa María, Chile
Eduardo Rodriguez-Tello   Civerstav – Tamaulipas, Mexico
Andrea Roli   Università di Bologna, Italy
Peter Ross   Edinburgh Napier University, UK
Frédéric Saubion   University of Angers, France
Patrick Siarry   University of Paris 12, France
Kevin Sim   Edinburgh Napier University, UK
Jim Smith   University of the West of England, UK
Giovanni Squillero   Politecnico di Torino, Italy
Thomas Stützle   Université Libre de Bruxelles, Belgium
El-ghazali Talbi   Université des Sciences et Technologies de Lille, France
Renato Tinós   University of Sao Paulo, Brazil
Nadarajen Veerapen   University of Stirling, UK
Sébastien Verel   Université du Littoral Côte d'Opale, France
Bing Xue   Victoria University of Wellington, New Zealand
Takeshi Yamada   NTT Communication Science Laboratories, Japan

Contents

A Computational Study of Neighborhood Operators for Job-Shop Scheduling Problems with Regular Objectives

Hayfa Hammami[✉] and Thomas Stützle

IRIDIA, Université libre de Bruxelles (ULB), Brussels, Belgium
{Haifa.Hammami,stuetzle}@ulb.ac.be

Abstract. Job-shop scheduling problems have received a considerable attention in the literature. While the most tackled objective in this area is makespan, job-shop scheduling problems with other objectives such as the minimization of the weighted or unweighted tardiness, the number of late jobs, or the sum of the jobs' completion times have been considered. However, the problems under the latter objectives have been generally less studied than makespan. In this paper, we study job-shop scheduling under various objectives. In particular, we examine the impact various neighborhood operators have on the performance of iterative improvement algorithms, the composition of variable neighborhood descent algorithms, and the performance of metaheuristics such as iterated local search in dependence of the type of local search algorithm used.

1 Introduction

Scheduling problems have received a great deal of attention in the research community both from an application side due to their practical relevance and from an algorithmic side due to the difficulty that poses their solution [13]. Job-shop scheduling problems concern the scheduling of jobs on machines where the order in which the jobs are to be processed on the various machines may differ from job to job. Among job-shop scheduling problems, various variants exist and one difference among these can be the objective function that is to be minimized. The most common variant is the minimization of the makespan, that is, the completion time of the last job [12,15]. However, in many practical situations, other objectives are more relevant. For example, if jobs have associated due dates, a common objective is to minimize the tardiness of the jobs, possibly weighted by their importance [4,14,17]. Other objectives may be to minimize the (weighted) sum of the completion times of jobs, or the weighted number of tardy jobs [9,13]. However, these alternative objective functions have received less attention than the makespan objective.

In this paper, we study the impact that various neighborhood operators have on the performance of local search algorithms for job-shop problems under three objectives, the minimization of (i) the total weighted tardiness, (ii) the total weighted computation, (iii) and the weighted number of late jobs. In particular,

© Springer International Publishing AG 2017
B. Hu and M. López-Ibáñez (Eds.): EvoCOP 2017, LNCS 10197, pp. 1–17, 2017.
DOI: 10.1007/978-3-319-55453-2_1

we consider six neighborhood structures and these types of algorithms: iterative improvement algorithms under the best- and first-improvement pivoting rules; extensions of these algorithms to variable neighborhood descent algorithms [3] that use two or three neighborhood structures; integrating the various neighborhood structures into simple iterated local search algorithms [8].

The studies in the literature closest to ours are those by Kuhpfahl and Bierwirth [6] and Mati et al. [9]. The former considers various neighborhoods for the job-shop scheduling problem under the minimization of the total weighted tardiness. Here, we implemented a subset of the neighborhoods used in [6], which comprises the most promising ones identified there. We extend the study in [6] considering additional objectives (weighted sum of completion times and weighted number of tardy jobs), pivoting rules (first-improvement), and additional algorithms (iterated local search). The latter study by Mati et al. proposes an iterated local search algorithm that tackles job-shop problems under different objective functions [9]. Here, we adopt their iterated local search algorithm by re-implementing its structure and extend it considering different additional neighborhoods, the usage of variable neighborhood descent, and an additional fine-tuning of the algorithm by using the irace software [7]. Our experimental study shows that the adoption of the first-improvement pivoting rule seems beneficial across all problems. The usefulness of considering a variable neighborhood descent algorithm depends on the particular objective considered; and the final iterated local search algorithms generally reach high-quality results.

The article is structured as follows. In the next section, we introduce the tackled problems more formally and present the disjunctive arc representation. In Sect. 3, we give details on the neighborhood structures we have considered and in Sect. 4 we present the experimental results. We conclude in Sect. 5.

2   The Job-Shop Scheduling Problem

The job-shop scheduling problem is defined by n jobs that are to be processed on m machines in a given order. Each job J_i, $i = 1, 2, \ldots, n$, can have a different number of operations and has its own processing order on the machines. The common assumptions of the job-shop scheduling problem are that all processing times of the jobs on the machines are fixed and known in advance. The processing of a job on a machine is called operation. Once started, preemption of an operation is not allowed. The machines are continuously available (no breakdowns), each machine can process at most one job at a time and each job can be processed on at most one machine at a time. Infinite in-process storage is allowed. The objective is to obtain a production sequence of the jobs on the machines so that the processing constraints are satisfied and a given criterion is optimized. Most scheduling criteria use the completion times of the jobs at the machines, which are denoted as C_{ij} ($i \in n, j \in m$); C_i is the completion time of job J_i on the last machine. Here, we focus on minimizing objectives related to the due dates of jobs and to the flow time that are less used in literature than makespan but might be more relevant in practical situations.

Table 1. An instance of the job shop scheduling problem with three machines and three jobs.

Job	Routing			Processing		
	M_1	M_2	M_3	M_1	M_2	M_3
J_1	2	3	1	3	6	1
J_2	1	3	2	8	10	5
J_3	3	2	1	8	4	5

The flow time of a job is the time between its release and completion time. Here we assume that all release times of jobs are zero. In this case, the flow time of a job J_i corresponds to its completion time C_i. Generally, minimizing the sum of completion times of all jobs, $\sum_i C_i$, might be more interesting than the maximum completion time of jobs C_{max}, especially in service-oriented environments. We can also minimize the total *weighted* flow time $\sum_i w_i C_i$ by introducing a weight w_i to job J_i that specifies its relative importance. Tardiness and lateness based objectives consider the due dates d_i of jobs, which denote the desired completion time of job J_i on the last machine. We aim at minimizing the total weighted tardiness $\sum_i w_i T_i$ related to the importance of each job. Let U_i be one if $T_i > 0$ and zero otherwise. Then, another relevant criterion is to minimize the number of tardy jobs $\sum_i U_i$ or the weighted number of tardy jobs $\sum_i w_i U_i$, which is related to satisfying customers on time or not.

The job-shop scheduling problem can be represented with a disjunctive graph noted $G = (V, C, D)$ as proposed by Singer and Pinedo [14]. The set of nodes V represent the operations of the jobs; to these are added a dummy node 0 that represents the *starting* node and a set of n sink nodes B_i $i = (1, ..., n)$, which represent the *ending* nodes of each job J_i. Each operation has a weight p_{ij}, which is equal to the processing time of job J_i on machine M_j. A set of conjunctive arcs C represent the precedence constraints between operations of each job. The undirected arcs are the set of disjunctive arcs D and represent machine constraints. Each pair of operations that requires the same machine cannot be executed simultaneously. Figure 1(a) shows an example of a disjunctive graph G for a 3-job, 3-machine instance described in Table 1. A feasible solution is obtained if and only if all the undirected arcs are turned into directed ones and the resulting graph G' is acyclic.

The length of the longest path from 0 to the sink node B_i represents the maximal completion time of job J_i. This path is the critical path, which is composed of critical blocks. Each critical block contains critical operations executed on the same machine without idle time. A critical arc connects two adjacent critical operations in a critical block. Figure 1(b) represents a feasible solution of the instance by randomly selecting one arc of each pair of disjunctive arcs. The length of the longest path of job J_1 starting from 0 to B_1 is $C_1 = 17$, $C_2 = 27$ for J_2 and $C_3 = 19$ for J_3.

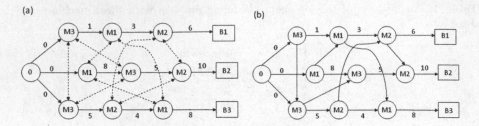

Fig. 1. Disjunctive graph model and a solution for the example from Table 1.

3   Neighborhood Structures for Job Shop Scheduling

A common approach for tackling job-shop scheduling problems is by local search algorithms. In local search, iteratively the current solution is replaced by a neighboring one. A basic local search algorithm is iterative improvement, where at each step an improving candidate solution is accepted until no more improvements can be found in the neighborhood of the current solution; in other words, the algorithm stops in a local optimum. In job-shop scheduling, a neighboring solution is commonly obtained (either improving or not) by some specific modification on its critical path. A modification defined by reversing critical arcs in the disjunctive graph representation of a feasible solution always yields another feasible one [15]. Some larger modifications lead to infeasible solutions, so that a feasibility test is needed. Well known operators are the transpose, insert and sequence moves. In this section, we present different neighborhoods used by Kuhpfahl and Bierwirth [6]. In their paper, they have evaluated existing and newly designed neighborhoods for the job-shop scheduling problem with total weighted tardiness objective (JSSP-WT). In their work, transpose-based neighborhoods and insertion-based neighborhoods are considered to be the most interesting ones regarding the average gap to the best known solutions. Here, we briefly review the six of these neighborhood structures that we consider in this paper.

CT: The *critical transpose* neighborhood [15] consists in reversing a pair of adjacent operations u and v assigned to the same machine on a critical path. It is shown by Van Laarhoven et al. [15] that any solution obtained by the critical transpose operator is feasible.

CET: A restricted version of the CT operator called *critical end transpose* which considers only the first or the last adjacent operations of a critical block to be swapped [12]. The feasibility of the new schedules provided by CET neighborhood always holds since it is a subset of the CT neighborhood operator.

CET+2MT: This perturbation affects multiple machines in a schedule, called *critical end transpose + 2-machine transpose*. It is an extension of the CET neighborhood operator by swapping two further arcs related to predecessors and successors of the critical arc (u, v). We denote by $SJ(i)$ and $PJ(i)$ the operation of the job succeeding or preceding operation i, respectively. $SM(i)$ and $PM(i)$

denote the successor and predecessor machine of operation i respectively. With the CET+2MT operator, two additional arcs associated to (u, v) are reversed, which are $(SJ(u), SM(SJ(u)))$ and $(PM(PJ(v)), PJ(v))$, provided that some conditions are satisfied, see [10]. The feasibility guarantee also holds for this operator since it requires no machine idle time in the processing of the adjacent operations corresponding to the three arcs.

ECET: This new neighborhood operator is proposed by Kuhpfahl and Bierwirth [6] called the *extended critical end transpose* neighborhood, which considers not only the first or the last arc of a critical block to be inverted but also both simultaneously. The feasibility guarantee holds in this perturbation, see [6].

CEI: This operator, proposed by Dell'Amico and Trubian [2], moves an operation inside a critical block to the first or the last position of the block to which it belongs. This operator is called *critical end insert* neighborhood. The neighboring solution obtained by this operator might be infeasible; therefore a feasibility test is required for each move and infeasible schedules are discarded.

CEI+2MT: This perturbation introduced in [5], is an extension of the CEI operator named *critical end insert + 2-machine transpose*. The principle of this neighborhood is like that of the CET+2MT operator; it consists of a CEI move with a reversal of two additional arcs related to predecessors and successors of the critical arc under consideration provided that some conditions hold, see [6]. This operator also does not ensure the feasibility of the solutions, thus a feasibility test is executed in the neighboring schedules as in the CEI operator.

The neighborhood operators described above have been shown by Kuhpfahl and Bierwirth [6] to be the best performing over a set of 53 instances using an iterative best-improvement algorithm as the local search algorithm. The types of experiments were considered to assess and compare the performance of neighborhood operators for the JSSP with total weighted tardiness as the only objective. In the subsequent section, we compare the performance of each neighborhood operator presented above within iterative first- or best-improvement local searches on different objectives of the job-shop scheduling problem.

4   Experimental Study

4.1   Experimental Evaluation

In this section, we compare the performance of the neighborhoods on different objectives using iterative best- and first-improvement algorithms. For different objectives based on weights and due dates, we consider the same procedure used by Singer and Pinedo [14] to define the job weights and the due dates. In particular, the first 20% of the jobs are very important and have weights $w_i = 4$, 60% of the jobs are of average importance and their weights are equal to $w_i = 2$ and the last 20% of jobs have a weight of $w_i = 1$. The due date of a job depends on the processing times of its operations and it is calculated using the following formula $d_i = f \cdot \sum_{j=1}^{m} p_{ij}$ where f is a tightness factor equal to 1.3

for all experiments. The computational tests are done over a set of 22 instances with 10 jobs and 10 machines proposed by Singer and Pinedo [14]. Notice that the last five jobs of instances LA21–24 are removed to make them 10×10.

All code is implemented in Java and experiments presented in this paper are run on an Intel Xeon E5-2680 CPU (2.5 GHz) with 16 MB cache, under Cluster Rocks Linux. If nothing else is said, we assess the statistical significance of possible differences with a paired Student t-test at significance level $\alpha = 0.05$.

4.2 Iterative Best Improvement Versus Iterative First Improvement for Single Neighborhoods

In the experimental evaluation, we assess the neighborhoods on different objectives by the two iterative best and first improvement algorithms. We start the iterative best improvement algorithm by generating a sequence of random start solutions $s_i = (s_i^1, s_i^2, ...)$ for each of the 22 problem instances. Each neighborhood operator of the six operators is executed in the following way to reach local optima for each instance. In the best-improvement case, starting from an initial solution, all neighboring solutions are generated by the neighborhood operator and the best neighboring one is accepted as the next one. This process continues iteratively until no improvement is found anymore, that is a local optimum is reached. This process is repeated by generating new local optima, starting a new local search process from the next random solution from the sequence s_i, until one of two termination criteria is reached. (This experimental setup with the two stopping criteria follows Kuhpfahl and Bierwirth [6].) The first termination criterion (experiment 1) is a fixed number of local optima reached by each neighborhood operator on each problem instance and the second termination criterion (experiment 2) is a fixed number of neighboring schedules evaluated for every operator and for every problem instance. The entire procedure is repeated for the iterative first improvement algorithm, where the local search immediately accepts a new candidate solution as soon as it is found. This procedure is tested on each one of the objective functions we consider in this paper: the total weighted tardiness, $\sum_i w_i T_i$; the total weighted flow time, $\sum_i w_i C_i$; and the weighted number of tardy jobs, $\sum_i w_i U_i$.

Experiment 1. In this experiment, we generated 100 local optima for each problem instance and each neighborhood operator and we compare for each operator the average of 100 local optima found in each instance for the best- and first-improvement pivoting rules minimizing the total weighted tardiness objective JSSP-WT. A first result is that for all neighborhood operators, the first-improvement algorithm is significantly better than best-improvement algorithm with the only exception being the CEI neighborhood operator, where the difference between first and best improvement is not statistically significant. This is shown in Fig. 2, which clearly shows the dominance of the first-improvement algorithm on all instances and for each operator (all instances start from the same initial solutions provided in sequence s_i): the diagonal is showing the line of equal performance and points below the line indicate better performance of

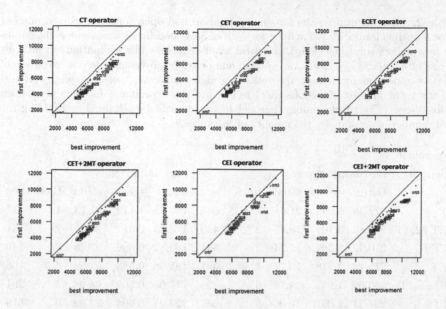

Fig. 2. Plots of the average solution cost per instance comparing first- (value on y-axis) and best-improvement (value on x-axis) of neighborhood operators with 100 local optima per instance for the JSSP-WT (Experiment 1)

the first-improvement operator. Only for the CEI operator most points are on the diagonal, indicating similar performance of the two algorithms.

For each operator and both best- and first-improvement, Table 2 presents the gap to the best known solutions, which is computed as

$$Gap(j) = \frac{1}{22} \sum_{i=1}^{22} \frac{BFS(i,j) - BKS(i)}{BKS(i)} \times 100 \qquad (1)$$

where BKS is the best known solution, and BFS the best found solution by operator j in the experiment, and other statistics explained in the table caption. It is clear that best-improvement consumes almost five times the number of evaluations used by the first-improvement algorithm and consequently its computation time is much higher. This is due to that best improvement algorithm requiring a complete evaluation of all neighbors in each search step. Concerning the solution quality indicated by the gap value, the first-improvement algorithm also performs better than the best-improvement algorithm, making it clearly preferable. First-improvement also requires a larger number of improving steps to reach local optima, but needs less scans of the entire neighborhood to do so, therefore also resulting in shorter computation times.

Summarizing, our implementations of first-improvement are superior to the best-improvement versions of the iterative improvement algorithms for all operators. Overall, the CT operator performed the best with the CEI operator giving overall the worst performance. The high computation times of the latter are in large part due to the additional feasibility tests it requires.

Table 2. Experimental results for single neighborhood operators with a fixed number of local optima for JSSP-WT. Given for each of best- and first-improvement the gap to the best known solutions (Gap), the total number of schedule evaluations in millions (across all instances) (#Eval), the average number of improvements in one execution of iterative improvement (#Imp), the rank of the algorithms w.r.t. solution quality (Rank) and the total number of seconds (Sec) to execute the experiment for the operator on all instances. (The total running time per instance could be obtained by dividing the total number of seconds by the number of instances.)

Experiment 1										
Operator	Best-improvement					First-improvement				
	Gap	#Eval	#Imp	Rank	Sec	Gap	#Eval	#Imp	Rank	Sec
CT	172.36	5.093	21.68	1	683	148.88	1.112	31.73	1	173
CET+2MT	202.89	3.101	18.85	2	435	175.23	0.61	28	2	131
ECET	210.82	3.309	17.74	3	534	180.82	0.690	27.43	3	140
CET	211.13	2.978	18.77	4	394	185.98	0.583	27.38	4	106
CEI+2MT	240.09	1.957	14.78	6	893	217.05	0.562	24.98	5	291
CEI	227.51	3.051	16.66	5	1686	227.97	0.664	24.02	6	519

Table 3. Experimental results for single neighborhood operators with a fixed number of evaluations for JSSP-WT. #Lopt gives the number of local optima generated. For the meaning of the other table entries, we refer to the caption of Table 2.

Experiment 2										
Operator	Best improvement					First improvement				
	Gap	#Lopt	#Imp	Rank	Sec	Gap	#Lopt	#Imp	Rank	Sec
CT	191.63	1008	21.78	1	241	141.98	4671	31.23	1	289
CET+2MT	208.67	1608	18.83	2	285	154.52	8230	27.8	2	355
ECET	220.55	1512	17.85	4	315	158.45	7316	27.28	4	385
CET	214.31	1669	18.78	3	266	158.4	8646	27.21	3	335
CEI+2MT	234.21	2612	14.63	5	880	165.7	9083	24.52	5	913
CEI	231.56	1660	16.55	6	1076	181.89	7795	23.75	6	1501

Experiment 2. In this experiment, the number of schedule evaluations per instance is fixed to a maximum number of $10 \times n^2 \times m^2$ evaluations, following [6]. In our experiments on Singer and Pinedo's instances this means 100 000 evaluations per instance. The results of this experiment are reported in Table 3. Due to the typically higher number of evaluations (corresponding to a total of 2.2×10^6 evaluations), especially for the first-improvement versions many more local optima than in experiment 1 could be seen, amplifying the advantage of the first-improvement versions over the best-improvement ones.

Next, we report the results of the same analysis for the job shop scheduling problems minimizing the total weighted flow time (JSSP-WC) and the total

Table 4. Experimental results for single neighborhood operators with a fixed number of local optima for JSSP-WC and JSSP-WU. For the meaning of the table entries, we refer to the caption of Table 2.

Experiment 1											
Criterion	Operator	Best improvement					First improvement				
		Gap	#Eval	#Imp	Rank	Sec	Gap	#Eval	#Imp	Rank	Sec
$\sum_i w_i C_i$	CT	15.98	5.093	21.93	1	691	12.96	1.115	31.7	1	177
	CET+2MT	17.66	3.098	18.95	2	442	15.73	0.607	27.73	2	121
	ECET	18.59	6.613	17.83	4	1084	15.96	0.688	27.23	3	144
	CET	18.21	2.97	18.83	3	401	16.17	0.581	27.19	4	109
	CEI+2MT	20.63	1.962	14.87	6	908	18.17	0.561	24.79	5	295
	CEI	19.08	3.062	16.77	5	1638	19.23	0.664	24	6	504
$\sum_i w_i U_i$	CT	112.31	0.291	0.14	1	71	112.31	0.264	0.15	1	69
	CET+2MT	115.28	0.216	0.14	4	59	115.28	0.196	0.14	3	55
	ECET	115.28	0.246	0.14	5	75	115.28	0.223	0.14	5	68
	CET	115.28	0.213	0.13	3	58	115.28	0.193	0.13	4	56
	CEI+2MT	115.23	0.17	0.13	6	108	115.23	0.185	0.13	6	117
	CEI	114.58	0.239	0.14	2	152	114.58	0.217	0.14	2	158

weighted number of late jobs (JSSP-WU). The evaluation of the operators is performed with first and best improvement on the same 22 instances. The experimental results are given in Table 4 for experiment 1 and in Table 5 for experiment 2. Given the lack of best known solutions in these instances for JSSP-WC and JSSP-WU, we have executed an iterated local search algorithm for a large number of iterations, taking the best solutions found as best-known ones.

The results for the JSSP-WC match relatively closely the main conclusions on the performance of the operators from JSSP-WT. First, the first-improvement versions are faster then the best-improvement versions of our implementation. Concerning the quality of the solutions the operators obtain, again CT is the best performing one and the ranking overall is almost the same as for JSSP-WT. The second experiment is confirming these conclusions. Considering JSSP-WU, the behavior of the algorithms is very different, mainly because the problem is dominated by very large plateaus in the search landscape, making it difficult to identify improvements. Hence, whether one uses a best-improvement or first-improvement local search does essentially not make any significant difference, as both require at least one full scan of the neighborhood. In the first experiment, CT appears to be the best performing one, while in the second experiment, this is not anymore the case. In fact, in the latter experiments the CEI+2MT and CET+2MT operators appear to be preferable as they probably can identify some improvements through the machine transpose moves.

4.3 First-Improvement Variable Neighborhood Descent

The performance of the local search with single neighborhood operator depends on the underlying neighborhood relation and, in particular on the size of the neighborhood. Generally using large neighborhoods rather than small ones can

Table 5. Experimental results for single neighborhood operators with a fixed number of schedule evaluations for JSSP-WC and JSSP-WU. For the meaning of the table entries, we refer to the caption of Table 2.

Experiment 2											
Criterion	Operator	Best improvement					First improvement				
		Gap	#Lopt	#Imp	Rank	Sec	Gap	#Lopt	#Imp	Rank	Sec
$\sum_i w_i C_i$	CT	17.68	1010	21.97	1	245	12.68	4647	31.18	1	298
	CET+2MT	18.35	1602	18.96	2	291	13.61	8219	27.46	2	372
	ECET	19.26	1510	17.98	4	320	14.18	7315	27	4	399
	CET	18.61	1674	18.84	3	270	14.17	8628	26.92	3	350
	CEI+2MT	20.24	2604	14.72	6	874	15.06	9047	24.42	5	940
	CEI	19.43	1654	16.67	5	1069	15.64	7782	23.68	6	1465
$\sum_i w_i U_i$	CT	89.85	17136	0.16	3	469	91.14	19168	0.16	5	501
	CET+2MT	87.82	22794	0.14	2	491	86.65	25398	0.14	2	508
	ECET	90.98	20118	0.14	4	523	89.12	22447	0.15	4	572
	CET	87.82	23199	0.14	2	508	87.3	25795	0.14	3	510
	CEI+2MT	86.6	29215	0.14	1	1043	86.16	27398	0.15	1	1004
	CEI	93.81	20823	0.14	5	1183	92.51	23251	0.15	6	1205

offer chances for finding improving search steps. Variable neighborhood descent (VND) [3] is a well-known approach of improving iterative improvement algorithms by considering various neighborhoods.

In this section, we study two different possible VNDs, considering a VND with two neighborhood operators (VND-2op) and a VND with three neighborhood operators. In our experiments here, we only consider VNDs where the neighborhoods are searched in a first-improvement order as the previous experimental results clearly indicate this as the better choice. In the case of VND-2ops there are a total of 30 possible combinations of how to define a VND. We have implemented all and run experiments following the setup of experiment 1 from the previous section. From these 30 possible combinations, we have selected six promising configurations as judged by the quality of the solutions generated and the computation time. In particular, we preferred combinations resulting in the best quality solutions taking into account that they result in short computation times. In particular, in the set of six VNDs are included the best four according to the solution quality they generate are included, complemented by others that are very fast but still reach good quality solutions. For instance, we did not combine the CEI operator with other operators due to the high running times of this operator. Table 6 shows the most promising combinations over 30 possible configurations using Experiment 1 as the evaluation process. Comparing the solution quality reached by the VNDs, one can notice a significant improvement over using only a single operator (see Table 2).

When we want to test a VND with three operators (VND-3op), the number of possible combinations increases to 120. To avoid evaluating all of them, we have used iterated F-race [1] as implemented in the irace automatic configuration

Table 6. Best configurations of VND-2op first improvement for JSSP-WT

	Operator 1	Operator 2	Gap	#Eval	#Imp	Rank	Sec
VND1	CT	CEI+2MT	135.24	1.434	33.59	1	311
VND2	CET+2MT	CT	146.81	1.368	32.08	4	194
VND3	CT	CET+2MT	147.79	1.273	31.98	5	205
VND4	CET	CT	145.92	1.351	31.46	3	194
VND5	ECET	CT	143.93	1.476	31.51	2	230
VND6	CET	CET+2MT	182.11	0.746	27.72	6	132

Table 7. Best configurations of VND-3op obtained by i-race with first improvement for JSSP-WT

	Operator 1	Operator 2	Operator 3	Gap	#Eval	#Imp	Rank	Sec
VND1	CT	CEI	CEI+2MT	134.13	1.748	34.79	1	463
VND2	CET	CT	CEI	138.17	1.848	33.73	2	380
VND3	ECET	CT	CEI	138.85	1.979	33.75	3	419
VND4	CT	CET+2MT	CEI	142.07	1.733	34.22	5	384
VND5	CT	CEI	CET+2MT	142.96	1.682	34.22	6	374
VND6	CET+2MT	CT	CEI	139.3	1.855	34.31	4	397

tool [7] to select the most promising configurations of a VND-3op. In order to discard the worst combinations of the operators, we apply irace using as initial candidates configurations all 120 possible VND-3ops and let irace select among these the 10 best ones. Among these ten, we have selected six of the most promising ones analogous to how we have done the selection for the VND-2op. In Table 7, these six configurations are evaluated according to the scheme of experiment 1. We observe that the CT operator is presented in each configuration since it consists on swapping every critical arc in a critical block and it is considered the best operator with the smallest gap in the previous experiments. These best combinations are based on CEI and CEI+2MT operators in the second or the third position of the configurations. When compared to the results of VND-2op, overall the quality of the solutions reached by the VND-3op algorithms is better, however, this comes at the cost of further increased computation times.

The variants of the VND-2op and VND-3op for the JSSP-WC and JSSP-WU have been chosen in a similar process as reported above. Considering JSSP-WC (see Tables 8 and 9), the high-level conclusions are similar to those of JSSP-WT; the VND-2op and VND-3op reach better quality solutions with those of VND-3op naturally being the best, however, at an increased computation time. For the JSSP-WU this same process led, as expected, only to minor differences among the solution quality of the VNDs, as the major improvement should come through the frequent execution of local searches. We nevertheless selected various VNDs, all making use of the CT operator as the first one, except for one VND-2op, where CT is used as the second operator.

Table 8. Best configurations of VND-2op first improvement for JSSP-WC

	Operator 1	Operator 2	Gap	#Eval	#Imp	Rank	Sec
VND1	CT	CET	12.96	1.242	31.7	5	209
VND2	CT	CEI	12.56	1.504	33.89	2	383
VND3	CT	CET+2MT	12.89	1.275	31.96	4	219
VND4	CT	CEI+2MT	12.09	1.433	33.58	1	330
VND5	CET	CET+2MT	15.68	0.742	27.54	6	136
VND6	CET+2MT	CT	12.82	1.371	32.03	3	224

Table 9. Best configurations of VND-3op obtained by irace for JSSP-WC

	Operator 1	Operator 2	Operator 3	Gap	#Eval	#Imp	Rank	Sec
VND1	CET+2MT	CEI	CT	12.14	1.842	34.27	2	421
VND2	CET+2MT	CT	CEI	11.98	1.845	34.21	1	372
VND3	CT	CEI	CEI+2MT	12.15	1.740	34.76	3	434
VND4	CT	CET+2MT	CEI	12.17	1.724	34.22	4	370
VND5	CT	CEI	CET+2MT	12.35	1.672	34.2	5	366
VND6	CT	CET	CEI	12.56	1.683	33.89	6	365

4.4   Iterated Local Search Algorithm

As a next step, we considered the various iterative improvement algorithms as
local searches inside an iterated local search (ILS) algorithm [8]. For a generic
outline of ILS, see Fig. 3. For the specific ILS algorithm, we followed the three-
step approach described by Mati et al. [9]. We made this ILS a parametrized
algorithm in which we can choose according to a parameter the iterative improvement algorithm to be used, different possibilities for the perturbation (either a
fixed number of perturbation steps or a randomly chosen number within some
interval) and different acceptance criteria.

The local search step in this ILS has two phases: the first phase is an improving phase starting from a given initial solution until a local optimum is reached.
In our implementation, we consider as possible alternative choices the first-
improvement algorithms using the six different operators, or our six candidate
VND-2op or VND-3op algorithms for each of the three objective functions we
consider. The second phase is an intermediate phase that starts from the solution
obtained by the improving step, but uses a second objective function to select a
best move without degrading the value of the local optimum. We use the sum of
completion times as the second objective function and the CT operator as the
neighborhood structure. These two phases are repeated until no improvement is
found in the first and second objective functions.

The perturbation step in the original algorithm by Mati et al. consists of a
number of steps at each of which a random neighbor in the CT neighborhood

$s_0 :=$ GenerateInitialSolution
$s^* :=$ LocalSearch(s_0)
while not stopping criteria met **do**
  $s' :=$ Perturbation (s^*)
  $s^{*\prime} :=$ LocalSearch(s')
  $s^* :=$ AcceptanceCriterion ($s^*, s^{*\prime}$)
end while

Fig. 3. Algorithmic scheme of the ILS

is accepted. The number of steps in Mati et al. is chosen uniformly at random in $t \in [t_{min}, t_{min} + \Delta]$. Here, we consider also the possibility of using a fixed number of steps in the perturbation. As the acceptance criterion, Mati et al. have accepted every new local optimum as the new incumbent solution; this acceptance criterion we call *accept_always*. Here, we also consider an acceptance criterion that accepts a worse solution with a probability given by the Metropolis condition [11], $exp(f(s^*) - f(s^{*\prime}))/T$ where $f(s^*)$ and $f(s^{*\prime})$ are the objective value of the current and the new solution, respectively. We consider the temperature T fixed in this algorithm (Fig. 3).

We consider three variants of ILS denoted by ILS1, ILS2 and ILS3. ILS1 considers only single neighborhood operators in the improving phase, ILS2 considers only VND-2op and ILS3 considers only VND-3op in the improving step. We fine-tuned the parameters for each variant ILS1, ILS2 and ILS3 on each of the JSSP variants, that is, JSSP-WT, JSSP-WC and JSSP-WU. The range of possible values for T was $[0.01, 10000]$, for t it was $t_{min} \in [0, 10]$ and $\Delta \in [1, 10]$ (if perturbation steps are variable) and $t \in [2, 10]$ (if perturbation steps are of fixed size) and for the acceptance criterion it was $[accept\_always, accept\_with\_Probability]$.

As for VND-3op, we used irace in its default setting for the automatic tuning [7]. For each ILS, we performed one run of the automatic tuner and allocated a limit of 1000 experiments for each irace run. Each experiment involving the execution of one algorithm configuration uses a time limit of 60 s. Table 10 describes the parameter settings of ILS variants that we obtained and that we used to compare the behaviour of each algorithm on the different objectives.

For the experimental analysis of the ILS variants, we did 5 runs of each ILS with a time limit of 60 s per run on the Singer and Pinedo instances and for the comparison of the algorithms we choose the median of the 5 results we obtained per algorithm and problem. Using the median is preferable over the averages, as the median is a more stable statistic than the mean, especially in case results can be rather variable. We then made the analysis based on the median gaps to the best-known solutions. Table 11 summarizes the average of these median gaps of the ILS variants for each objective function. The number in parenthesis indicates the number of optimal or best-known solutions reached. Best results of ILS are clearly better than just restarting the LS from random initial solutions for 60 s per run and the best results with this latter approach are the percentage deviation of 92.11% and 9.51% away from the optimum with

Table 10. Parameter settings found by automatic tuning for different ILS algorithms on different JSSP variants.

ILS	Neighborhood	perturbFixed	Acceptance	t_{min}	t	Δ	T
ILS1-WT	CET	False	always	7	-	9	-
ILS2-WT	VND2	True	with_prob	-	9	-	479.69
ILS3-WT	VND6	False	always	8	-	10	-
ILS1-WC	CT	False	always	2	-	9	-
ILS2-WC	VND6	True	always	-	8	-	-
ILS3-WC	VND4	True	with_prob	-	9	-	5393.17
ILS1-WU	CET+2MT	False	always	6	-	3	-
ILS2-WU	VND4	False	with_prob	4	-	2	426.02
ILS3-WU	VND2	False	always	9	-	5	-

VND-2op for JSSP-WT and JSSP-WC respectively. With VND-3op, the best gaps are 96.46% and 9.61% for JSSP-WT and JSSP-WC respectively.

Table 11. Comparison between different ILS algorithms using a single (ILS1) neighborhood operator, a VND with two operators (ILS2) or a VND with three operators (VND3) for the three variants of the job-shop scheduling problem we consider. Given is the average gap of the algorithms measured across our benchmark set of 22 instance; for each instance, we measured the median gap to the optimal or best-known solutions.

Objective function	ILS1	ILS2	ILS3	P-values		
				ILS1-ILS2	ILS1-ILS3	ILS2-ILS3
$\sum_i w_i T_i$	14.16(2)	12.66(5)	20.1(4)	0.3202	0.04198	0.014523
$\sum_i w_i C_i$	1.44(3)	1.69(2)	2.02(2)	0.3956	0.11172	1
$\sum_i w_i U_i$	10.41(10)	16.56(8)	33.92(2)	0.1308	0.000573	0.0009168

We assess the statistical significance of the differences using the Wilcoxon paired test at a significance level $\alpha = 0.05$. We compared the results obtained by ILS1, ILS2 and ILS3 on the three JSSP variants. Multiple comparisons are carried out between ILS1-ILS2, ILS1-ILS3 and ILS2-ILS3, using Holm's correction. The results are given in Table 11.

JSSP-WT: The difference between ILS1-WT and ILS2-WT are not statistically significant, while the differences of ILS1-WT and ILS2-WT w.r.t. ILS3-WT are statistically significant. Overall, taking the average performance, ILS2 is the best performing algorithm with a gap 12.66 to the optimal solutions.

JSSP-WC: The results obtained by the test show that there is no statistically significant differences between the variants. Anyway, ILS1-WC obtains the lowest averages, followed rather closely by ILS2-WC.

JSSP-WU: There is no statistically significant difference between ILS1-WU and ILS2-WU, but the differences between both algorithms and ILS3-WU are statistically significant.

Overall, the ILS algorithms improve strongly w.r.t. a repeated application of local search operators, as can be observed when comparing the gaps in Table 11 to any of the other experiments. In general, there is no clear trend whether ILS1 or ILS2 is preferable and possibly further tests on larger instances would be necessary to differentiate more clearly between them. However, the experimental results also clearly indicate that the ILS3 variants are inferior to ILS1 or ILS2, which probably is due to the fact that the slight improvement in solution quality reached by concatenating the search according to three operators does not pay off the additional computation time it requires.

5   Conclusions

In this paper, we have studied six neighborhood operators for three job-shop scheduling problems that differ in the objective functions to be minimized. The neighborhoods have been studied within iterative first- and best-improvement algorithms, and within variable neighborhood descent algorithms. The experimental results showed first-improvement local search is preferable over the best-improvement one due to shorter computation times and often better quality solutions. Among variable neighborhood descent algorithms we considered ones where either two or three neighborhoods are integrated. Generally, the critical transpose neighborhood was found to be important for all three job-shop scheduling problems considered, even though the best VNDs had different shapes, thus, depended on the specific problem. Once either the single neighborhood iterative improvement or the two- and three-neighborhood VNDs are integrated into an iterated local search algorithm, the VND with three neighborhoods was generally found to be not cost effective and the faster local searches to be preferable.

We intend to extend this study in a number of directions. First, we may implement even more neighborhood structures in addition to those studied here. We have done so with CT+2MT, which seemed particularly promising given the performance of CT and CET+2MT, and obtained actually improved solution quality when compared to CT. However, we do not necessarily expect the addition of further neighborhoods to change the other conclusions taken here. We may also improve the iterated local search algorithm by considering also other operators than the critical transpose one for generating the solution perturbations. Interesting is probably to extend the work considering more variants of the job-shop scheduling problem that do not only differ in the objectives but also in other features such as sequence-dependent setup times, constraints on inventory and buffer sizes, or more general job routing possibilities. In fact, providing a generalized job-shop environment in the style of what done by Vidal et al. for vehicle routing [16], should provide interesting insights and results, as well as increasing the practical value of the research.

Acknowledgments. This research and its results have received funding from the COMEX project (P7/36) within the Interuniversity Attraction Poles Programme of the Belgian Science Policy Office. Thomas Stützle acknowledges support from the Belgian F.R.S.-FNRS, of which he is a Senior Research Associate.

References

1. Birattari, M., Yuan, Z., Balaprakash, P., Stützle, T.: F-race and iterated F-race: an overview. In: Bartz-Beielstein, T., Chiarandini, M., Paquete, L., Preuss, M. (eds.) Experimental Methods for the Analysis of Optimization Algorithms, pp. 311–336. Springer, Heidelberg (2010)
2. Dell'Amico, M., Trubian, M.: Applying tabu search to the job-shop scheduling problem. Ann. Oper. Res. **41**(3), 231–252 (1993)
3. Hansen, P., Mladenović, N., Brimberg, J., Pérez, J.A.M.: Variable neighborhood search. In: Gendreau, M., et al. (eds.) Handbook of Metaheuristics. International Series in Operations Research and Management Science, vol. 146, 2nd edn, pp. 61–86. Springer, New York (2010)
4. Kreipl, S.: A large step random walk for minimizing total weighted tardiness in a job shop. J. Sched. **3**(3), 125–138 (2000)
5. Kuhpfahl, J., Bierwirth, C.: A new neighbourhood operator for the job shop scheduling problem with total weighted tardiness objective. In: Proceedings of International Conference on Applied Mathematical Optimization and Modelling, pp. 204–209 (2012)
6. Kuhpfahl, J., Bierwirth, C.: A study on local search neighborhoods for the job shop scheduling problem with total weighted tardiness objective. Comput. Oper. Res. **66**, 44–57 (2016)
7. López-Ibáñez, M., Dubois-Lacoste, J., Pérez Cáceres, L., Stützle, T., Birattari, M.: The irace package: iterated racing for automatic algorithm configuration. Oper. Res. Perspect. **3**, 43–58 (2016)
8. Lourenço, H.R., Martin, O., Stützle, T.: Iterated local search. In: Glover, F., et al. (eds.) Handbook of Metaheuristics, pp. 321–353. Kluwer Academic Publishers, Norwell (2002)
9. Mati, Y., Dauzère-Pèrés, S., Lahlou, C.: A general approach for optimizing regular criteria in the job-shop scheduling problem. Eur. J. Oper. Res. **212**(1), 33–42 (2011)
10. Matsuo, H., Suh, C.J., Sullivan, R.S.: A controlled search simulated annealing method for the general jobshop scheduling problem. Graduate School of Business, University of Texas, Austin, TX (1988)
11. Metropolis, N., Rosenbluth, A.W., Rosenbluth, M.N., Teller, A., Teller, E.: Equation of state calculations by fast computing machines. J. Chem. Phys. **21**, 1087–1092 (1953)
12. Nowicki, E., Smutnicki, C.: A fast taboo search algorithm for the job shop problem. Manag. Sci. **42**(6), 797–813 (1996)
13. Pinedo, M.L.: Scheduling: Theory, Algorithms, and Systems, 4th edn. Springer, New York (2012)
14. Singer, M., Pinedo, M.: A computational study of branch and bound techniques for minimizing the total weighted tardiness in job shops. IIE Trans. **30**(2), 109–118 (1998)
15. Van Laarhoven, P.J., Aarts, E.H., Lenstra, J.K.: Job shop scheduling by simulated annealing. Oper. Res. **40**(1), 113–125 (1992)

16. Vidal, T., Crainic, T.G., Gendreau, M., Prins, C.: A unified solution framework for multi-attribute vehicle routing problems. Eur. J. Oper. Res. **234**(3), 658–673 (2014)
17. Yang, Y., Kreipl, S., Pinedo, M.: Heuristics for minimizing total weighted tardiness in flexible flow shops. J. Sched. **3**(2), 89–108 (2000)

A Genetic Algorithm for Multi-component Optimization Problems: The Case of the Travelling Thief Problem

Daniel K.S. Vieira[1,4]([⊠]), Gustavo L. Soares[2], João A. Vasconcelos[3], and Marcus H.S. Mendes[4]

[1] Instituto de Ciências Exatas, Federal University of Minas Gerais, Belo Horizonte, Brazil
danielv@dcc.ufmg.br
[2] Pontifical Catholic University of Minas Gerais, Belo Horizonte, Brazil
gsoares@pucminas.br
[3] Evolutionary Computation Laboratory, PPGEE, Federal University of Minas Gerais, Belo Horizonte, Brazil
jvasconcelos@ufmg.br
[4] Instituto de Ciências Exatas e Tecnológicas, Universidade Federal de Viçosa, Florestal, Brazil
marcus.mendes@ufv.br

Abstract. Real-world problems are often composed of multiple interdependent components. In this case, benchmark problems that do not represent that interdependence are not a good choice to assess algorithm performance. In recent literature, a benchmark problem called Travelling Thief Problem (TTP) was proposed to better represent real-world multi-component problems. TTP is a combination of two well-known problems: 0-1 Knapsack Problem (KP) and the Travelling Salesman Problem (TSP). This paper presents a genetic algorithm-based optimization approach called Multi-Component Genetic Algorithm (MCGA) for solving TTP. It aims to solve the overall problem instead of each sub-component separately. Starting from a solution for the TSP component, obtained by the Chained Lin-Kernighan heuristic, the MCGA applies the evolutionary process (evaluation, selection, crossover, and mutation) iteratively using different basic operators for KP and TSP components. The MCGA was tested on some representative instances of TTP available in the literature. The comparisons show that MCGA obtains competitive solutions in 20 of the 24 TTP instances with 195 and 783 cities.

Keywords: Genetic Algorithm · Multi-component problem · Travelling Thief Problem

1 Introduction

Classic problems in computer science have been proposed and have recently been studied to define strategies to obtain a good solution for real-world problems.

© Springer International Publishing AG 2017
B. Hu and M. López-Ibáñez (Eds.): EvoCOP 2017, LNCS 10197, pp. 18–29, 2017.
DOI: 10.1007/978-3-319-55453-2_2

Many of these problems belong to the NP-hard class, which means that they are combinatorial problems to which we do not have algorithms that can find the best solution in a polynomial time nowadays. Therefore, it is not possible to obtain the best solution in large instances due to the necessary time to process each possible solution. For this reason, heuristics have been developed to obtain a satisfactory solution in an acceptable time.

According to Bonyadi et al. (2013), real-world problems often have interdependent components and these should not be solved separately. This correlation should be considered for the achievement of better solutions to the overall problem.

In order to cover the complexity of a real-world problem, a new problem called Travelling Thief Problem (TTP) was proposed (Bonyadi et al. 2013). It is a combination of two well-known problems: the 0-1 Knapsack Problem (KP) and the Travelling Salesman Problem (TSP).

Some strategies have been proposed to solve TTP. Faulkner et al. (2015) presents algorithms that focus on manipulating the KP component and obtaining the TSP component from the Chained Lin-Kernighan algorithm (CLK) (Applegate et al. 2003). Bonyadi et al. (2014) use a co-evolutionary approach called CoSolver where different modules are responsible for each component of the TTP, which in turn communicates with others and combines solutions to obtain an overall solution to the problem. In this way, the CoSolver attempts to solve the TTP by manipulating both components at the same time, instead of obtaining a solution for one component using the solution of other component. Mei et al. (2014) seek large-scale TTP instances proposing complexity reduction strategies for TTP with fitness approximation schemes and applying these techniques in a Memetic Algorithm, which outperforms the Random Local Search and Evolutionary Algorithm proposed by Polyakovskiy et al. (2014). Wagner (2016) proposes a swarm intelligence approach based ant colony that builds efficient routes for TTP instead of TSP, and combines it with a packing heuristic. This strategy outperforms state-of-the-art heuristics on instances with up to 250 cities. Finally, an Evolutionary Algorithm that tackles both sub-problems (KP and TSP) at the same time is presented by Lourenço et al. (2016). The computational experiments in this last study used six instances with a number of cities ranging from 51 to 100.

This paper proposes an algorithm based on Genetic Algorithm (GA) concept called Multi-component Genetic Algorithm (MCGA) to solve small and medium instances of TTP. MCGA has four basic steps in each iteration: evaluation of the solution (individual) to know how good this solution is; selection of solutions based on their performance (fitness); crossover the solutions to create new solutions (children) based on the features (chromosome) of existing solutions (parents); disturbance of the solutions by applying some mutation operator that changes their features (change some alleles of the genes of their chromosome). After some iterations (generations) the algorithm tends to achieve good solutions to the problem.

The article is organized as follows. Section 2 describes the TTP and its specifications. In Sect. 3, the Multi-component Genetic Algorithm (MCGA) is defined. Section 4 explains the methodology. Finally, we present the conclusion in Sect. 5, as well as the possibilities for further research.

2   Travelling Thief Problem

According to Polyakovskiy et al. (2014), TTP is defined as having a set of cities $N = \{1, \ldots, n\}$ where the distance d_{ij} between each pair of cities i and j is known, with $i, j \in N$. Every city i, except the first, has a set of items $M_i = \{1, \ldots, m_i\}$. Each item k in a city i is described by its value (profit) p_{ik} and weight w_{ik}. The candidate solution must visit each city only once and return to the starting city.

Additionally, items can be collected in cities while the sum of the weight of the collected items does not exceed the knapsack maximum capacity W. A rent rate R must be paid by each time unit that is used to finish the tour. v_{max} and v_{min} describes the maximum and minimum speed allowed along the way, respectively.

$y_{ik} \in \{0, 1\}$ is a binary variable equals to 1 if the item k is collected in the city i. W_i specifies the total weight of the collected items when the city i is left. Hence, the objective function for a tour $\Pi = (x_1, \ldots, x_n)$, $x_i \in N$ and a picking plan $P = (y_{21}, \ldots, y_{nm_i})$ is defined as:

$$Z(\Pi, P) = \sum_{i=2}^{n} \sum_{k=1}^{m_i} p_{ik} y_{ik} - R \left(\frac{d_{x_n x_1}}{v_{max} - \nu W_{x_n}} + \sum_{i=1}^{n-1} \frac{d_{x_i x_{i+1}}}{v_{max} - \nu W_{x_i}} \right) \quad (1)$$

where $\nu = \frac{v_{max} - v_{min}}{W}$. The aim is to **maximize** $Z(\Pi, P)$. The equation is summarized in penalize the profit gains from collected items with a value that represents the total travel time multiplied by the renting rate R.

Figure 1 shows an example of TTP where the number of items for each city is equal to the test instances provided by Polyakovskiy et al. (2014). This case has 2 items to each city (except the first city, that does not have items) and 3 cities in total. Each item $I_{ij}(p_{ij}, w_{ij})$ is described as being the item j of the city

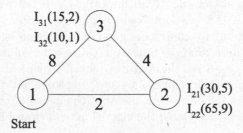

Fig. 1. TTP example.

i that has a value p_{ij} and a weight w_{ij}. Let us assume the knapsack capacity $W = 10$, the renting rate $R = 1$, $v_{max} = 1$ and $v_{min} = 0.1$. A feasible solution for this problem is $P = (0, 1, 0, 1)$ and $\Pi = (1, 2, 3)$, which describes that the items I_{22}, I_{32} which are in the cities 2 and 3, respectively, will be collected, making a tour starting from the city 1 to the city 2, and then to the city 3, and finally returning to the city 1. This solution results in $Z(\Pi, P) \approx -28.053$.

Note that, in this example, if only the TSP component of the problem is considered, all the possible solutions have the same cost, since all connections between the cities will be used anyway. However, considering the overall problem, it can be seen that the order of the tour affects the solution cost, due to the variation of the time that an item is kept in the knapsack. In other words, a heavy item picked at the start of the tour affects the travelling speed for a longer time, compared to the same item picked at the end of the tour, which slows the solution and increases the cost.

3   Multi-component Genetic Algorithm

The proposed MCGA[1] has a different encoding type for each component of the TTP. The TSP component is encoded by enumerating the city indices in order, starting from the city 1 and ending the tour at the same city. The KP component is encoded using a binary array with the size of existing items. The items are ordered according to the city to which they belong. If the problem has 5 items per city, the first five genes make reference to the items of the city two (since the first city, by definition, does not have items). Figure 2 shows a graphical representation of the chromosomes that compose the individual.

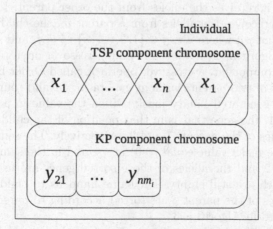

Fig. 2. Graphical representation of the MCGA individual encoding.

---

[1] The source code can be found at https://github.com/DanielKneipp/Genetic AlgorithmTravelingThiefProblem.

The initial population is obtained using the Chained Lin-kernighan (Applegate et al. 2003) (CLK) for the TSP component and all items unpacked (KP component). After generating the initial population, an iterative process starts with the selection of the individuals who will pass through an evolutionary process, based on their fitness. In this step, it was used the tournament method, in which, given a tournament size k, k individuals are selected to be compared with each other and the best individual is selected to the next step. This procedure is repeated until the size of the selected population reaches the size of the original population. Algorithm 1 shows a k-tournament where s^* individuals are selected from population H.

Algorithm 1. Tournament (H, s^*, k)

1: $H\_subset \leftarrow \{\}$
2: **for** $i := 0$ to $s^* - 1$ **do**
3:     $H\_subset \leftarrow k$ random individuals from H
4:     add to H^* the best individual of H_subset
5: **end for**
6: **return** H^*

The crossover step creates new individuals (children) based on the chromosomes of the existing individuals (parents). Due to the multi-component characteristic of the TTP, a different crossover method was used on each component of the problem. For the KP component, a crossover operator called N-point was used. This operator combines the picking plan of two parents in two children in a way that one child receives the alleles of one parent until a certain point is reached. Then, it receives the alleles from the other parent. In the crossover, while the child c_1 receives the alleles from a parent p_i, the child c_2 receives the alleles from a parent p_j, where $\forall\ i,\ j \in \{1, 2, \ldots, s\} : i \neq j$, and s is the number of parents. The Algorithm 2 shows the N-point crossover operator.

For the TSP component, a crossover operator called Order-based is used to combine the tours of two parents without generating invalid tours (Davis 1985). It uses a random-generated binary mask. When the value in a position of the mask is equal to 1, the gene that is in that position of the children 1 and 2 are filled with the genes of the parents 1 and 2, respectively. The remaining genes in parent 1 (with the mask value equal to 0) are sorted in the same order as they appear in parent 2 and the alleles of those sorted genes are used to fill in the child 1 genes which are still empty. The same happens to child 2, although in this case, the genes of the parent 2 are sorted according to parent 1 order. This operator is detailed in Algorithm 3.

After the application of the crossover operators, the population size doubles due to the creation of two children for each two parents. Hence, a selection procedure is applied to decrease the population back to its original value. This second selection step also uses the Tournament method, as shown in Algorithm 1, using the number of individuals before the crossover step as the population size.

Algorithm 2. N-point crossover (P_{p_1}, P_{p_2})

1: $s \leftarrow$ size of the chromosome.
2: $Points \leftarrow n$ different loci (positions) in the chromosome sorted in increasing order
3: $j \leftarrow 0$
4: **for** $i := 0$ to $s - 1$ **do**
5:   **if** $j < n, i = Points[j]$ **then**
6:     $j \leftarrow j + 1$
7:   **end if**
8:   **if** j is an even number **then**
9:     $P_{c_1}[i] \leftarrow P_{p_1}[i]$
10:    $P_{c_2}[i] \leftarrow P_{p_2}[i]$
11: **else**
12:    $P_{c_2}[i] \leftarrow P_{p_1}[i]$
13:    $P_{c_1}[i] \leftarrow P_{p_2}[i]$
14: **end if**
15: **end for**
16: **return** P_{c_1}, P_{c_2}

Algorithm 3. Order-based crossover (Π_{p_1}, Π_{p_2})

1: $s \leftarrow$ size of the chromosome.
2: $Mask \leftarrow$ random array of bits with size s
3: **for** $i := 0$ to $s - 1$ **do**
4:   **if** $Mask[i] = 1$ **then**
5:     $\Pi_{c_1}[i] \leftarrow \Pi_{p_1}[i]$
6:     $\Pi_{c_2}[i] \leftarrow \Pi_{p_2}[i]$
7:   **end if**
8: **end for**
9: $L_1 \leftarrow$ list of genes in Π_{p_1} which are in a position i that the $Mask[i] = 0$
10: $L_2 \leftarrow$ list of genes in Π_{p_2} which are in a position i that the $Mask[i] = 0$
11: Sort L_1 so that the genes appear in the same order as in Π_{p_2}
12: Sort L_2 so that the genes appear in the same order as in Π_{p_1}
13: Fill in the still empty genes of Π_{c_1} with the L_1
14: Fill in the still empty genes of Π_{c_2} with the L_2
15: **return** Π_{c_1}, Π_{c_2}

Similarly to the crossover step, the mutation step has a different operator for each component of the problem. To mutate the KP component, a simple Bit-flip operator was used. Each item i has a probability p of being removed in case of i already be selected, or added to the knapsack, otherwise.

2-OPT was used to mutate the TSP component (Watson et al. 1998). This mutation operator swaps a pair of edges in the tour. A way to do that is given two vertexes v_1 and v_2 (they cannot be the first or the last vertex, which reference the same city), it made a copy of the tour until v_1, then the sub-tour $[v_1, v_2]$ is copied in reverse order. Next, the remainder of the tour is copied. It can be seen in the Algorithm 4, assuming that the index range of the chromosome starts from 0.

Algorithm 4. 2-OPT mutation (Π_p)

1: $s \leftarrow$ size of the chromosome.
2: $v_1 \leftarrow$ random integer number $\in [1, s-3]$
3: $v_2 \leftarrow$ random integer number $\in [v_1 + 1, s-2]$
4: **for** $i := 0$ to $v_1 - 1$ **do**
5:    $\Pi_c[i] \leftarrow \Pi_p[i]$
6: **end for**
7: **for** $i := v_1$ to v_2 **do**
8:    $\Pi_c[i] \leftarrow \Pi_p[v_2 - (i - v_1)]$
9: **end for**
10: **for** $i := v_2 + 1$ to $s - 1$ **do**
11:    $\Pi_c[i] \leftarrow \Pi_p[i]$
12: **end for**
13: **return** Π_c

In the TSP, the order of the tour does not affect the solution cost, since the distance from a city c_1 to another city c_2 is equal to that from c_2 to c_1, for example. However, in TTP, this change affects the time that an item will be carried, and consequently, the cost. Therefore, the 2-OPT operator can dramatically change the individual. For this reason, the mutation operators are used with parsimony. The 2-OPT has a 17.5% chance of being chosen, Bit-flip has 65% and both operators have 17.5%. Note that the probability of 82.5% (65% + 17.5%) of Bit-flip being used do not necessarily mean that several items are going to be removed or added because there is the configuration parameter p of the Bit-flip. For example: with $p = 0.2\%$, an item i has a $0.002 \times 0.825 = 0.00165 = 0.165\%$ chance of being added or removed.

We use elitism to maintain a set of the best individuals (also called elites) in each generation to the next generation. In the mutation step, the best individuals are not included in the procedure. In the crossover step, before the procedure starts, a set of the best individuals is ensured to be part of H^*. Note that these best individuals are included in the tournament.

The Eq. 1 is used to evaluate the individuals. If an individual is invalid (the sum of the weight of picked items exceeds the knapsack capacity), a correction procedure is applied and removes the worst items of the individual until it becomes valid. The worst items are those with the lowest values of $\frac{p_i}{w_i}$, where p_i is the value of the item i and w_i its weight.

The MCGA combines all these operators as shown in the Algorithm 5, where K and Q stores the MCGA configuration and the stop conditions, respectively. Inside K, it is specified the population size s; two different tournament sizes t_s, t_c, for the selection step and the tournament after the crossover, respectively; three different sizes of elite set e_s, e_c, e_m, for the selection step, tournament procedure after the crossover and the mutation step, respectively; a number of points n_c for the N-point crossover operator; the probability p for Bit-flip mutation operator.

Algorithm 5. Multi-component Genetic Algorithm (K, Q)

1: $O \leftarrow$ GenerateInitialPopulation$(K.s)$
2: EvaluateIndividuals(O)
3: **while** $\neg Q$ **do**
4:    $O \leftarrow$ Select$(O, K.s, K.t_s, K.e_s)$
5:    $\Theta \leftarrow$ Crossover$(O, K.n_c)$
6:    EvaluateIndividuals(Θ)
7:    $O \leftarrow \{O, \Theta\}$
8:    $O \leftarrow$ Select$(O, K.s, K.t_c, K.e_c)$
9:    $O \leftarrow$ Mutate$(O, K.p, K.e_m)$
10:    EvaluateIndividuals(O)
11: **end while**
12: **return** BestIndividual(O)

---

4   Methodology and Results

A subset of 9720 TTP benchmark instances, proposed by Polyakovskiy et al. (2014), was used in the experiments. These 9720 instances have the number of cities ranging from 51 to 85,900 (81 different sizes); the number of items per city $F \in \{1, 3, 5, 10\}$ (called item factor); ten capacity categories (varying knapsack capacity); three KP types: *uncorrelated* (the values of the items are not correlated with their weights), *uncorrelated with similar weights* (same as the uncorrelated, but the items have similar weights) and *bounded strongly correlated* (items with values strongly correlated with their weights and likely presence of multiple items with the same characteristics).

According to Faulkner et al. the 72 instances selected by them are a representative subset to cover small, medium, and large size instances with different characteristics of the original set of 9720 instances. In this paper, we selected a subset including 36 instances of those 72 selected by Faulkner et al. (2015). This subset consists of 3 different numbers of cities n between 195 and 3038, two item factors $F \in 3, 10$, all three types of knapsacks t and two capacity categories C being equal to 3 or 7.

Different configurations for the MCGA were defined (based on a simple empirical study) due to the variety of problem sizes, namely:

C1: 200 individuals, Tournament size of both selection procedures equal to 2, number of elites of both selection steps and mutation step equal to 12, number of points for the N-point crossover operator equal to 3, and probability p for the Bit-flip mutation operator equal to 0.2%. The stop condition is 10 min of runtime. The CLK heuristic has no runtime limit;

C2: 80 individuals, Tournament size of both selection procedures equal to 2, number of elites of both selection steps and mutation step equal to 6, number of points for the N-point crossover operator equal to 3, and probability p for the Bit-flip mutation operator equal to 0.2%. The stop conditions are 50 min of runtime. The CLK heuristic has runtime limit $t_{CLK} = \frac{0.6 \times t_{MCGA}}{s}$, where

Table 1. MCGA performance compared to the S5 heuristic and MIP approach on a subset (36 instances) of the 72 representative instances used by Faulkner et al. (2015).

n	m	t	C	MCGA	S5	MIP
195	582	bsc	3	84328.5 ($\pm$2169.3)	86516.9	**86550.9**
			7	**121166** ($\pm$5597.43)	110107	110555
		unc	3	**61666.25** ($\pm$1629.41)	56510.6	56518
			7	**76813** ($\pm$1180.9)	70583.8	70728
		usw	3	**31049.1** ($\pm$721.32)	28024.7	28061.5
			7	**53808** ($\pm$1110.14)	48023	48332
	1940	bsc	3	**229924** ($\pm$2217.35)	227063	227965
			7	**385503.5** ($\pm$3464.83)	359614	359527
		unc	3	**174049.5** ($\pm$999.1)	157297	157532
			7	**250635** ($\pm$996.26)	227502	227637
		usw	3	**115036** ($\pm$830.88)	102568	103417
			7	**193773** ($\pm$1814.72)	168931	169168
783	2346	bsc	3	**288460** ($\pm$7089.17)	263725	263691
			7	**478978** ($\pm$11832.2)	435157	433814
		unc	3	**207616** ($\pm$3640.77)	189949	189671
			7	**287819** ($\pm$4739.09)	263367	263258
		usw	3	**140840.5** ($\pm$2420.11)	130409	130901
			7	**233716** ($\pm$3593.83)	213893	213943
	7820	bsc	3	**942002** ($\pm$7588.29)	940002	940141
			7	**1501090** ($\pm$16182.4)	1425821	1424650
		unc	3	581237 ($\pm$5916.8)	**637793**	637487
			7	905915 ($\pm$9955.25)	**910032**	909343
		usw	3	399076 ($\pm$9569.79)	434180	**435368**
			7	**707502** ($\pm$15864.72)	698730	699101
3038	9111	bsc	3	798247 ($\pm$19108.8)	**1217786**	1214427
			7	1312630 ($\pm$35161.7)	**1864413**	1858773
		unc	3	37385.3 ($\pm$24576.2)	**782652**	780574
			7	362986 ($\pm$31487.3)	**1093259**	1090977
		usw	3	93402.2 ($\pm$16183.7)	**568509**	567102
			7	298075 ($\pm$25109.7)	**873670**	869420
	30370	bsc	3	1042960 ($\pm$81288.8)	**4023124**	4006061
			7	1784080 ($\pm$201566)	**5895031**	5859413
		unc	3	-1576520 ($\pm$44662.3)	**2595328**	2589287
			7	-897125 ($\pm$117563)	**3603613**	3600092
		usw	3	-729530 ($\pm$40322.8)	1800448	**1801927**
			7	-679838 ($\pm$94382.3)	**2863437**	2856140

t_{MCGA} is the total runtime (50 min in this case) and s is the population size (80 in this case).

Note that the MCGA runtime includes the generation of the initial population procedure. Therefore, the time spent by the CLK is taken into account.

In the computational experiments we compare MCGA with two heuristics proposed by Faulkner et al. (2015): (1) a local search algorithm called S5, which runs CLK and then an iterative greedy heuristic (called PackIterative), until the time is up; and (2) a mixed integer programming (MIP) based approach. The computer used has two Intel Xeon E5-2630 v3 totaling 32 cores (only one was used for each run), 128 GB of RAM and Ubuntu 14.04 LTS OS.

The results presented in Table 1, using C1 configuration in MCGA, represent the mean of 30 independent runs (with standard deviation, in case of MCGA). It can be seen that MCGA outperforms the other algorithms in 83.33% (20 of the 24) instances derived from the TSP problem component rat with 195 and 783 cities. On the other hand, for the problem with 3038 cities, it has clearly underperformed and demonstrated that the MCGA performance is visibly linked with the TSP component characteristics, which can be the distance pattern or the problem size or maybe both. The overall result shows that MCGA obtained better averages in 55.56% (20 of the 36) of the presented instances; S5 outperforms the other algorithms in 36.11% (13 of the 36, mainly due to its results in the instances with 3038 cities) and the MIP based approach was the best in 8.33% (3 of the 36) of the benchmark problems.

The algorithm evolution was analyzed for further investigation of the MCGA behavior on instances with 3038 cities. We discovered that, for many instances such as pcb3038_n30370_uncorr_07 (3038 cities, 10 items per city, uncorrelated type and capacity category 7), the performance was limited by time restriction. Figure 3 shows that MCGA convergence was not attained. Probably, a better solution can be found with more time available. In face of this, a configuration with more time available (C2) was used to analyze the MCGA behavior while solving the instance (pcb3038_n30370_uncorr_07). We limited the CLK

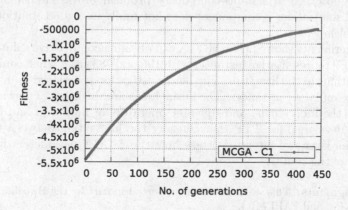

Fig. 3. MCGA with C1 configuration in the instance pcb3038_n30370_uncorr_07.

heuristic runtime (set to 37.5 s to generate each individual) in order to guarantee considerable runtime to MCGA.

Figure 4 shows that MGCA achieves significantly higher fitness value if more time is available. However, this value still remains below the S5 and MIP results. Therefore, the amount of time is not the only MCGA feature that requires improvement.

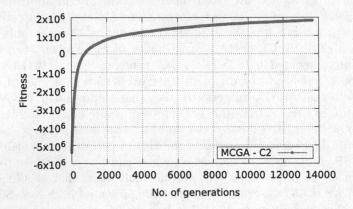

Fig. 4. MCGA with C2 configuration in the instance `pcb3038_n30370_uncorr_07`.

5   Conclusion and Future Work

In this paper, a Genetic Algorithm approach called Multi-component Genetic Algorithm (MCGA) was proposed in an attempt to solve a new multi-component combinatorial problem called Travelling Thief Problem (TTP). It applies basic mutation and crossover operators for each component of the problem. The inter-dependence observed in a multi-component problem proves challenging, since the optimal solution for a component does not imply in a good solution for the overall problem.

The experiments showed that MCGA can obtain a competitive solution in 20 of the 24 TTP representative instances with 195 and 783 cities when compared to other algorithms in the literature. Further investigation is required to improve its performance on larger instances. The initial ideas are the using of new strategies to evaluate the individuals, and different operators of mutation and crossover specifically developed for TTP that consider the interaction between the components. Moreover, we intended assess how the MCGA parameters influences the quality of the results.

Acknowledgments. This work has been supported in part by the Brazilian agencies CAPES, CNPq, and FAPEMIG.

References

Applegate, D., Cook, W., Rohe, A.: Chained Lin-Kernighan for large traveling salesman problems. INFORMS J. Comput. **15**(1), 82–92 (2003)

Bonyadi, M.R., Michalewicz, Z., Przybyŏek, M.R., Wierzbicki, A.: Socially inspired algorithms for the travelling thief problem. In: Proceedings of the 2014 Conference on Genetic and Evolutionary Computation, pp. 421–428. ACM (2014)

Bonyadi, M., Michalewicz, Z., Barone, L.: The travelling thief problem: the first step in the transition from theoretical problems to realistic problems. In: 2013 IEEE Congress on Evolutionary Computation (CEC), pp. 1037–1044, June 2013

Davis, L.: Applying adaptive algorithms to epistatic domains. IJCAI **85**, 162–164 (1985)

Faulkner, H., Polyakovskiy, S., Schultz, T., Wagner, M.: Approximate approaches to the traveling thief problem. In: Proceedings of the 2015 on Genetic and Evolutionary Computation Conference, pp. 385–392. ACM (2015)

Lourenço, N., Pereira, F.B., Costa, E.: An evolutionary approach to the full optimization of the traveling thief problem. In: Chicano, F., Hu, B., García-Sánchez, P. (eds.) EvoCOP 2016. LNCS, vol. 9595, pp. 34–45. Springer, Heidelberg (2016). doi:10.1007/978-3-319-30698-8_3

Mei, Y., Li, X., Yao, X.: Improving efficiency of heuristics for the large scale traveling thief problem. In: Dick, G., et al. (eds.) SEAL 2014. LNCS, vol. 8886, pp. 631–643. Springer, Heidelberg (2014). doi:10.1007/978-3-319-13563-2_53

Polyakovskiy, S., Bonyadi, M.R., Wagner, M., Michalewicz, Z., Neumann, F.: A comprehensive benchmark set and heuristics for the traveling thief problem. In: Proceedings of the 2014 Conference on Genetic and Evolutionary Computation, GECCO 2014, pp. 477–484. ACM, New York (2014). http://doi.acm.org/10.1145/2576768.2598249

Wagner, M.: Stealing items more efficiently with ants: a swarm intelligence approach to the travelling thief problem. In: Dorigo, M., Birattari, M., Li, X., López-Ibáñez, M., Ohkura, K., Pinciroli, C., Stützle, T. (eds.) ANTS 2016. LNCS, vol. 9882, pp. 273–281. Springer, Heidelberg (2016). doi:10.1007/978-3-319-44427-7_25

Watson, J., Ross, C., Eisele, V., Denton, J., Bins, J., Guerra, C., Whitley, D., Howe, A.: The traveling salesrep problem, edge assembly crossover, and 2-opt. In: Eiben, A.E., Bäck, T., Schoenauer, M., Schwefel, H.-P. (eds.) PPSN 1998. LNCS, vol. 1498, pp. 823–832. Springer, Heidelberg (1998). doi:10.1007/BFb0056924

A Hybrid Feature Selection Algorithm Based on Large Neighborhood Search

Gelareh Taghizadeh$^{(\boxtimes)}$ and Nysret Musliu

Database and Artificial Intelligence Group,
Vienna University of Technology, Vienna, Austria
{gtaghiza,musliu}@dbai.tuwien.ac.at

Abstract. Feature selection aims at choosing a small number of relevant features in a data set to achieve similar or even better classification accuracy than using all features. This paper presents the first study on Large Neighborhood Search (LNS) algorithm for the feature selection problem. We propose a novel hybrid Wrapper and Filter feature selection method using LNS algorithm (WFLNS). In LNS, an initial solution is gradually improved by alternately destroying and repairing the solution. We introduce the idea of using filter ranking method in the process of destroying and repairing to accelerate the search in identifying the core feature subsets. Particularly, WFLNS either adds or removes features from a candidate solution based on the correlation based feature ranking method. The proposed algorithm has been tested on twelve benchmark data sets and the results have been compared with ten most recent wrapper methods where WFLNS outperforms other methods in several the data sets.

Keywords: Feature selection · Large Neighborhood Search · Classification

1 Introduction

One of the aims of the feature selection is to remove irrelevant and redundant features from a set of original features to improve the classification performance. Generally, recent interest in feature selection has been increased due to challenges which are caused by sheer volume of data that is expected to have a rapid growth over next years. Such data volume would not only increase the demand of computational resources but also effect on the quality of several data mining tasks such as classification. Moreover, learning from large data sets would be a more complex task when it includes irrelevant, redundant and noisy features. Feature selection techniques address such challenges by decreasing the dimensionality, reducing the amount of data needed for the learning process, shortening the running time, and improving the performance of the learnt classifiers [1].

Throughout the literature, feature selection approaches are mainly categorized into three main groups: filter, wrapper and hybrid approaches. Filter

© Springer International Publishing AG 2017
B. Hu and M. López-Ibáñez (Eds.): EvoCOP 2017, LNCS 10197, pp. 30–43, 2017.
DOI: 10.1007/978-3-319-55453-2_3

approaches select and evaluate subsets of features based on the general characteristics of the training data without involving a learning model while wrapper approaches find subset of features by considering a learning model. Wrapper methods have better quality subsets than filter methods, however they impose a high computational cost as a result of constructing a learning model when evaluating every single subset while filter methods are more computationally efficient. Hybrid approaches aim at benefiting from advantages of both filter and wrapper methods, where filter method can search through the feature space efficiently while the wrapper method provides good accuracy by providing higher quality subset of features for a classifier.

Wrapper methods guide the search through the search space of feature subset by a learning algorithm where a search algorithm is wrapped around the classification model. To this end a common open question that all wrapper techniques try to deal with is to develop an efficient search algorithm for finding an optimal feature subset. An exhaustive search algorithm may be employed, as a possible option to find the optimal solution, however it is often impractical for most data sets. Alternatively, metaheuristic methods have been developed with varying degrees of success (see the next section) to find near-optimal solutions in a reasonable amount of time.

This paper proposes a novel hybrid Wrapper and Filter algorithm designed based on LNS (WFLNS) to solve the feature selection problem. In this study we aim at developing a LNS with improved capabilities of destroy, repair and acceptance methods that has leading-edge performance over the most recent metaheuristics algorithms for feature selection. Our idea is based on hybridizing a filter ranking method as a heuristic in the process of destroying and repairing of the LNS to transform a current solution into a different solution by constantly adding or removing features based on the correlation based feature ranking method [2]. Moreover, we proposes a new acceptance method in WFLNS by incorporating the Simulated Annealing acceptance probability. It is worth mentioning that to the best of our knowledge no systematic studies have been carried out to investigate the capability of LNS algorithms for the feature selection problem so far. The performance of the proposed algorithm is compared with ten state of the art metaheuristic algorithms which all were employed for solving the feature selection problem in [3]. Moreover, we used the same high dimensional and real-valued benchmark data sets as [3], which were originally collected from UCI repository [4]. We were able to conclude that our WFLNS outperforms other algorithms in terms of classification accuracy, particularly for large size data set. The rest of the paper is organized: Sect. 2 reviews the prior studies in metaheuristic algorithms for feature selection problem. Section 3 proposes Large Neighborhood Search for the feature selection problem and experimental results and discussions are presented in Sect. 4. Finally, Sect. 5 concludes the paper.

2   Literature Review

Feature selection problem is the problem of choosing the best subset of features out of the whole feature set. Searching for the best subset in the feature space requires checking all possible combinations of features in the search space. To this end, feature selection problem could be considered as a NP-hard problem [5]. Metaheuristic algorithms are introduces into feature selection as an appropriate methods as they have been demonstrated to be superior methodologies in many NP hard problems. As the core concept of WFLNS is close to the concepts of both wrapper and hybrid approaches, here we concentrated on the literature that has focused on wrapper and hybrid feature selection by employing metaheuristic algorithms. Different feature selection algorithms have been developed based on different sort of optimization techniques such as metaheuristic algorithms. We explored and categorized the existing metaheuristic-based feature selection algorithms into three groups: trajectory-based feature selection, population-based feature selection and hybrid feature selection. A trajectory-based algorithm typically uses one solution at a time, which will gradually improve the current solution by constantly changing the current solution as the iterations continue. The trajectory-based algorithms developed for feature selection problem include Tabu Search (TS) [6,7], Simulated Annealing (SA) [8] and Harmony Search (HS) [9]. In [10], a hybrid approach based on TS and probabilistic neural networks is proposed for the feature selection problem. In contrast with other TS based approached for the feature selection problem, this approach employed long-term memory to avoid the necessity of tuning the memory length and decrease the risk of trapping into local optimal solutions. A TS was employed in [11] to select effective emotion states from the physiological signals based on K-nearest neighbor classifier. In [12], a SA-SVM approach is developed for tuning the parameter values of SVM and finding the best subset of features in a way that maximize the classification accuracy of SVM. Another SA-based algorithm is proposed in [13], where feature selection applied on marketing data to build large scale regression model. In the proposed approach, SA was compared with stepwise regression [14] (as a typical example of an iterative improvement algorithm) and the results have shown the superiority of SA in providing a better predictive model by escaping the local optimum that stepwise regression is fall into. In [15], hybrid combination of SA and GA proposed to combine the capability of SA and GA to avoid being trapped in a local minimum and benefit from the high rate of convergence of the crossover operator of genetic algorithms in the search process. In [16] HS is employed for feature selection in email classification task, where HS was incorporated with the fuzzy support vector machine and Naive Bayesian classifiers. In [17], the authors propose a hybrid feature selection algorithm based on the filter and HS algorithm for gene selection in micro-array data sets.

Population-based approaches rely on a set of candidate solutions rather than on one current solution. There exist two main population based algorithms, which are developed for feature selection problem, evolutionary algorithms and swarm intelligence algorithms. Former involves Genetic Algorithms (GA) [18], Memetic Algorithms (MA) [19] and Artificial Immune Algorithms [20] while later referring to Particle Swarm Optimization (PSO) [21], Ant Colony Optimization (ACO) [22], and Artificial Bee Colony (ABC) [23] optimization. In [24] a GA-base feature selection was proposed in which a feature subset is represented by a binary string of length of the total number of features, called a chromosome. A population of such chromosomes is randomly initialized and maintained, and

those with higher classification accuracy are propagated into the later generations. In [25], a new hybrid algorithm of GA and PSO was developed for classification of hyper spectral data set based on SVM classifier. [26] developed a feature
selection algorithm based on MA for multi-label classification problem which
prevent premature convergence by employing a local search to refine the feature subsets found through a GA search. A hybrid filter-wrapper method based
on memetic framework was proposed in [27], which employes the filter-ranking
method at each iteration of the MA to rank the features and reduce the neighborhood size, then local search operators are applied. Clonal Search Algorithm
(CSA) is inspired from Artificial Immune Algorithm, which was developed for
the feature selection problem by Shojaei and Moradi in [28] where the proposed
approach enables both feature selection and parameter tuning of the SVM classifier. Among swarm intelligence metaheuristic algorithms, the ACO-based feature
selection approach proposed in [29] that each ant has a binary vector where 0 and
1 represents deselected and selected features respectively. [30] presents a variation
of ACO for the feature selection problem, which is called enriched ACO. It aims
at considering the previously traversed edges in the earlier executions to adjust
the pheromone values appropriately and prevent premature convergence of the
ACO. [31] proposed a rough set-based binary PSO algorithm to perform feature
selection. In the algorithm, each particle represents a potential solution, and
these are evaluated using the rough set dependency degree. An ABC algorithm
was proposed in [32] to solve multi-objective feature selection problem where the
number of features should be minimized while the classification accuracy should
be maximized. The results where evaluated based on mutual information, fuzzy
mutual information and the proposed fuzzy mutual information.

In order to improve the searching ability of proposed feature selection algorithms, hybrid algorithms were proposed. The most effective forms of the combination is to use both filter and wrapper approach at the same time. Generally,
hybrid approaches involve two main steps. At the first step a filter method applies
to reduce the number of features and consequently the searching space. The second step is a wrapper method that explores the subsets, which were built on the
first step. In [33] a combination of information gain as a filter method and GA
as a wrapper method was proposed. The K-nearest neighbor (KNN) classifier
with leave-one-out cross-validation (LOOCV) employed as an evaluator of the
proposed algorithm. Another hybrid approach is developed for developing short
term forecasting in [34], which first uses partial mutual information based filter
method to remove most irrelevant features, and subsequently applies a wrapper
method through firefly algorithm to further reduce the redundant features without degrading the forecasting accuracy. Another hybrid approach was developed
in [35] for feature selection in DNA micro-arrays data set. The proposed algorithm employed both univariate and multivariate filter methods along with the
GA as a wrapper method where the reported results show that a multivariate
filter outperforms a univariate filter in filter-wrapper hybrid. [36] presents the
application of rough set method on the outcome of Principal Components Analysis (PCA) for the feature selection problem on neural network classifiers for a face

image data set. Greedy Randomized Adaptive Search Procedure (GRASP) is a multi-start two-phase algorithm, construction and local search phase. A feasible solution is built in the construction phase, and then its neighborhood is explored by the local search. [37] proposes a GRASP-based algorithm for the feature selection problem, where the proposed algorithm uses some random part of the features subset at each iteration, and then selects features based on cooperation between all the previously found non-dominated solutions. [38] is investigated binary classification high dimension data sets while employing GRASP algorithm for feature selection to reduce the computation time. In [39] a feature selection approach is proposed based on a linear programming model with integer variables on biological application. To deal with such approach the authors presents a metaheuristic algorithm based on GRASP procedure which is extended with the adoption of short memory and a local search strategy. The reported results show that the method performs well on a very large number of binary or categorical features. In [3] authors carried out a comprehensive review on the most recent metaheuristic algorithms for the feature selection problem. The performance of developed algorithms were examined and compared with each other based on twelve benchmark data set from UCI repository. We considered this paper as a base line for evaluating our proposed method, and compare our experiments with their reviewed algorithms with the same setting and data sets.

3   WFLNS: A Wrapper Filter Feature Selection Based on LNS

LNS algorithm [40] is a metaheuristic search algorithm, which aims at finding a near-optimal solution by iteratively transforming a current solution to an alternative solution in its neighborhood. The notion of the neighborhood in LNS algorithm refers to a set of similar candidate solutions, which is achieved by applying destroy and repair methods over the current solution [41]. Using large neighborhoods makes it possible to find better candidate solutions at each iteration of the algorithm and hence explore more promising part of the search space. LNS-based algorithms have been applied on many optimization problems, including the traveling salesman problem space [42], timetabling problems [43,44] and capacitated vehicle routing problem [44] because of their capabilities to explore a wide samples of the search space and escape from local optima by means of destroy and repair techniques along with an appropriate acceptance criterion. To the best of our knowledge no systematic studies have been carried out to investigate the capability of LNS-based algorithms for the feature selection problem. Here we present WFLNS, a hybrid Wrapper and Filter feature selection algorithm based on the LNS framework. The novelty of the WFLNS includes embedding new problem-specific destroy, repair and acceptance methods in the LNS which let the algorithm search the feature space more efficiently with improved intensification and diversification. In the following, the algorithmic flow of the WFLNS is explained. Moreover, the Pseudo-code of the WFLNS is presented at Algorithm 1, where defined functions and variables are as follows: The functions

$des()$, $rep()$ and $acc()$ define destroy, repair and acceptance methods respectively. The variable xb is the best observed solution during the search, x is the current solution, xd is the outcome of the $des()$ function which would be served as the input to the $rep()$ function, and xt is the temporary solution.

> **input** : an initial solution x;
> xb=x;
> **while** *Stop criterion is not met* **do**
>> $xd = \mathbf{des}(x)$: Select K percentage of the lowest ranked selected features (ones) to remove from x ;
>> $xt = \mathbf{rep}(xd)$: Select R percentage of the highest ranked features from the removed features and insert them to xt ;
>> **if** *acc(xt , xb)* **then**
>>> $xb = xt$;
>> **end**
> **end**
> **Output** : xb;

Algorithm 1. Procedure of WFLNS

3.1   Encoding Representation and Initialization

A candidate solution in WFLNS is encoded as a binary string with D digits, where D is the total number of features and we aim at choosing a string with d digits out of it, where d is the subset size. Each binary digit represents a feature values 1 and 0, in which 1 indicates a selected feature and 0 an unselected feature. As an example, string 010100 means that the second and fourth features are selected. When prior knowledge about the optimal number of features is available, we may limit d to no more than the predefined value; otherwise, d is equal to D. To make our results comparable with [3], we considered d equal to D in all the experiments. We initialized the initial solution at random and tried to minimize the randomness effects by repeating the experiments for five independent runs. The maximum number of iterations in [3], as our baseline for comparison, is set to the very large value, 5000, to allow all of the studied algorithms to fully converge. In our study, the number of iterations is set to 500 as in most of our experiments the solution was not improved after around 350 iterations.

3.2   Destroy and Repair Methods

The notion of neighborhood in LNS is defined by employed strategies for destructing and rebuilding the current solution to transform it to another solution. Thus both destroy and repair methods have significant impact on the quality of the final solution. Typically, the employed strategies for destroying different parts of the solution are applied randomly and the neighborhood of a solution is then defined as the set of solutions that can be reached by the repair method. The main drawback is caused by destructing the large part of the solution which leads to a neighborhood containing a large amount of candidate solutions that need to be explored. In other words, for each destruction choice there are many possible repairing solutions. In WFLNS, we incorporate a CFS filter method [45] for both destroy and repair methods. CFS is a filter method in which features are ranked based on their correlation with each other as well as the classifier. The main hypothesis behind CFS is that a good feature subsets contain features highly correlated with the classification, yet uncorrelated to each other. Thus, irrelevant features would be ignored as they typically show low correlation with the classifier. Suppose a given current solution x, we define K as a degree of destruction, which would remove K percentage of the solution (K is selected randomly and it is between 0 and 100). In order to destruct the suitable part of the solution, we employed CFS to rank features and remove lowest rank K percentage of selected features which were set to one. Afterwards, we rebuild the destructed part of the solution based on R, where R is the size of the neighborhood (R is selected randomly and it is between 1 and K). In other words, in the destroy process the lowest rank features (K percentage) would be removed and in the repair process the highest ranked features (R percentage) from the destroyed set would be considered as selected features. Given a current solution x, we define the functionality of destroy and repair methods by $des()$ and $rep()$ respectively. As mentioned before, function $des()$ selects K percentage of the lowest ranked selected features (ones) from x, using the CFS ranking selection and move it to xd. Function $rep()$ selects R percentage of the highest ranked features from destroyed set, using the CFS ranking selection and inserts them.

3.3   Objective Function

The objective function is defined by the classification accuracy, i.e.

$$ObjF(x) = Accuracy(Sx) \tag{1}$$

where Sx denotes the corresponding selected feature subset encoded in the current solution x, and the feature selection criterion function $Accuracy(Sx)$ evaluates the significance for the given feature subset Sx. As we considered paper [3] as the baseline for this work, so to make our results comparable with their results, we needed to use the same classifiers as our objective functions: a tree based classification (C4.5) [46] and the probabilistic Bayesian classifier [47] with naive independence assumptions.

3.4   Acceptance Method

We propose a new acceptance method for WFLNS, which inspired from acceptance probably of Simulated Annealing algorithm. In our proposed method xt is always accepted as xb if $ObjF(xt) > ObjF(xb)$, and accepted with probability $e^{-(ObjF(xb)-ObjF(xt))/T}$ if $ObjF(xb) > ObjF(xt)$. Here we set the initial temperature to $T = 1$. By choosing this temperature non improving solutions would be allowed to be accepted. Within the search progress, T decreases and towards the end of the search only a few or non improving solutions would be accepted. It is worth noting that based on the original LNS in [40], the acceptance method evaluates solutions of the neighborhood and then allows the solution with the highest objective function value to be considered as a current solution if and only if it was improved. The particularity of the proposed acceptance method against its original counterpart is that it allows WFLNS move to a new solution, which might make the objective function worse in the hope of not trapping in to local optimal solutions.

4 Experimental Results and Discussion

We considered the paper [3] as a baseline for conducting and comparing our experiments because to the best of our knowledge this paper has been the most recent comparison of existing wrapper methods. We used the same classifiers, Naive Bayesian (NB) and C4.5, with the same setting to measure the classification accuracy. Then, we employed the same twelve real value data sets (Table 1) with [3], which were originally chosen from the UCI repository [4]. Data sets are both high and low dimension to present reasonable challenges to the proposed algorithm. Finally, we compared the performance of the WFLNS with all state of the art employed metaheuristic algorithms in [3], such as Swarm Intelligence algorithms, Evolutionary algorithms and Local Search algorithms. Table 1 illustrates the accuracy on all data sets for both employed classifiers with no feature selection. For the sake of performance analysis, we categorized these data sets into three main categories: small size data sets (number of features below 20), medium size data sets (number of features between 20 and 100) and large size data sets (number of features more than 100). The small sized data sets include

Table 1. Data set information

Data set	Feature	Instance	Class	C4.5 (%)	NB (%)
Heart	14	270	2	77.56	84.00
Cleveland	14	297	5	51.89	55.36
Ionosphere	35	230	2	86.22	83.57
Water	39	390	3	81.08	85.40
Waveform	41	699	2	75.49	79.99
Sonar	60	208	2	73.59	67.85
Ozone	73	2,534	2	92.70	67.66
Libras	91	360	15	68.24	63.635
Arrhythmia	280	452	16	65.97	61.40
Handwritten	257	1,593	10	75.74	86.21
Secom	591	1,567	2	89.56	30.04
Mutifeat	650	2000	10	94.54	95.30

Cleveland and Hearth. The next six data sets include Ionosphere, Water, Waveform, Ozone, Sonar and Libras are categorized as medium sized. The last four data sets Libras, Arrhythmia, Handwritten, Secom and multifeat are considered as large size data sets. The obtained results have been discussed in terms of classification accuracy, destruction degree and acceptance criteria. For each data set, the final accuracy was obtained by averaging out the accuracy of WFLNS over five independent runs with an identical initial solution. Also, the number of iterations and initial temperature (T) were set to 500 and 1 respectively.

4.1 Classification Accuracy

As Table 2 shows, for both classifiers the proposed method achieved the highest accuracy in comparison with other algorithms in most data sets. WFLNS achieved the highest accuracy for seven and eight out of twelve data sets for C4.5 and NB respectively. More specifically, in case of C4.5 classifier: for small size data set WFLNS achieved the highest accuracy among all other algorithms. For medium size data set (Ionosphere, Water, Waveform, Sonar and Ozone), the best subset was found by WFLNS for both Ionosphere and Sonar with 89.9 and 76.2 respectively while for Water data set GA achieved the highest accuracy with 83.4 and for Waveform data set CSA and SA achieved the same highest result with 77.6. In case of large data set, WFLNS reached the best classification accuracy for Arrhythmia data set with 68.1. For Secom and Multifeat data set, ABC and SA gain the highest accuracy with 93.4 and 95.1 respectively.

Table 2. C4.5 (left) and NB (right) classification accuracies

Data set	ABC		ACO		CSA		FF		GA		HS		MA		PSO		SA		TS		WFLNS	
Heart	81.7	83.0	81.2	84.1	81.7	83.0	82.2	83.0	80.6	85.0	80.6	85.0	80.6	85.0	80.7	85.0	82.1	82.3	81.7	82.9	83.2	85.5
Cleveland	55.8	56.3	55.3	56.5	55.8	56.3	55.8	56.6	56.4	56.9	56.4	56.9	55.0	55.7	56.4	56.9	55.0	55.7	56.3	57.0	57.8	55.3
Ionosphere	82.2	86.6	87.6	86.3	87.8	86.8	88.4	86.9	88.4	88.9	88.3	88.9	88.3	86.9	88.0	86.4	87.4	87.0	87.3	87.0	89.9	88.3
Water	81.9	84.9	82.7	85.8	82.8	85.9	83.4	85.9	83.4	85.9	83.3	85.9	83.3	85.9	82.1	85.2	83.0	85.6	82.9	85.6	81.4	85.9
Waveform	76.9	79.8	77.4	80.5	77.6	80.7	76.8	79.5	75.5	80.2	77.5	80.2	77.5	80.2	76.9	79.7	77.6	86.9	77.5	80.5	76.6	82.0
Sonar	72.3	66.6	73.2	66.3	73.0	66.6	73.4	65.9	73.1	66.6	73.2	66.6	73.3	66.5	72.8	66.3	72.3	66.6	74.1	66.9	76.2	67.3
Ozone	93.4	75.8	93.4	77.2	93.1	74.8	93.3	76.1	93.3	73.9	93.2	73.9	93.4	73.7	93.3	73.5	93.4	78.4	93.1	74.0	95.1	79.2
Libras	65.6	60.7	62.1	57.3	67.0	61.6	65.5	61.0	67.6	62.1	67.3	61.6	68.2	61.8	66.8	61.4	65.9	61.4	66.9	61.3	68.3	63.0
Arrhythmia	63.0	63.2	66.8	67.0	67.1	68.5	66.8	67.0	66.8	66.8	66.9	68.9	66.9	67.4	63.5	63.3	67.4	69.0	67.2	69.0	68.1	70.2
Handwritten	75.0	83.7	70.2	72.9	75.9	84.7	74.3	82.0	75.5	85.2	75.9	85.3	75.4	85.5	75.3	85.3	76.0	83.8	78.1	84.9	77.1	86.1
Secom	93.4	88.7	92.5	82.6	92.5	84.2	92.7	75.1	90.7	84.1	92.1	71.2	91.2	88.7	92.7	74.2	92.5	84.7	92.4	83.7	89.5	86.7
Mutifeat	94.3	95.7	93.4	95.2	94.9	97.1	92.8	95.7	94.6	96.3	94.9	96.8	94.6	95.8	94.7	95.9	95.1	97.2	94.9	97.2	92.8	95.8

In case of NB classifier, WFLNS achieved the highest accuracy for Heart data set by 85.5 while for Cleveland data set, TS achieved the highest accuracy with 57.0. For all medium size data sets, the proposed algorithm outperforms other algorithms by achieving 85.9, 82.0, 67.3, 79.2 and 63 for Ionosphere, Water, Waveform, Sonar, Ozone and Libras data sets respectively. For large size data sets, apart from Secom and Multifeat data sets, WFLNS gain the highest accuracy for Arrhythmia and Handwritten with 70.2 and 86.1 classification accuracy. Moreover, Table 3 shows the number of selected features along with the selected features by WFLNS for medium size data sets, where $d(C4.5)$ and $d(NB)$ represent the number of selected features by C4.5 and NB classifiers respectively.

Table 3. Selected features by WFLNS

Data Set	d(C4.5)	d(NB)	Selected Features by C4.5	Selected Features by NB
Waveform	13	17	{2,4,6,7,10,11,13,15,17,18, 19,20,21}	{1,2,3,4,5,6,7,8,10,11, 12,13,14,18,19,20,21}
Sonar	32	51	{1,2,3,4,6,7,8,9,14,17,18, 19,21,22,23,24,26,28,29,34,37,38, 40,41,42,45,46,47,49,52,55,58}	{1,3,4,5,6,7,8,9,10,11,13,14,15,16, 17,18,19,22,23,24,25,26,27,28, 29,30,31,32,33,34,35,36,37,38,39, 40,41,42,43,45,48,50,51,53,54,55, 56,57,58,59,60}
Ozone	43	40	{1,4,5,6,7,8,9,12,16,17, 18,19,21,22,23,24,25,27,28,30, 32,33,37,38,39,40,41,44,45,46, 47,48,52,54,55,56,57,58,61,64, 68,72,73}	{1,2,5,7,9,10,11,12,13,15, 17,18,19,20,22,23,30,31,32,37, 38,40,41,42,45,48,49,52,55,56, 57,58,60,62,65,66,67,70,71,72}
Libras	48	71	{1,3,4,7,8,15,16,17,20,23, 24,25,26,29,32,33,34,35,37,39, 40,41,46,47,48,50,55,56,57,59, 60,61,63,64,65,67,70,71,73,76, 77,80,83,84,85,86,88,90}	{1,2,3,4,5,6,7,8,9,11, 12,13,14,15,17,18,19,20,21,22, 28,29,30,32,33,35,36,37,38,39, 40,41,42,43,45,46,47,48,49,50, 51,52,53,54,55,57,59,60,61,63, 65,66,67,68,69,71,72,73,74,75, 79,81,82,83,84,85,86,87,88,89, 90}

4.2 Effect of Destruction Degree Parameter

Choosing an appropriate degree of destruction(K), for destroy method has an impact on the quality of the search process and consequently on the quality of classification accuracy. We chose C4.5 classifier along with Ionosphere, Sonar and Libras data sets as a representative of small size, medium size and large size data set respectively to investigate the effect of destruction degree parameter on the quality of the classification accuracy.

As illustrates in Fig. 1, the best classification accuracy were achieved by 0.5, 0.6 and 0.6 destruction degree in Ionosphere, Sonar and Libras data sets respectively. So, we can conclude that selecting either small or too large destruction degree lead to undesirable effects because if the small percentage of the solution is destructed then the effect of a large neighborhood search would be lost in WFLNS as it explores the smaller part of the search space and subsequently it failed to achieve the highest classification accuracy. On the other hand, if the large percentage of the solution is selected to be destructed then WFLNS turns in to random search.

4.3 Effect of Acceptance Criteria

The acceptance criterion has an important role in both diversification and intensification of the search process of WFLNS. Our proposed acceptance method for

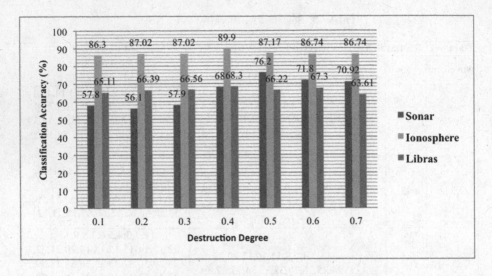

Fig. 1. Effect of destruction degree on C4.5 classifier

WFLNS, inspired from the acceptance probability in Simulated Annealing algorithm, in which even non-improving temporary solutions would have the chance to be considered as an accepted solution in a hope of finding better solution in following iterations. Based on the acceptance method in LNS algorithm, only the best improving temporary solution would be considered as an accepted solution (best solution) for each iteration.

Table 4. Effect of acceptance criteria

Data set	IS-C4.5	SA-C4.5	IS-NB	SA-NB
Heart	**83.2**	**83.2**	**85.5**	**85.5**
Ionosphere	88.8	**89.9**	**89.2**	88.3
Sonar	74.1	**76.2**	66.5	**67.3**
Libras	66.2	**68.3**	62.2	**63.0**
Arrhythmia	67.3	**68.1**	68.6	**70.2**

To evaluate the efficiency of the proposed acceptance method, we compare the results of both methods with one another based on both classification accuracy and on different data sets size such as: Heart, Ionosphere, Sonar, Libras and Arrhythmia data sets. Two acceptance methods in Table 4 refer to as IS (Improvement strategy) and SA, which represent the LNS and WFLNS acceptance method respectively. Our experiments show that for C4.5 classifier, the proposed acceptance method achieved better results than IS, apart from the Heart data set which both methods achieved the same result. For NB classifier, the proposed method proved its superiority over IS with one exception in Ionosphere data set that achieved 89.2 by IS.

5   Conclusion

In this paper, we presented the first study on applying LNS on the problem of feature selection by proposing a new hybrid algorithm called WFLNS. The core idea of WFLNS is based on designing a problem-specific destroy, repair and acceptance methods for the LNS algorithm to deal with the problem of feature selection. We incorporated the idea of filter ranking method into the destroy and repair methods, in which the algorithm was guided into the most promising part of the search space by adding and removing proper features iteratively. Furthermore, we introduced a new acceptance method for our WFLNS, which is inspired from the Simulated Annealing acceptance probability. The particularity of the proposed acceptance method is to let both improving and non-improving solutions be considered as the best found solution. The performance of the proposed algorithm is evaluated based on C4.5 and NB classifiers and twelve real value data sets used to test the algorithms based on the paper [3], which were originally chosen from the UCI repository. Experimental results show the proposed algorithm outperforms other metaheuristic search algorithms in most data sets.

References

1. Guyon, I., Elisseeff, A.: An introduction to variable and feature selection. J. Mach. Learn. Res. **3**, 1157–1182 (2003)
2. Hall, M.A.: Correlation-based feature selection of discrete and numeric class machine learning. In: Proceedings of 17th International Conference on Machine Learning (2000)
3. Diao, R., Shen, Q.: Nature inspired feature selection meta-heuristics. Artif. Intell. Rev. **44**(3), 311–340 (2015)
4. Lichman, M.: UCI machine learning repository (2013)
5. Amaldi, E., Kann, V.: On the approximability of minimizing nonzero variables or unsatisfied relations in linear systems. Theoret. Comput. Sci. **209**(1), 237–260 (1998)
6. Glover, F.: Tabu search-part I. ORSA J. Comput. **1**(3), 190–206 (1989)
7. Glover, F.: Tabu search-part II. ORSA J. Comput. **2**(1), 4–32 (1990)
8. Brooks, S.P., Morgan, B.J.: Optimization using simulated annealing. Statistician **44**, 241–257 (1995)
9. Geem, Z.W., Kim, J.H., Loganathan, G.: A new heuristic optimization algorithm: harmony search. Simulation **76**(2), 60–68 (2001)
10. Wang, Y., Li, L., Ni, J., Huang, S.: Feature selection using tabu search with long-term memories and probabilistic neural networks. Pattern Recogn. Lett. **30**(7), 661–670 (2009)
11. Wang, Y., Mo, J.: Emotion feature selection from physiological signals using Tabu search. In: 2013 25th Chinese Control and Decision Conference (CCDC) (2013)
12. Lin, S.W., Lee, Z.J., Chen, S.C., Tseng, T.Y.: Parameter determination of support vector machine and feature selection using simulated annealing approach. Appl. Soft Comput. **8**(4), 1505–1512 (2008)

13. Meiri, R., Zahavi, J.: Using simulated annealing to optimize the feature selection problem in marketing applications. Eur. J. Oper. Res. **171**(3), 842–858 (2006)
14. Miller, A.: Subset Selection in Regression. CRC Press, Boca Raton (2002)
15. Gheyas, I.A., Smith, L.S.: Feature subset selection in large dimensionality domains. Pattern Recogn. **43**(1), 5–13 (2010)
16. Wang, Y., Liu, Y., Feng, L., Zhu, X.: Novel feature selection method based on harmony search for email classification. Knowl.-Based Syst. **73**, 311–323 (2015)
17. Shreem, S.S., Abdullah, S., Nazri, M.Z.A.: Hybrid feature selection algorithm using symmetrical uncertainty and a harmony search algorithm. Int. J. Syst. Sci. **47**(6), 1312–1329 (2016)
18. Whitley, D.: A genetic algorithm tutorial. Stat. Comput. **4**(2), 65–85 (1994)
19. Moscato, P., et al.: On evolution, search, optimization, genetic algorithms and martial arts: towards memetic algorithms. Caltech Concurrent Computation Program, C3P Report **826** (1989)
20. Gao, X.Z.: Artificial immune systems and their applications. In: NICSO 2006, p. 7 (2006)
21. Eberhart, R.C., Kennedy, J., et al.: A new optimizer using particle swarm theory. In: Proceedings of 6th International Symposium on Micro Machine and Human Science, New York, NY, vol. 1, pp. 39–43 (1995)
22. Colorni, A., Dorigo, M., Maniezzo, V., et al.: Distributed optimization by ant colonies. In: Proceedings of 1st European Conference on Artificial Life, Paris, France, vol. 142, pp. 134–142 (1991)
23. Karaboga, D.: An idea based on honey bee swarm for numerical optimization. Technical report, tr06, Engineering Faculty, Computer Engineering Department, Erciyes University (2005)
24. Oh, I.S., Lee, J.S., Moon, B.R.: Hybrid genetic algorithms for feature selection. IEEE Trans. Pattern Anal. Mach. Intell. **26**(11), 1424–1437 (2004)
25. Ghamisi, P., Benediktsson, J.A.: Feature selection based on hybridization of genetic algorithm and particle swarm optimization. IEEE Geosci. Remote Sens. Lett. **12**(2), 309–313 (2015)
26. Lee, J., Kim, D.W.: Memetic feature selection algorithm for multi-label classification. Inf. Sci. **293**, 80–96 (2015)
27. Zhu, Z., Ong, Y.S., Dash, M.: Wrapper-filter feature selection algorithm using a memetic framework. IEEE Trans. Syst. Man Cybern. Part B: Cybern. **37**(1), 70–76 (2007)
28. Shojaie, S., Moradi, M.: An evolutionary artificial immune system for feature selection and parameters optimization of support vector machines for ERP assessment in a p300-based GKT. In: Cairo International Biomedical Engineering Conference, CIBEC 2008, pp. 1–5. IEEE (2008)
29. Kashef, S., Nezamabadi-pour, H.: An advanced ACO algorithm for feature subset selection. Neurocomputing **147**, 271–279 (2015)
30. Forsati, R., Moayedikia, A., Jensen, R., Shamsfard, M., Meybodi, M.R.: Enriched ant colony optimization and its application in feature selection. Neurocomputing **142**, 354–371 (2014)
31. Wang, X., Yang, J., Teng, X., Xia, W., Jensen, R.: Feature selection based on rough sets and particle swarm optimization. Pattern Recogn. Lett. **28**(4), 459–471 (2007)
32. Hancer, E., Xue, B., Zhang, M., Karaboga, D., Akay, B.: A multi-objective artificial bee colony approach to feature selection using fuzzy mutual information. In: IEEE Congress on Evolutionary Computation (CEC), pp. 2420–2427. IEEE (2015)

33. Chuang, L.Y., Yang, C.H., Yang, C.H., et al.: IG-GA: a hybrid filter/wrapper method for feature selection of microarray data. J. Med. Biol. Eng. **30**(1), 23–28 (2010)
34. Hu, Z., Bao, Y., Xiong, T., Chiong, R.: Hybrid filter-wrapper feature selection for short-term load forecasting. Eng. Appl. Artif. Intell. **40**, 17–27 (2015)
35. Fahy, C., Ahmadi, S., Casey, A.: A comparative analysis of ranking methods in a hybrid filter-wrapper model for feature selection in DNA microarrays. In: Bramer, M., Petridis, M. (eds.) Research and Development in Intelligent Systems XXXII, pp. 387–392. Springer, Cham (2015)
36. Swiniarski, R.W., Skowron, A.: Rough set methods in feature selection and recognition. Pattern Recogn. Lett. **24**, 833–849 (2003)
37. Bermejo, P., Gámez, J.A., Puerta, J.M.: A grasp algorithm for fast hybrid (filter-wrapper) feature subset selection in high-dimensional datasets. Pattern Recogn. Lett. **32**(5), 701–711 (2011)
38. Bertolazzi, P., Felici, G., Festa, P., Lancia, G.: Logic classification and feature selection for biomedical data. Comput. Math. Appl. **55**(5), 889–899 (2008)
39. Bertolazzi, P., Felici, G., Festa, P., Fiscon, G., Weitschek, E.: Integer programming models for feature selection: new extensions and a randomized solution algorithm. Eur. J. Oper. Res. **250**(2), 389–399 (2016)
40. Shaw, P.: Using constraint programming and local search methods to solve vehicle routing problems. In: Maher, M., Puget, J.-F. (eds.) CP 1998. LNCS, vol. 1520, pp. 417–431. Springer, Heidelberg (1998). doi:10.1007/3-540-49481-2_30
41. Pisinger, D., Ropke, S.: Large neighborhood search. In: Gendreau, M., Potvin, J.-Y. (eds.) Handbook of Metaheuristics, vol. 146, pp. 399–419. Springer, Berlin (2010)
42. Lin, S., Kernighan, B.W.: An effective heuristic algorithm for the traveling-salesman problem. Oper. Res. **21**(2), 498–516 (1973)
43. Demirovic, E., Musliu, N.: Maxsat based large neighborhood search for high school timetabling. Computers Oper. Res. **78**, 172–180 (2017)
44. Meyers, C., Orlin, J.B.: Very large-scale neighborhood search techniques in timetabling problems. In: Burke, E.K., Rudová, H. (eds.) PATAT 2006. LNCS, vol. 3867, pp. 24–39. Springer, Heidelberg (2007). doi:10.1007/978-3-540-77345-0_2
45. Hall, M.A.: Correlation-based feature selection for machine learning. Ph.D. thesis, The University of Waikato (1999)
46. Quinlan, J.R.: C4.5: Programs for Machine Learning. Elsevier, Amsterdam (2014)
47. John, G.H., Langley, P.: Estimating continuous distributions in Bayesian classifiers. In: Proceedings of 11th Conference on Uncertainty in Artificial Intelligence, pp. 338–345. Morgan Kaufmann Publishers Inc. (1995)

A Memetic Algorithm to Maximise the Employee Substitutability in Personnel Shift Scheduling

Jonas Ingels and Broos Maenhout[✉]

Faculty of Economics and Business Administration, Ghent University,
Tweekerkenstraat 2, 9000 Gent, Belgium
{jonas.ingels,broos.maenhout}@ugent.be

Abstract. Personnel rosters are typically constructed for a medium-term period under the assumption of a deterministic operating environment. However, organisations usually operate in a stochastic environment and are confronted with unexpected events in the short term. These unexpected events affect the workability of the personnel roster and need to be resolved efficiently and effectively. To facilitate this short-term recovery, it is important to consider robustness by adopting proactive scheduling strategies during the roster construction. In this paper, we discuss a proactive strategy that maximises the employee substitutability value in a personnel shift scheduling context. We propose a problem-specific population-based approach with local and evolutionary search heuristics to solve the resulting non-linear personnel shift scheduling problem and obtain a medium-term personnel shift roster with a maximised employee substitutability value. Detailed computational experiments are presented to validate the design of our heuristic procedure and the selection of the heuristic operators.

Keywords: Personnel shift scheduling · Employee substitutability · Memetic algorithm

1 Introduction

The personnel management and planning process is widely studied in the operational research literature [1–4]. This process generally consists of three hierarchical phases where the higher phases constrain the lower phases in terms of decision freedom. We distinguish the strategic staffing phase, the tactical scheduling phase and the operational allocation phase [1,5]. In the staffing phase, the personnel mix and budget required to meet the service demand in the long term is determined. In this phase, the personnel characteristics, including the competencies and degree of employment, are determined. During the tactical scheduling phase, a personnel roster is constructed for a medium-term period based on predictions and/or assumptions about the service demand and employee availability. However, the actual service demand and employee availability may be different in

© Springer International Publishing AG 2017
B. Hu and M. López-Ibáñez (Eds.): EvoCOP 2017, LNCS 10197, pp. 44–59, 2017.
DOI: 10.1007/978-3-319-55453-2_4

the short-term operational allocation phase [4], which may require unexpected adjustments to restore the workability of the original roster.

A proactive approach builds in a certain degree of robustness in the original roster constructed during the tactical scheduling phase to protect and facilitate its workability in the operational allocation phase. This built-in robustness improves the absorption and/or adjustment capability of the original roster to achieve stability and/or flexibility. A stable roster has a high absorption capability and therefore requires a small number of changes when operational disruptions arise [6]. A flexible roster has a high adjustment capability such that an appropriate number of possibilities for schedule changes are available to recover from unexpected events [7].

In general, substitutability provides an indication of robustness in crew and aircraft scheduling [6]. Indeed, a crew and/or aircraft substitution enables the recovery of operational disruptions [7–11]. In this respect, Shebalov and Klabjan [12] proactively maximise the number of move-up crews, i.e. crews that can be substituted to overcome operational disruptions. Ionescu and Kliewer [7] also proactively improve the flexibility of a roster by introducing substitution possibilities in the scheduling phase. In contrast to Shebalov and Klabjan [12], the authors prioritise the utility of substitutions rather than the number of substitutions.

In this paper, we investigate the personnel shift scheduling problem, which entails the assignment of employees to cover the staffing requirements for specific skills and shifts for a medium-term period. We focus on proactively improving the short-term adjustment capability of the tactical personnel shift roster by maximising the employee substitutability. In this respect, we consider a non-linear optimisation problem that is solved with a problem-specific memetic algorithm, which comprises a population-based approach with local and evolutionary search heuristics, and provides a personnel shift roster for a medium-term period with a maximised value of personnel substitution possibilities.

The remainder of this paper is organised as follows. In Sect. 2, we define the tactical scheduling phase and describe different types of employee substitution possibilities. We formulate the personnel shift scheduling problem to maximise the value of these substitution possibilities, i.e. the employee substitutability value. In Sect. 3, we define and explain the building blocks of our algorithm to obtain a medium-term personnel shift roster. The test design, test instances and computational experiments are discussed in Sect. 4. We provide conclusions and directions for future research in Sect. 5.

2   Problem Definition and Formulation

In this paper, we study a personnel shift scheduling problem with common personnel and shift characteristics [1,4]. This problem encompasses the assignment of employees, who possess one or multiple categorical skills, to shifts during the tactical scheduling phase. Categorical skills cannot be hierarchically ranked and each employee is either perfectly capable to work shifts requiring an individual skill or not at all [13].

We extend this problem to consider employee substitutability. More specifically, we formulate a bi-criteria objective function to simultaneously minimise the costs and maximise the value of substitution possibilities. A substitution actually arises when an employee can be reassigned to take over the working assignment of another employee on the same day. In this respect, we distinguish three substitution types, i.e.

- A *between-skill substitution* comprises the possible reassignment of a working employee to another skill during another or the same shift on the same day.
- A *within-skill substitution* indicates the potential to reassign an employee from one shift to another shift within the same skill category.
- A *day-off-to-work substitution* is the potential conversion of a day off assignment to a working assignment.

This definition of employee substitutability leads to a non-linear problem, which we linearise in the following mathematical model, i.e.

Notation
Sets

G	set of skills (index m)
N	set of employees (index i)
D	set of days (index d)
S	set of shifts (index j)
T'_{dj}	set of shifts that cannot be assigned the day before day d and shift assignment j (index s)
T''_{dj}	set of shifts that cannot be assigned the day after day d and shift assignment j (index s)

Parameters

M	a large number
b_{im}	1 if employee i possesses skill m, 0 otherwise
c^w_{imdj}	wage cost of assigning an employee i to skill m and shift j on day d
c^{wu}_{mdj}	cost of understaffing shift j on day d for skill m
p_{idj}	preference penalty cost if an employee i receives a shift assignment j on day d
γ_{imdj}	the benefit value of a substitution possibility to employee i with a shift assignment j for skill m on day d
R^w_{mdj}	staffing requirements on day d for shift j and skill m
$\eta^{w,min}_i$	minimum number of assignments for employee i
$\eta^{w,max}_i$	maximum number of assignments for employee i

Variables

x^w_{imdj}	1 if employee i receives a shift assignment j for skill m on day d, 0 otherwise
x^v_{id}	1 if employee i receives a day off on day d, 0 otherwise
x^{wu}_{mdj}	the number of employees short on day d for shift j and skill m

f_{idj} 1 if a minimum rest violation would arise if employee i is reassigned to work on day d during shift j, 0 otherwise

χ_{imdj} the number of substitution possibilities to employee i with a shift assignment j for skill m on day d

Mathematical Formulation

$$\min \sum_{i \in N} \sum_{m \in G} \sum_{d \in D} \sum_{j \in S} (c^w_{imdj} + p_{idj}) x^w_{imdj} + \sum_{m \in G} \sum_{d \in D} \sum_{j \in S} c^{wu}_{mdj} x^{wu}_{mdj} \qquad (1a)$$

$$- \sum_{i \in N} \sum_{m \in G} \sum_{d \in D} \sum_{j \in S} \gamma_{imdj} \chi_{imdj} \qquad (1b)$$

In this study, we minimise the personnel assignment costs, i.e. the wage and preference penalty cost, and the cost of understaffing (Eq. (1a)). Simultaneously, we maximise the value of substitution possibilities (Eq. (1b)).

$$\sum_{i \in N} b_{im} x^w_{imdj} + x^{wu}_{mdj} \geq R^w_{mdj} \qquad \forall m \in G, \forall d \in D, \forall j \in S \quad (2)$$

$$\sum_{m \in G} \sum_{j \in S} x^w_{imdj} + x^v_{id} = 1 \qquad \forall i \in N, \forall d \in D \qquad (3)$$

$$\sum_{m \in G} x^w_{imdj} + \sum_{m \in G} \sum_{s \in T''_{dj}} x^w_{im(d+1)s} \leq 1 \qquad \forall i \in N, \forall d \in D, \forall j \in S \qquad (4)$$

$$\sum_{m \in G} \sum_{d \in D} \sum_{j \in S} x^w_{imdj} \leq \eta^{w,max}_i \qquad \forall i \in N \qquad (5)$$

$$\sum_{m \in G} \sum_{d \in D} \sum_{j \in S} x^w_{imdj} \geq \eta^{w,min}_i \qquad \forall i \in N \qquad (6)$$

The number of employees required to satisfy the personnel demand for every skill category, day and shift is defined in Eq. (2), which is relaxed by allowing understaffing.

The time-related constraints require an employee to either receive a shift assignment or a day off (Eq. (3)), and a minimum rest period between consecutive shift assignments (Eq. (4)). The other time-related constraints include the maximum (Eq. (5)) and minimum (Eq. (6)) number of assignments for every employee.

$$\chi_{imdj} \leq MR^w_{mdj} x^w_{imdj} \qquad \forall i \in N, \forall m \in G, \forall d \in D, \forall j \in S \qquad (7)$$

A substitution possibility to an employee, shift and skill category on a day can only exist if that employee has an assignment during that day, for this shift and in that skill category. Moreover, a substitution possibility is only valuable if employees are required, i.e. $R^w_{mdj} > 0$ (Eq. (7)).

$$\chi_{imdj} \le \sum_{i_1 \in N \setminus \{i\}} \left(\sum_{m_1 \in G \setminus \{m\}} \sum_{j_1 \in S} b_{i_1 m} x^w_{i_1 m_1 dj_1} + \sum_{j_1 \in S \setminus \{j\}} x^w_{i_1 mdj_1} + b_{i_1 m} x^v_{i_1 d} \right)$$

(8a)

$$- \sum_{i_1 \in N \setminus \{i\}} b_{i_1 m} f_{i_1 dj} \qquad \forall i \in N, \forall m \in G, \forall d \in D, \forall j \in S$$

(8b)

$$M f_{idj} \ge \sum_{m \in G} \left(\sum_{s \in T''_{dj}} x^w_{im(d+1)s} + \sum_{s \in T'_{dj}} x^w_{im(d-1)s} \right) \qquad \forall i \in N, \forall d \in D, \forall j \in S$$

(8c)

In Eq. (8), we count the number of substitution possibilities to an employee i that works a shift j in skill category m on day d. The *between-skill substitutions, within-skill substitutions* and *day-off-to-work substitutions* are defined in Eq. (8a). Note that we define a correction variable f_{idj} to exclude substitution possibilities that would violate the minimum rest period (Eq. (8b)). This variable is defined in Eq. (8c).

$$
\begin{aligned}
x^w_{imdj} &\in \{0,1\} & \forall i \in N, \forall m \in G, \forall d \in D, \forall j \in S \\
x^v_{id} &\in \{0,1\} & \forall i \in N, \forall d \in D \\
x^{wu}_{mdj} &\ge 0 \, and \, integer & \forall m \in G, \forall d \in D, \forall j \in S \\
f_{idj} &\in \{0,1\} & \forall i \in N, \forall d \in D, \forall j \in S \\
\chi_{imdj} &\ge 0 \, and \, integer & \forall i \in N, \forall m \in G, \forall d \in D, \forall j \in S
\end{aligned}
$$

(9)

Constraints (9) embody the integrality conditions.

3   A Memetic Algorithm to Maximise the Employee Substitutability

We propose a memetic algorithm to obtain personnel shift rosters for a medium-term period that have a maximised employee substitutability value, i.e. we solve model (1)–(9). Figure 1 illustrates the building blocks and flowchart of this algorithm.

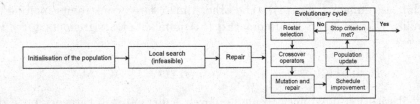

Fig. 1. Algorithm to obtain medium-term personnel shift rosters

First, we initialise the optimisation procedure by constructing initial personnel shift rosters for a simplified model (Sect. 3.1). Second, we improve the personnel shift rosters in the population by applying local search heuristics in order to increase the employee substitutability value in a cost-efficient way (Sect. 3.2). The local search heuristics include the selection of an employee-based and/or day-based guiding order and a neighbourhood structure. Note that we allow infeasibilities in terms of Eqs. (4)–(6), which we repair in a third step (Sect. 3.3). Fourth, we apply an evolutionary cycle in a repetitive manner in order to improve the overall objective function value (Sect. 3.4). We repeat this evolutionary cycle until a stop criterion is reached, i.e. the number of iterations without improvement. Thus, whenever a number of cycles is executed without an improvement in the objective function value of the best found solution (Eq. (1)), the algorithm is terminated.

3.1   Population Initialisation

In the initialisation phase, we construct a population of initial personnel shift rosters using the commercial optimisation software Gurobi [14]. The population consists of n_{pop} initial personnel shift rosters and we relax the maximisation of employee substitutability by omitting the objective (Eq. (1b)) and corresponding constraints (Eqs. (7) and (8)) during the construction. In this phase, we construct personnel shift rosters that minimise the costs (Eq. (1a)) and satisfy the staffing requirements (Eq. (2)), all time-related constraints (Eqs. (3)–(6)) and the integrality conditions (Eq. (9)).

We consider different population sizes and construct a diverse set of personnel shift rosters by randomising the personnel assignment costs, i.e. the wage and preference penalty cost.

3.2   Local Search (LS)

We select an initial personnel shift roster from the population and employ local search heuristics to improve the objective function value (Eq. (1)). These heuristics include the determination of a single search guiding order to manage the search process and the selection of a single neighbourhood structure to characterise the neighbourhood move. We reiterate this process for every personnel shift roster in the population.

Neighbourhood Search Guiding Order. The guiding order determines the search process, i.e. the sequence in which parts of the search area are revised during the local search [15–17]. This sequence is typically based on the objective function improvement potential [18]. In this paper, we calculate this potential for each cell in the personnel shift roster. Hence, we reconsider every daily assignment for each employee and evaluate whether a change in this assignment can improve the objective function value (Eq. (1)). In this respect, the objective function improvement potential can be utilised in two problem-specific guiding orders, i.e. the employee-based and day-based guiding orders.

Employee-based Guiding Orders (EGO). The employee-based guiding orders determine the order in which the employee schedules are revisited. Since skills can have a significant impact on the number of substitution possibilities, we divide employees into groups based on the number of skills they possess. Note that we do not make a distinction based on the type of skills because we consider categorical skills. The skill groups can be ordered according to two types, i.e.

- The *skill-based group order* either orders these groups according to a decreasing or an increasing number of skills. These orders are denoted as *EGO1-1* and *EGO1-2*, respectively.
- The *random group order* determines the sequence of the employee groups randomly (*EGO1-3*).

The employees within these groups can also be ordered, i.e.

- A first order arranges employees based on their objective function improvement potential (Eq. (1)). We define the *decreasing employee potential order* (*EGO2-1*) and the *increasing employee potential order* (*EGO2-2*).
- The *random employee order* (*EGO2-3*) randomly determines the sequence of the employees within each group.

Day-based Guiding Orders (DGO). This order determines the sequence in which the days in the planning horizon are revisited. We distinguish three types, i.e.

- The objective function improvement potential defines a *decreasing day potential order* (*DGO1*) and an *increasing day potential order* (*DGO2*).
- The days can also be searched in a *random day order* (*DGO3*).

Note that these guiding orders can be applied separately or combined. Therefore, we establish a horizontal, vertical or no guiding priority to reveal if priority is given to the employee-based guiding order, the day-based guiding order or a combination of both. As such, the guiding priority and orders determine the next search area in the personnel shift roster (Fig. 2).

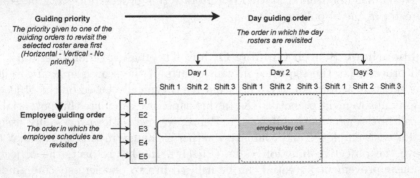

Fig. 2. Guiding priority and order

Neighbourhood Structure. The neighbourhood structure determines the characteristics of a neighbourhood move [15,16,19–22]. In this study, we propose a problem-specific *cell-by-cell search* that focuses on a single employee and day, i.e. one cell of the personnel shift roster. We distinguish three different types of search mechanisms based on the guiding priority (cf. Fig. 2), i.e.

- The *horizontal cell-by-cell search (N1)* investigates the complete schedule of an employee on a day-by-day basis before the schedule of another employee is explored, i.e. the guiding priority is horizontal. This mechanism first selects an employee schedule according to the chosen employee-based guiding order, after which the day-based guiding order determines the sequence in which the daily assignments are reconsidered.
- The *vertical cell-by-cell search (N2)* considers a complete day roster on an employee-by-employee basis and then moves to another day roster, i.e. the guiding priority is vertical. A day roster is selected according to the chosen day-based guiding order after which the individual employee assignments are reconsidered based on the employee-based guiding order.
- *Combined cell-by-cell search (N3)* visits cells in a consecutive order based on a combination of the employee-based and day-based guiding orders as no guiding priority is imposed. This means that the consecutively visited cells are not necessarily located next to each other.

Note that we investigated other neighbourhood structures such as a horizontal schedule and vertical roster search, which reoptimise a complete employee schedule and day roster, respectively. However, these structures did not provide significant benefits in the objective function value (Eq. (1)) and required more CPU-time.

3.3 Repair (R)

In order to ensure the satisfaction of the time-related constraints (Eqs. (4)–(6)), we apply a destroy and repair neighbourhood [20]. First, we check the minimum rest constraint (Eq. 4) and the maximum number of working assignments (Eq. (5)), and delete all assignments that violate these constraints. Second, we ensure that every employee receives their minimum imposed number of working assignments (Eq. (6)) in a greedy manner. Therefore, we maximise the improvement of changing a day off assignment to a working assignment in terms of the objective function value (Eq. (1)). Finally, we add assignments to employee schedules until it is no longer feasible or it is no longer possible to improve the objective function value (Eq. (1)).

After the repair of the time-related constraints, there may be some residual understaffing. This is resolved by applying the shift chain neighbourhood [23] to shift overstaffed assignments to shifts with understaffing. Again, we use the objective function value (Eq. (1)) to optimise the value of substitution possibilities and the cost.

3.4 Evolutionary Cycle

In this section, we discuss the building blocks of the evolutionary cycle (cf. Fig. 1). First, we randomly select two personnel shift rosters by defining their selection probabilities based on the objective function value (Eq. (1)). As such, personnel shift rosters with a high value corresponding to substitution possibilities and a low cost have an increased selection probability. Second, these selected personnel shift rosters are combined to generate new personnel shift rosters based on problem-specific crossover operators (CO). Third, we apply mutation (M) on the newly generated personnel shift rosters to diversify the search with a given probability (r_{mut}). Our mutation procedure entails an exchange of the complete day rosters of two random days. Given the application of crossover and mutation, we may need to repair the feasibility of the time-related constraints (Eqs. (4)–(6)) using a repair function (cf. Sect. 3.3). Fourth, we perform a schedule-based improvement function (SBI) to improve the personnel assignment costs, and more specifically the preference penalty cost. Finally, we update the population based on the principle of elitism, i.e. the personnel shift roster with the worst objective function value is replaced with the newly improved personnel shift roster if its objective function value is better.

In the remainder of this section, we discuss the crossover operators (CO) and schedule-based improvement (SBI) in more detail.

Crossover (CO). Maenhout and Vanhoucke [24] give an overview of crossover operators applied for different personnel shift scheduling problems. We focus on the employee-based and day-based crossover operators to generate a new personnel shift roster.

We distinguish three employee-based crossover operators, i.e. *randomly selected crossover (ECO1)*, *one-point crossover (ECO2)* and *crossover with tournament selection (ECO3)*. The *crossover with tournament selection (ECO3)* starts from the personnel shift roster with the highest value of substitution possibilities (Eq. (1b)) and copies the employee schedules for the multi-skilled employees to the new personnel shift roster. Hereafter, the single-skilled employees receive their corresponding schedule from the first or second original personnel shift roster depending on the value of substitution possibilities.

Similarly, we consider three day-based crossover operators, i.e. *randomly selected crossover (DCO1)*, *one-point crossover (DCO2)* and *crossover with tournament selection (DCO3)*. The latter provides a personnel shift roster that is the best combination of days in terms of employee substitutability. For each day, the two original personnel shift rosters are compared based on the value of substitution possibilities. The day roster with the highest value becomes part of the newly generated personnel shift roster. Note that the application of day-based crossover operators may result in infeasible personnel shift rosters, which need to be repaired (cf. Sect. 3.3).

Schedule-Based Improvement (SBI). We aim to reduce the preference penalty cost for all employees by exchanging their respective schedules based on the logic of the schedule-based local search proposed by Maenhout and Van-houcke [22]. To avoid infeasibilities, this exchange of schedules is limited to employees that possess exactly the same skill(s). First, we determine the preference penalty cost of all employee schedules for each individual employee. Second, employees are assigned their cheapest schedule in terms of their preference penalty cost in a random order.

4   Computational Experiments

In this section, we provide insight into our methodology to improve the value of employee substitutability. In Sect. 4.1, we describe our test design and discuss the parameter settings of the problem instances. In Sect. 4.2, we gain insight into the impact of different optimisation strategies in the proposed heuristic framework and we evaluate the performance of the suggested algorithm. All tests were carried out on an Intel Core processor 2.5 GHz and 4 GB RAM.

4.1   Test Design

In this section, we provide detailed information on the test instances and the parameter settings. All test instances have a planning horizon of 7 days.

Personnel Characteristics. We generate test instances with 10, 20 and 40 employees and the employees have a maximum of 2 skills. In total, we distinguish 11 skill possession settings according to the triplet (m_1% - m_2% - m_3%), which indicates the percentage of employees that uniquely possesses skill 1 (m_1%), skill 2 (m_2%) or both skills (m_3%). This triplet varies between (50%, 50%, 0%) and (0%, 0%, 100%) with intervals of 5% for m_1 and m_2, and an interval of 10% for m_3.

Note that these percentages can lead to fractional values. In this case, we round the number of employees for skill 1 down ($N_1 = \lfloor N \times m_1\% \rfloor$) and round the number of employees for skill 2 up ($N_2 = \lceil N \times m_2\% \rceil$). The remainder of employees possess both skills, i.e. $N_3 = N - N_1 - N_2$.

Shift Characteristics. The problem instances are characterised by three different 8-hour shifts with specific start and end times. We make a distinction between an easy test set and a hard test set based upon the definition of these start and end times. The easy test set assumes that the start and end times of the shifts are identical, and hence, the three shifts embody in fact different tasks that are carried out simultaneously on one particular day. The hard test set assumes non-overlapping shifts that start at 6 a.m., 2 p.m. and 10 p.m.

Staffing Requirements. The staffing requirements are generated based on three indicators proposed by Vanhoucke and Maenhout [25], that determine the demand profile as follows, i.e.

- The *Total Coverage Constrainedness* (TCC) provides an indication of the total staffing requirements over the planning horizon. As the TCC-value increases, the total staffing requirements rise. We generate demand profiles with a TCC-value of 0.30, 0.40 and 0.50 and distribute these staffing requirements evenly over the two skill categories.
- The *Day Coverage Distribution* (DCD) and the *Shift Coverage Distribution* (SCD) reflect the variability of the staffing requirements over the days of the planning horizon and over the shifts for a single day, respectively. As these values augment, the variability over the days and shifts increases. We investigate test instances with a value of 0.00, 0.25 and 0.50 for both indicators.

We generate 10 instances that are combined with 27 ($3 \times 3 \times 3$) (TCC, DCD, SCD)-combinations and 11 skill possession settings for 10, 20 and 40 employees, which implies that the computational experiments are based on $8910 (= 10 \times 27 \times 11 \times 3)$ instances.

Time-Related Constraints. All personnel members need to work a single shift per day or receive a day off (Eq. (3)) and need a minimum rest period of 11 hours between two consecutive working shifts (Eq. (4)). Moreover, each employee can work a maximum ($\eta_i^{w,max}$) of 5 shifts (Eq. (5)) and should work a minimum ($\eta_i^{w,min}$) of 4 shifts (Eq. (6)).

The small number of constraints provides extra assignment flexibility and improves the fit between the available employees and the staffing requirements. This is important to get an unbiased idea about the number and value of substitution possibilities.

Objective Function. The objective function weights include for each employee a wage cost (c_{imdj}^w) of 10 and a preference penalty cost (p_{idj}) randomly generated in the range of 1 to 5. The cost of understaffing (c_{mdj}^{wu}) is fixed at 20 and each substitution possibility has a benefit value of 1 (γ_{imdj}). Note that this benefit value can be easily adapted to consider the level of uncertainty of capacity for employee i and the level of uncertainty of demand during shift j on day d for skill m.

4.2   Validation of the Proposed Procedure

In this section, we validate the individual building blocks and the complete heuristic procedure of the proposed memetic algorithm based on the following parameters, i.e.

OF_δ the solution value of objective δ

$\%Dev_{\delta,\lambda}$ the percentage deviation in objective δ for the algorithmic solution compared to solution λ
   with $\delta = 1.1$, the total personnel assignment and understaffing cost (cf. Eq. (1a))

$\delta = 1.2$, the total value of substitution possibilities (cf. Eq. (1b))

$\delta = 1$, the total objective function value (cf. Eq. (1))

with $\lambda = Init$, the initial personnel shift roster with minimum personnel assignment and understaffing cost

$\lambda = BestMA$, the best found solution using the complete algorithm

$\lambda = IP$, the best found solution using commercial optimisation software

$CPU(s)$ the required time to obtain the final personnel shift roster in seconds

#cycles   the number of executed evolutionary cycles

These parameters are utilised to determine the contribution of the different building blocks of the memetic algorithm (cf. Sect. 3). In order to determine the contribution of the complete memetic algorithm, we benchmark our performance against a non-dedicated IP optimisation algorithm. Unless otherwise stated, we discuss the average results for the hard test instances over 10, 20 and 40 employees for every (TCC, DCD, SCD)-combination and all skill possession settings.

Note that computational experiments have shown that the best results are obtained for the proposed procedure with a stop criterion of 250 iterations without improvement in the objective function value.

Contribution of the Building Blocks of the Memetic Algorithm. Table 1 shows the contribution of the best performing strategy of each individual building block through a gradual assembly of our memetic algorithm. We incrementally add each building block until the complete algorithm is composed. In the table, we start from '(1) Initial roster', which represents the personnel shift roster with the minimum personnel assignment and understaffing cost (Eq. (1a)). The next columns show the results for the gradual introduction of the other building blocks, which also consider the maximisation of the employee substitutability value (Eq. (1b)). We successively add the following building blocks, i.e. multiple personnel shift rosters in a population ('(2) population'), local search and repair ('(3) LS+R') and an evolutionary cycle (4), which consists of a crossover operator ('(4.1) CO'), mutation and repair ('(4.2) M+R') and schedule-based improvement heuristics ('(4.3) SBI'). This last column displays the results corresponding to the personnel shift roster obtained by the complete algorithm.

The *initial roster* embodies a personnel shift roster that is optimal in terms of costs with a total cost of 1,065.97 and an employee substitutability value of 1,067.54 (*Init*). The construction of multiple personnel shift rosters results in a *population* for which the best personnel shift roster improves the employee substitutability with 2.78% at a cost increase of 0.68%. This *population* consists of 20 (n_{pop}) different personnel shift rosters obtained by marginally randomising the personnel assignment costs, i.e. c^w_{imdj} and p_{idj}. In this respect, we subtract a random number ranging between 0 and 5 from the original assignment costs (cf. Sect. 4.1). This marginal randomisation outperforms a complete

Table 1. Computational contribution of the individual building blocks of the memetic algorithm

	(1) Initial roster	(1)+ (2) Population	(1)–(2)+ (3) LS+R	(4) Evolutionary cycle		
				(1)–(3)+ (4.1) CO	(1)–(4.1)+ (4.2) M+R	(1)-(4.2)+ (4.3) SBI
$OF_{1.1}$	1,065.97	1,073.19	1,130.32	1,139.68	1,136.24	1,121.94
$\%Dev_{1.1,Init}$	0.00%	+0.68%	+6.04%	+6.91%	+6.59%	+5.25%
$OF_{1.2}$	1,067.54	1,097.23	1,247.08	1,298.01	1,297.41	1,303.52
$\%Dev_{1.2,Init}$	0.00%	+2.78%	+16.82%	+21.59%	+21.53%	+22.10%
OF_1	−1.57	−24.04	−116.76	−158.33	−161.17	−181.58
$CPU(s)$	0.05	1.07	1.63	1.91	2.29	3.98
$\#cycles$	0.00	0.00	0.00	427.19	447.05	702.42

randomisation strategy, which results in personnel shift rosters that are too expensive considering the small gain in employee substitutability value.

The introduction of the *local search heuristics and repair function* further ameliorates the value of substitution possibilities (16.82%) at the expense of a cost rise with 6.04%. A detailed analysis of the guiding orders and priorities and the neighbourhood structures reveals that the most significant contributors to this improvement, comprise a *decreasing skill-based group order (EGO1-1)* and a *random day order (DGO3)*. This means that it is best to prioritise the schedules of multi-skilled employees given that these employees inherently offer more flexibility and therefore employee substitutability. This can lead to local optima but is avoided through the application of the *random day order*.

The expansion of the algorithm with an *evolutionary cycle* to diversify the search further improves the total objective function value. The combination of two personnel shift rosters by means of a *crossover operator (CO)* leads to an increase in the value of substitution possibilities (21.59%) and the costs (6.91%). Nevertheless, it is important to apply the appropriate type of crossover operators. The day-based crossover operators significantly outperform the employee-based crossover operators. Hence, it is very important to preserve the day rosters to facilitate the transfer of cost-efficient substitution possibilities from the two original personnel shift rosters to the newly generated personnel shift roster. However, it is interesting to note that not all substitution possibilities can be transferred to the new personnel shift roster with a day-based crossover. This is due to the fact that this new personnel shift roster is a combination of the day rosters of two personnel shift rosters, which changes the variable f_{idj} in Eq. (8c) and thus the number of substitution possibilities (cf. Eqs. (8a) and (8b)). This explains the poor performance of the *crossover with tournament selection (DCO3)*. In this respect, the best results were obtained with *randomly selected crossover (DCO1)* to diversify the search space.

The introduction of additional randomness through *mutation and repair* *(M+R)* does not provide significant benefits for varying mutation probabilities (r_{mut}).

The *schedule-based improvement (SBI)* method simply swaps assignments between personnel members and is able to reduce the costs and in particular the personnel preference penalty cost while also increasing the value of substitution possibilities.

Overall, Table 1 shows that every step in the algorithm has a positive impact on the objective function value while the CPU-time and the number of executed evolutionary cycles rise. The execution of the complete algorithm (*BestMA*) provides a personnel shift roster that exhibits a rise of 22.10% in the value of substitution possibilities at a cost increase of 5.25% compared to the initial personnel shift roster (*Init*).

Comparison with an Exact Optimisation Technique. We provide insight in the comparison of our algorithm (*MA*) with the exact branch-and-bound algorithm (*IP*) in Fig. 3 for both the easy and hard test set. The solution quality (OF_1) is displayed on the Y-axis as a function of the CPU-time on the X-axis. Note that the scale of the X-axis is biased to denote the behaviour of the proposed heuristic algorithm and the IP exact approach in the early search phase more accurately. Figure 3 shows the reference solutions *Init*, *BestMA* and *IP*. The heuristic approach is able to find high-quality solutions very rapidly while the exact optimisation approach only finds solutions of similar quality after 100 and 450 seconds for the easy and hard test set, respectively.

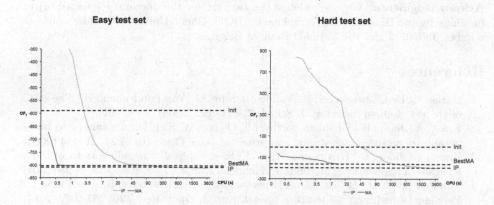

Fig. 3. The solution quality as a function of CPU-time

5   Conclusions

In this paper, we consider a bi-objective personnel shift scheduling model that not only minimises the costs but also maximises the employee substitutability to improve the robustness of a personnel shift roster. We incorporate three types of substitution possibilities between personnel members, which gives rise to a non-linear optimisation problem.

Due to the heavy computational burden, we propose a memetic algorithm to solve the problem under study. This algorithm is initialised by constructing a population of personnel shift rosters that does not consider the objective of employee substitutability but only the costs. The other problem-specific operators in the algorithm, i.e. the local search and the evolutionary cycle, aim to gradually increase employee substitutability in balance with the additional cost.

We validate the proposed algorithm with different heuristic optimisation strategies and a non-dedicated exact algorithm. The computational experiments show that individual personnel schedules should be improved in such an order that the personnel members whose schedule may yield the highest potential improvement are explored first. This implies that employees are grouped based on their skills and that multi-skilled employees are considered first. Moreover, sufficient attention should be dedicated to the day structure of a personnel shift roster in order to preserve the substitutability between employees. Finally, we can conclude that the algorithm is able to increase the value of substitution possibilities at the expense of a small cost increase.

Future research should primarily focus on the validation of the robustness of personnel shift rosters with maximised employee substitutability. In this respect, specific strategies should be developed to consider the level of uncertainty of capacity and demand in the definition of the benefit values of substitution possibilities.

Acknowledgments. We acknowledge the support for the doctoral research project fundings by the Bijzonder Onderzoekfonds (BOF, Ghent University) under contract number 01N00712 and the National Bank of Belgium.

References

1. Burke, E., De Causmaecker, P., Vanden Berghe, G., Van Landeghem, H.: The state of the art of nurse rostering. J. Sched. **7**, 441–499 (2004)
2. Ernst, A., Jiang, H., Krishnamoorthy, M., Owens, B., Sier, D.: An annotated bibliography of personnel scheduling and rostering. Ann. Oper. Res. **127**, 21–144 (2004)
3. Ernst, A., Jiang, H., Krishnamoorthy, M., Sier, D.: Staff scheduling and rostering: a review of applications, methods and models. Eur. J. Oper. Res. **153**, 3–27 (2004)
4. Van den Bergh, J., Beliën, J., De Bruecker, P., Demeulemeester, E., De Boeck, L.: Personnel scheduling: a literature review. Eur. J. Oper. Res. **226**, 367–385 (2013)
5. Abernathy, W., Baloff, N., Hershey, J.: A three-stage manpower planning and scheduling model - a service sector example. Oper. Res. **21**, 693–711 (1973)
6. Dück, V., Ionescu, L., Kliewer, N., Suhl, L.: Increasing stability of crew and aircraft schedules. Transp. Res. Part C: Emerg. Technol. **20**(1), 47–61 (2012)
7. Ionescu, L., Kliewer, N.: Increasing flexibility of airline crew schedules. Proc. - Soc. Behav. Sci. **20**, 1019–1028 (2011)
8. Abdelghany, A., Ekollu, G., Narasimhan, R., Abdelghany, K.: A proactive crew recovery decision support tool for commercial airlines during irregular operations. Ann. Oper. Res. **127**(1–4), 309–331 (2004)

9. Abdelghany, K., Abdelghany, A., Ekollu, G.: An integrated decision support tool for airlines schedule recovery during irregular operations. Eur. J. Oper. Res. **185**(2), 825–848 (2008)
10. Eggenberg, N., Salani, M., Bierlaire, M.: Constraint-specific recovery network for solving airline recovery problems. Comput. Oper. Res. **37**(6), 1014–1026 (2010)
11. Gao, C., Johnson, E., Smith, B.: Integrated airline fleet and crew robust planning. Transp. Sci. **43**(1), 2–16 (2009)
12. Shebalov, J., Klabjan, D.: Robust airline crew pairing: move-up crews. Transp. Sci. **40**, 300–312 (2006)
13. De Bruecker, P., Van den Bergh, J., Beliën, J., Demeulemeester, E.: Workforce planning incorporating skills: state of the art. Eur. J. Oper. Res. **243**(1), 1–16 (2015)
14. Gurobi Optimization Inc.: Gurobi optimizer reference manual (2015). http://www.gurobi.com/documentation/
15. Bellanti, F., Carello, G., Della Croce, F.: A greedy-based neighborhood search approach to a nurse rostering problem. Eur. J. Oper. Res. **153**, 28–40 (2004)
16. Burke, E., Curtois, T., Post, G., Qu, R., Veltman, B.: A hybrid heuristic ordering and variable neighbourhood search for the nurse rostering problem. Eur. J. Oper. Res. **188**, 330–341 (2008)
17. Burke, E.K., Li, J., Qu, R.: A pareto-based search methodology for multi-objective nurse scheduling. Ann. Oper. Res. **196**(1), 91–109 (2012)
18. Maenhout, B., Vanhoucke, M.: An evolutionary approach for the nurse rerostering problem. Comput. Oper. Res. **38**, 1400–1411 (2011)
19. Beddoe, G., Petrovic, S.: Enhancing case-based reasoning for personnel rostering with selected tabu search concepts. J. Oper. Res. Soc. **58**(12), 1586–1598 (2007)
20. Bilgin, B., De Causmaecker, P., Rossie, B., Vanden Berghe, G.: Local search neighbourhoods for dealing with a novel nurse rostering model. Ann. Oper. Res. **194**(1), 33–57 (2012)
21. Burke, E.K., Li, J., Qu, R.: A hybrid model of integer programming and variable neighbourhood search for highly-constrained nurse rostering problems. Eur. J. Oper. Res. **203**(2), 484–493 (2010)
22. Maenhout, B., Vanhoucke, M.: An electromagnetic meta-heuristic for the nurse scheduling problem. J. Heuristics **13**, 359–385 (2007)
23. Dowsland, K.: Nurse scheduling with tabu search and strategic oscillation. Eur. J. Oper. Res. **106**, 393–407 (1998)
24. Maenhout, B., Vanhoucke, M.: Comparison and hybridization of crossover operators for the nurse scheduling problem. Ann. Oper. Res. **159**, 333–353 (2008)
25. Vanhoucke, M., Maenhout, B.: On the characterization and generation of nurse scheduling problem instances. Eur. J. Oper. Res. **196**, 457–467 (2009)

Construct, Merge, Solve and Adapt Versus Large Neighborhood Search for Solving the Multi-dimensional Knapsack Problem: Which One Works Better When?

Evelia Lizárraga[1], María J. Blesa[1], and Christian Blum[2(✉)]

[1] Computer Science Deptartment, Universitat Politècnica
de Catalunya – BarcelonaTech, Barcelona, Spain
{evelial,mjblesa}@cs.upc.edu
[2] Artificial Intelligence Research Institute (IIIA-CSIC),
Campus UAB, Bellaterra, Spain
christian.blum@iiia.csic.es

Abstract. Both, Construct, Merge, Solve and Adapt (CMSA) and Large Neighborhood Search (LNS), are hybrid algorithms that are based on iteratively solving sub-instances of the original problem instances, if possible, to optimality. This is done by reducing the search space of the tackled problem instance in algorithm-specific ways which differ from one technique to the other. In this paper we provide first experimental evidence for the intuition that, conditioned by the way in which the search space is reduced, LNS should generally work better than CMSA in the context of problems in which solutions are rather large, and the opposite is the case for problems in which solutions are rather small. The size of a solution is hereby measured by the number of components of which the solution is composed, in comparison to the total number of solution components. Experiments are conducted in the context of the multi-dimensional knapsack problem.

1 Introduction

The development and the application of hybrid metaheuristics has enjoyed an increasing popularity in recent years [1,2]. This is because these techniques often allow to combine the strengths of different ways of solving optimization problems in a single algorithm. Especially the combination of heuristic search with exact techniques—a field of research often labelled as matheuristics [3]—has been quite

This work was funded by project TIN2012-37930-C02-02 (Spanish Ministry for Economy and Competitiveness, FEDER funds from the European Union) and project SGR 2014-1034 (AGAUR, Generalitat de Catalunya). Evelia Lizárraga acknowledges funding from the Mexican National Council for Science and Technology (CONACYT, doctoral grant number 253787).

B. Hu and M. López-Ibáñez (Eds.): EvoCOP 2017, LNCS 10197, pp. 60–74, 2017.
DOI: 10.1007/978-3-319-55453-2_5

fruitful. One of the most well known, and generally applicable, algorithms from this field is called Large Neighborhood Search (LNS) [4], which is based on the following general idea. Given a valid solution to the tackled problem instance, first, destroy selected parts of it, resulting in a partial solution. Then apply some other, possibly exact, technique to find the best valid solution on the basis of the given partial solution, that is, the best valid solution that contains the given partial solution. Thus, the destruction step defines a *large neighborhood*, from which a best (or nearly best) solution is determined, not by naive enumeration but by the application of a more effective alternative technique. Apart from LNS, the related literature offers algorithms that make use of alternative ways of defining large neighborhoods, such as the so-called Corridor Method [5], POPMUSIC [6], and Local Branching [7].

One of the latest algorithmic developments in the line of LNS is labelled Construct, Merge, Solve and Adapt (CMSA) [8]. Just like LNS, the main idea of CMSA is to iteratively apply a suitable exact technique to reduced problem instances, that is, sub-instances of the original problem instances. Note that the terms *reduced problem instance* and *sub-instance* refer, in this context, to a subset of the set of solutions to the tackled problem instance which is obtained by a reduction of the search space. The idea of both algorithms—LNS and CMSA—is to identify substantially reduced sub-instances of a given problem instance such that the sub-instances contain high-quality solutions to the original problem instance. This might allow the application, for example, of an exact technique with reasonable computational effort to the sub-instance in order to obtain a high-quality solution to the original problem instance. In other words, both algorithms employ techniques for reducing the search space of the tackled problem instances.

1.1   Our Contribution

Although both LNS and CMSA are based on the same general idea, the way in which the search space is reduced differs from one to the other. Based on this difference we had the intuition that LNS would (generally) work better than CMSA for problems for which solutions are rather large, and the opposite would be the case in the context of problems for which solutions are rather small. The size of solutions is hereby measured by the number of solution components (in comparison to the total number) of which they are composed. For example, in the case of the travelling salesman problem, the complete set of solution components is composed of the edges of the input graph. Moreover, solutions consist of exactly n components, where n is the number of vertices of the input graph. The above-mentioned intuition is based on the consideration that, for ending up in some high-quality solution, LNS needs to find a path of over-lapping solutions from the starting solution to the mentioned high-quality solution. The smaller the solutions are, the more difficult it should be to find such a path. A theoretical validation of our intuition seems, a priori, rather difficult to achieve. Therefore, we decided to study empirical evidence that would support (or refute) our intuition. For this purpose, we used the multi-dimensional knapsack problem

(MDKP). As will be outlined later, for this problem it is possible to generate both, problem instances for which solutions are small and problem instances for which solutions are large. We implemented both LNS and CMSA for the MDKP and performed an empirical study of the results of both algorithms for problem instances over the whole range between small and large solutions. The outcome of the presented study is empirical evidence for the validity of our intuition.

1.2   Outline of the Paper

The remainder of this paper is organized as follows. Section 2 provides a general, problem-independent, description of both LNS and CMSA, whereas Sect. 3 describes the application of both algorithms to the MDKP. The empirical study in the context of the MDKP is presented in Sect. 4, and the conclusions and an outline of future work is given in Sect. 5.

2   General Description of the Algorithms

In the following we provide a general description of both LNS and CMSA in the context of problems for which the exact technique used to solve sub-instances is a general-purpose integer linear programming (ILP) solver. For the following discussion we assume that a problem instance I is characterized by a complete set C of solution components. In the case of the well-known travelling salesman problem, for example, C consists of all edges of the input graph. Moreover, solutions are represented as subsets of C. Finally, any sub-instance in the context of CMSA—denoted by C'—is also a subset of C. Solutions to C' may only be formed by solution components from C'.

2.1   Large Neighborhood Search

The pseudo-code of a general ILP-based LNS is provided in Algorithm 1. First, in line 2 of Algorithm 1, an initial incumbent solution S_{cur} is generated in function GenerateInitialSolution(I). Solution S_{cur} is then partially destroyed at each iteration, depending on the *destruction rate* D_r. The way in which the incumbent solution is destroyed (randomly versus heuristically) is a relevant design decision. The resulting partial solution $S_{partial}$ is fed to the ILP solver; see function ApplyILPSolver($S_{partial}$, t_{max}) in line 7 of Algorithm 1. Apart from $S_{partial}$, this function receives a time limit t_{max} as input. Note that the complete solver is forced to include $S_{partial}$ in any considered solution. This means that the corresponding sub-instance comprises all solutions that contain $S_{partial}$. As a result, the complete solver provides the best valid solution found within the available computation time t_{max}. This solution, denoted by S'_{opt}, may or may not be the optimal solution to the tackled sub-instance. This depends on the given computation time limit t_{max} for each application of the complete solver. Finally, in the LNS version used in this paper, the better solution between S'_{opt} and S_{cur} is carried over to the next iteration. This seems, at first sight, restrictive. In

---

Algorithm 1. Large Neighborhood Search (LNS)

1: **input:** problem instance $\mathcal{I}$, values for parameters D^l, D^u, D^{inc}, and $t_{\max}$
2: $S_{\mathrm{cur}} :=$ GenerateInitialSolution($\mathcal{I}$)
3: $S_{\mathrm{bsf}} := S_{\mathrm{cur}}$
4: $D_r := D^l$
5: **while** CPU time limit not reached **do**
6:     $S_{\mathrm{partial}} :=$ DestroyPartially(S_{cur}, D_r)
7:     $S'_{\mathrm{opt}} :=$ ApplyILPSolver($S_{\mathrm{partial}}, t_{\max}$)
8:     **if** S'_{opt} is better than S_{bsf} **then** $S_{\mathrm{bsf}} := S'_{\mathrm{opt}}$
9:     **if** S'_{opt} is better than S_{cur} **then**
10:       $S_{\mathrm{cur}} := S'_{\mathrm{opt}}$
11:       $D_r := D^l$
12:     **else**
13:       $D_r := D_r + D^{\mathrm{inc}}$
14:       **if** $D_r > D^u$ **then** $D_r := D^l$
15:     **end if**
16: **end while**
17: **return** S_{bsf}

---

particular, other—more probabilistic—ways of selecting between S'_{opt} and S_{cur} would be possible. However, in turn the algorithm is equipped with a variable destruction rate D_r, which may vary between a lower bound D^l and an upper bound D^u. Hereby, D^l and D^u are parameters of the algorithm. A proper setting of these parameters enables the algorithm to escape from local minima. Note that the adaptation of D_r is managed in the style of variable neighborhood search algorithms [9]. In particular, if S'_{opt} is better than S_{cur}, the value of D_r is set back to the lower bound D^l. Otherwise, the value of D_r is incremented by D^{inc}, which is also a parameter of the algorithm. If the value of D_r—after this update—exceeds the upper bound D^u, it is set back to the lower bound D^l.

2.2 Construct, Merge, Solve and Adapt

The pseudo-code of an ILP-based CMSA algorithm is provided in Algorithm 2. Each algorithm iteration consists of the following actions. First, the best-so-far solution S_{bsf} is initialized to $\emptyset$, indicating that no such solution exists yet. Moreover, the restricted problem instance C', which is—as mentioned before—a subset of the complete set C of solutions components, is initialized to the empty set. Then, at each iteration, the restricted problem instance C' is augmented in the following way (see lines 5 to 11): n_a solutions are probabilistically generated in function ProbabilisticSolutionGeneration(C). The components found in the constructed solutions are added to C'. Hereby, the so-called age of each of these solution components ($age[c]$) is set to zero. Once C' was augmented in this way, a complete solver is applied in function ApplyILPSolver(C') to find a possibly optimal solution S'_{opt} to the restricted problem instance C'. If S'_{opt} is better than the current best-so-far solution S_{bsf}, solution S'_{opt} is taken as the

Algorithm 2. Construct, Merge, Solve and Adapt (CMSA)

1: **input:** problem instance $\mathcal{I}$, values for parameters n_a, age_{max}, and t_{max}
2: $S_{bsf} := \emptyset$; $C' := \emptyset$
3: $age[c] := 0$ for all $c \in C$
4: **while** CPU time limit not reached **do**
5: 　　**for** $i := 1, \ldots, n_a$ **do**
6: 　　　　$S := \mathsf{ProbabilisticSolutionGeneration}(C)$
7: 　　　　**for** all $c \in S$ and $c \notin C'$ **do**
8: 　　　　　　$age[c] := 0$
9: 　　　　　　$C' := C' \cup \{c\}$
10: 　　　**end for**
11: 　　**end for**
12: 　　$S'_{opt} := \mathsf{ApplyILPSolver}(C', t_{max})$
13: 　　**if** S'_{opt} is better than S_{bsf} **then** $S_{bsf} := S'_{opt}$
14: 　　$\mathsf{Adapt}(C', S'_{opt}, age_{max})$
15: **end while**
16: **return** s_{bsf}

---

new best-so-far solution. Next, sub-instance C' is adapted on the basis of solution S'_{opt} in conjunction with the age values of the solution components; see function $\mathsf{Adapt}(C', S'_{opt}, age_{max})$ in line 14. This is done as follows. First, the age of each solution component in $C' \setminus S'_{opt}$ is incremented while the age of each solution component in $S'_{opt} \subseteq C'$ is re-initialized to zero. Subsequently, those solution components from C' with an age value greater than age_{max}—which is a parameter of the algorithm—are removed from C'. This causes that solution components that never appear in solutions derived by the complete solver do not slow down the solver in subsequent iterations. On the other side, components which appear in the solutions returned by the complete solver should be maintained in C'.

2.3　Search Space Reduction in LNS and CMSA

The way in which the search space of the tackled problem instance is reduced by LNS, respectively CMSA, can be summarized as follows. LNS keeps an incumbent solution which, at each iteration, is partially destroyed. This results in a partial solution. The reduced search space consists of all solutions to the original problem instance that contain this partial solution. This is graphically illustrated in Fig. 1a. CMSA, on the other side, reduces the search space as follows: at each iteration, solutions to the original problem instance are constructed in a probabilistic way, using a greedy function as bias. The solution components found in these solutions are joined, forming a subset C' of the complete set of solution components. The set of solutions to the original problem instance that can be generated on the basis of the components in C' form the reduced search space in CMSA. This is graphically presented in Fig. 1b.

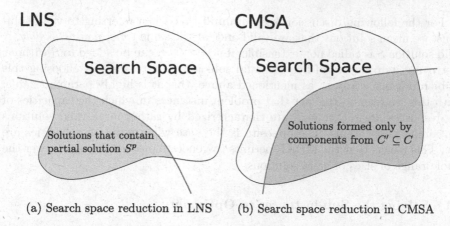

(a) Search space reduction in LNS   (b) Search space reduction in CMSA

Fig. 1. The way in which the search space is reduced in LNS, respectively CMSA. The term *search space* refers to the set of all valid solutions to the tackled problem instance. The grey-colored sub-spaces indicate the search spaces of the tackled sub-instances.

3   Application to the MDKP

For the aim of finding empirical evidence for the intuition phrased in the introduction of this work, we make use of the so-called *multi-dimensional knapsack problem* (MDKP), a well studied NP-hard combinatorial optimization problem and a popular test case for new algorithmic proposals (see, for example, [10–12]). The reason for choosing the MDKP is that it is parametrizable, as we will outline in more detail below.

The MDKP is defined as follows. Given is a set $C = \{1, \ldots, n\}$ of n items, and a set $K = \{1, \ldots, m\}$ of m different resources. Each resource $k \in K$ is available in a certain quantity (*capacity*) $c_k > 0$, and each item $i \in C$ requires from each resource $k \in K$ a given amount $r_{i,k} \geq 0$ (*resource consumption*). Moreover, each item $i \in C$ has associated a profit $p_i > 0$. Note that, in the context of the MDKP, the set C of items corresponds to the complete set of solution components.

A feasible solution to the MDKP is a selection (subsets) of items $S \subseteq C$ such that for each resource k the total consumption over all selected items $\sum_{i \in S} r_{i,k}$ does not exceed the resource's capacity c_k. The objective is to find a feasible item selection S of maximum total profit $\sum_{i \in S} p_i$. The MDKP can be stated in terms of an ILP as follows:

$$\text{maximize} \sum_{i \in C} p_i \cdot x_i \tag{1}$$

$$\text{s.t.} \sum_{i \in C} r_{i,k} \cdot x_i \leq c_k \qquad \forall k \in K \tag{2}$$

$$x_i \in \{0, 1\} \qquad \forall i \in C \tag{3}$$

Hereby, inequalities (2) limit the total consumption for each resource and are called *knapsack constraints*.

For the following discussion keep in mind that when referring to valid solutions, we mean solutions that are valid and, at the same time, *non-extensible*. A valid solution S is called non-extensible, if no $i \in C \setminus S$ can be added to S without destroying its property of being a valid solution. The reasons for choosing this problem for our study is, as mentioned above, that it is highly parametrizable. With this we refer to the fact that problem instances in which the capacities of the resources are rather high are characterized by rather large valid solutions containing many items. The opposite is the case when resource capacities are low. This means that the MDKP permits to generate problem instances over the whole range of sizes of valid solutions.

3.1   Solving the Sub-instances to Optimality

For solving a sub-instance determined by a partial solution S_{partial} in the context of LNS to optimality, the following constraints must be added to the ILP model for the MDKP that was outlined above:

$$x_i = 1 \quad \forall i \in S_{\text{partial}} \tag{4}$$

Similarly, for solving a sub-instance C' in the context of CMSA to optimality we simply have to apply the ILP model using set C' instead of the complete set C.

3.2   Constructing Solutions for the MDKP

Apart from solving the sub-instances to optimality, we require a way for generating the initial solution in the case of LNS and for generating solutions at each iteration of CMSA in a probabilistic way. For both purposes we used the greedy heuristic outlined in the following. Henceforth it is assumed that the items in C are ordered w.r.t. the following *utility values* in a non-increasing way:

$$u_i \leftarrow \frac{p_i}{\sum_{k \in K} r_{i,k}/c_k} \quad \forall\, i \in C. \tag{5}$$

That is, the items in C are ordered such that $u_1 \geq u_2 \geq \ldots \geq u_n$. This means that an item $i \in C$ has position/index i due to its utility value. The utility values are used as a static greedy weighting function in the heuristic described in Algorithm 3. This heuristic simply adds items in the order determined by the utility values to an initially empty partial solution S until no further item fits w.r.t. the remaining resource capacities.

The probabilistic way of constructing a solution employed in CMSA (function ProbabilisticSolutionGeneration(C) in line 6 of Algorithm 2) also adds one item at a time until no further item can be added without violating the constraints. At each solution construction step, let S denote the current partial solution and let l denote (the index of) the last item added to S. Remember that item l has index l. In case $S = \emptyset$, let $l = -1$. In order to choose the next item to be added to S, the first up to l_{size} items starting from item $l + 1$ that fit w.r.t. all resources are collected in a set L. Hereby, L is commonly called the *candidate*

Algorithm 3. Greedy Heuristic for the MDKP

1: **input:** a MDKP instance $\mathcal{I}$
2: $S \leftarrow \emptyset$
3: **for** $i \leftarrow 1, \ldots, n$ **do**
4:    **if** $\left(\sum_{j \in S} r_{j,k} \right) + r_{i,k} \leq c_k, \; \forall k = 1, \ldots, m$ **then**
5:       $S \leftarrow S \cup \{i\}$
6:    **end if**
7: **end for**
8: **return** S

---

list and l_{size}, which is an important parameter, is called the *candidate list size*. In order to choose an item from L, a number $\nu \in [0,1)$ is chosen uniformly at random. In case $\nu \leq d_{\text{rate}}$, the item $i^* \leftarrow \min\{i \in L\}$ is chosen and added to S. Otherwise—that is, in case $\nu > d_{\text{rate}}$—an item i^* from L is chosen uniformly at random. Just like l_{size}, the *determinism rate* d_{rate} is an input parameter of the algorithm for which a well-working value must be found.

3.3 Partial Destruction of Solutions in LNS

The last algorithmic aspect that must be specified is the way in which solutions in LNS are partially destroyed. Two variants were considered. In both variants, given the incumbent solution S, $\max\{3, \lfloor D_r \cdot |S| \rfloor\}$ items are chosen at random and are then deleted from S. However, while this choice is made uniformly at random in the first variant, the greedy function outlined above is used in the second variant in an inverse-proportional way in order to bias the random choice of items to be deleted. However, as we were not able to detect any benefit from the biased random choice, we decided to use the first variant for the final experimental evaluation.

4 Empirical Study

Both, LNS and CMSA, were coded in ANSI C++ using GCC 4.7.3 for compilation. The experimental evaluation was performed on a cluster of computers with "Intel® Xeon® CPU 5670" CPUs of 12 nuclei of 2933 MHz and (in total) 32 Gigabytes of RAM. Moreover, all ILPs in LNS and CMSA were solved with IBM ILOG CPLEX V12.1 (single-threaded mode).

In the following we describe the set of benchmark instances generated to test the two algorithms. Then, we describe the tuning experiments in order to determine a proper setting for the parameters of LNS and CMSA. Finally, the experimental results are presented.

4.1 Problem Instances

The following set of benchmark instances was created using the methodology described in [10,13]. In particular, we generated benchmark instances of

$n \in \{100, 500, 1000, 5000, 10000\}$ items and $m \in \{10, 30, 50\}$ resources. The so-called *tightness* of problem instances refers hereby to the size of the capacities. In the way of generating instances that we used—first described in [10,13]—the tightness of an instance can be specified by means of a parameter α which may take values between zero and one. The lower the value of α, the tighter is the resulting problem instance and the smaller are the solutions to the respective problem instance. In order to generate instances over the whole range of tightness values we chose $\alpha \in \{0.1, 0.2, \ldots, 0.8, 0.9\}$. More specifically, for our experiments we generated 30 random instances for each combination of values of the three above-mentioned parameters (n, m and α). For all instances, the resource requirements $r_{i,j}$ were chosen uniformly at random from $\{1, \ldots, 1000\}$. In total, the generated benchmark set consist of 4050 problem instances.

4.2   Tuning

We made use of the automatic configuration tool irace [14] for both algorithms. irace was applied for each combination of n (number of items) and α (the tightness value). More specifically, for each combination of n and α we generated three random instances for each $m \in \{10, 30, 50\}$, that is, in total nine tuning instances were generated for each application of irace. The budget of irace was set to 1000. Moreover, the following computation time limits were chosen for both LNS and CMSA: 60 CPU seconds for instances with $n = 100$, 120 CPU seconds for those with $n = 500$, 210 CPU seconds for those with $n = 1000$, 360 CPU seconds for those with $n = 5000$, and 600 CPU seconds for those with $n = 10000$.

Parameters of CMSA. The important parameters of CMSA that are considered for tuning are the following ones: (1) the number of solution constructions per iteration (n_a), (2) the maximum allowed age (age_{max}) of solution components, (3) the determinism rate (d_{rate}), (4) the candidate list size (l_{size}), and (5) the maximum time in seconds allowed for CPLEX per application to each sub-instance (t_{max}). The following parameter value ranges were chosen concerning the five parameters of CMSA.

- $n_a \in \{10, 30, 50\}$
- $age_{max} \in \{1, 5, 10, \inf\}$, where inf means that no solution component is ever removed from the sub-instance.
- $d_{rate} \in \{0.0, 0.3, 0.5, 0.7, 0.9\}$, where a value of 0.0 means that the selection of the next solution component to be added to the partial solution under construction is always done randomly from the candidate list, while a value of 0.9 means that solution constructions are nearly deterministic.
- $l_{size} \in \{3, 5, 10\}$
- $t_{max} \in \{1.0, 2.0, 4.0, 8.0\}$ (in CPU seconds) for all instances with $n \in \{100, 500\}$, and $t_{max} \in \{2.0, 4.0, 8.0, 16.0, 32.0\}$ for all larger instances.

Parameters of LNS. The parameters of LNS considered for tuning are the following ones: (1) the lower and upper bounds—that is, D^l and D^u—of the destruction rate, (2) the increment of the destruction rate (D^{inc}), and (3) the maximum time t_{max} (in seconds) allowed for CPLEX per application to a sub-instance. The following parameter value ranges were chosen concerning the five parameters of CMSA.

- $(D^l, D^u) \in \{(0.1, 0.1), (0.2, 0.2), (0.3, 0.3), (0.4, 0.4), (0.5, 0.5), (0.6, 0.6), (0.7, 0.7), (0.8, 0.8), (0.9, 0.9), (0.1, 0.3), (0.1, 0.5), (0.3, 0.5), (0.3, 0.7), (0.3, 0.9), (0.1, 0.9)\}$. Note that when $D^l = D^u$, the destruction rate D_r is fixed.
- $D^{inc} \in \{0.01, 0.02, \dots, 0.08, 0.09\}$
- The value range for t_{max} was chosen in the same way as for CMSA (see above).

The results of the tuning processes are shown in the three sub-tables of Fig. 4 in Appendix A.

4.3   Results

Both LNS and CMSA were applied to all problem instances exactly once, with the computation time limits as outlined at the beginning of Sect. 4.2. The results are shown graphically by means of boxplots in Fig. 2. Note that there is one graphic per combination of n (the number of items) and m (the number of resources). The x-axis of each graphic ranges from the 30 instances of tightness value $\alpha = 0.1$ to the 30 instances of tightness value $\alpha = 0.9$, that is, from left to right we move from instances with small solutions—that is, solutions containing few components—to instances with large solutions—that is, solutions containing many components. The boxes in these boxplots show the improvement of CMSA over LNS (in percent). This means that when data points have a positive sign (that is, greater than zero), CMSA has obtained a better result than LNS, and vice versa. In order to improve the readability of these figures, the area of data points with positive signs has a shaded background.

The following main observation can be made: In accordance with our intuition that CMSA should have advantages over LNS in the context of problems with small solutions, it can be observed that CMSA generally has advantages over LNS when the tightness values of instances are rather small. This becomes more and more clear with growing instance size (n) and with a decreasing number of resources (m). In turn, LNS generally has advantages over CMSA for instances with a high tightness value, that is, for instances with large solutions.

In order to shed some further light on the differences between CMSA and LNS, we also measured the percentage of items—that is, solution components—that appeared in at least one of the solutions visited by the algorithm within the allowed computation time. This information is provided in the graphics of Fig. 3 by means of barplots. Again, we present one graphic per combination of n and m. The following observations can be made:

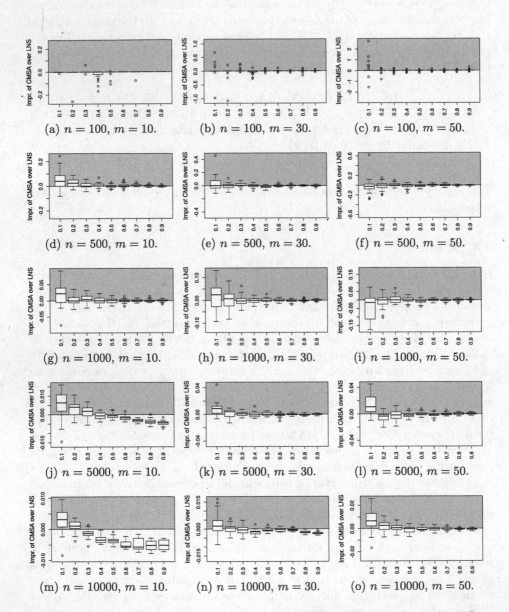

Fig. 2. Improvement of CMSA over LNS (in percent). Each box shows the differences for the corresponding 30 instances. Note that negative values indicate that LNS obtained a better result than CMSA, and vice versa.

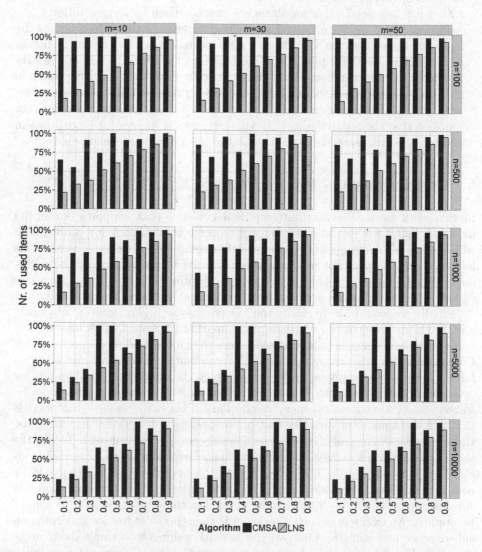

Fig. 3. Percentage of the items—that is, solution components—that were used in at least one visited solution. Each bar shows the average over the respective 30 problem instances.

- First of all, the percentage of used items is always much higher for CMSA than for LNS. This means that, with the optimized parameter setting as determined by irace, CMSA is much more *explorative* than LNS. This, apparently, pays off in the context of problems with small solutions. On the downside, this seems rather not beneficial when problems are characterized by large solutions.
- Additionally, it can be observed that the difference in the percentage of the usage of items between CMSA and LNS decreases with growing instances size (n). This can be explained by the fact that the absolute size of the sub-instances that are generated by CMSA naturally grows with growing problem size, and—as the efficiency of CPLEX for solving these sub-instances decreases with growing sub-instance size—the parameter values as determined by irace are such that the relative size of the sub-instances is smaller for large problem instances, which essentially means that the algorithm is less explorative.

5   Conclusions and Future Work

In this work we have given first empirical evidence that supports our initial intuition that LNS should generally work better than CMSA for problems in which solutions contain rather many solution components, and vice versa. This has been shown by means of experimental results in the context of the multi-dimensional knapsack problem. In the near future we intent to confirm this empirical evidence by the application to additional optimization problems.

Finally, we would like to clarify the following aspect. Our intuition obviously only holds for problems for which, a priori, neither LNS nor CMSA have advantages over the other one. In fact, it is not very difficult to find problems for which CMSA generally has advantages over LNS, no matter if solutions are small or large. Consider, for example, problems for which the number of variables and/or constraints in the respective ILP model are so large that the problem cannot be solved simply because of memory restrictions. This is the case in ILP models in which the number of variables and/or constraints are super-linear concerning the input parameters of the problem. Due to its specific way of reducing the search space, CMSA tackles sub-instances that correspond to reduced ILP models. This is not the case of LNS. Even though parts of the solution are fixed, the complete original ILP model must be built in order to solve the corresponding sub-instance. Therefore, CMSA can be applied in these cases, while LNS cannot be applied. An example of such a problem is the repetition-free longest common subsequence problem [15]. Contrarily, it is neither difficult to think about problems for which LNS generally has advantages over CMSA. Consider, for example, a problem where the main difficulty is not the size of ILP model but rather the computational complexity. Moreover, let us assume that when fixing a part of the solution, the sub-instance becomes rather easy to be solved, which is—for example—the case in problems with strong symmetries. In such a case LNS will most probably have advantages over CMSA. An example of such a problem is the most strings with few bad columns problem [16].

A   Appendix: Tuning results

See Fig. 4.

α	n_a	age_{max}	d_{rate}	l_{size}	t_{max}	n_a	age_{max}	d_{rate}	l_{size}	t_{max}	n_a	age_{max}	d_{rate}	l_{size}	t_{max}
			$n=100$					$n=500$					$n=1000$		
0.1	30	1	0.3	5	1.0	50	1	0.3	3	4.0	10	1	0.9	10	2.0
0.2	30	1	0.5	3	2.0	50	10	0.9	5	1.0	50	1	0.5	5	8.0
0.3	30	10	0.7	5	8.0	30	1	0	3	4.0	30	1	0.5	3	4.0
0.4	30	10	0.7	5	8.0	10	10	0.9	5	2.0	50	1	0.7	3	8.0
0.5	50	1	0.7	5	2.0	10	1	0.5	3	4.0	10	1	0.5	3	8.0
0.6	30	5	0.7	3	8.0	30	1	0.9	5	2.0	10	5	0.9	5	4.0
0.7	10	5	0.7	3	4.0	30	5	0.9	3	4.0	30	5	0.7	3	8.0
0.8	50	1	0.9	3	4.0	30	1	0.9	3	8.0	30	5	0.9	3	4.0
0.9	10	1	0.3	3	8.0	10	1	0.5	10	2.0	10	1	0.3	10	2.0

(a) Tuning CMSA for instances with $n \in \{100, 500, 1000\}$.

α	n_a	age_{max}	d_{rate}	l_{size}	t_{max}	n_a	age_{max}	d_{rate}	l_{size}	t_{max}
			$n=5000$					$n=10000$		
0.1	10	1	0.5	3	4.0	30	1	0.7	5	8.0
0.2	10	1	0.9	5	8.0	50	1	0.9	5	8.0
0.3	10	1	0.9	5	8.0	50	5	0.9	5	16.0
0.4	10	1	0.5	10	4.0	10	inf	0.5	3	16.0
0.5	10	1	0	10	8.0	50	10	0.9	5	32.0
0.6	50	inf	0.9	3	16.0	50	inf	0.9	3	32.0
0.7	30	1	0.9	3	16.0	10	1	0.7	10	32.0
0.8	10	5	0.9	3	32.0	30	inf	0.9	3	32.0
0.9	10	1	0	10	16.0	10	10	0.9	3	32.0

(b) Tuning CMSA for instances with $n \in \{5000, 10000\}$.

α	D^l	D^u	D^{inc}	t_{max}	D^l	D^u	D^{inc}	t_{max}	D^l	D^u	D^{inc}	t_{max}	D^l	D^u	D^{inc}	t_{max}	D^l	D^u	D^{inc}	t_{max}
			$n=100$				$n=500$				$n=1000$				$n=5000$				$n=10000$	
0.1	0.8	0.8	0.08	4.0	0.9	0.9	0.05	1.0	0.8	0.8	0.06	4.0	0.8	0.8	0.04	8.0	0.9	0.9	0.08	8.0
0.2	0.9	0.9	0.06	4.0	0.9	0.9	0.03	1.0	0.9	0.9	0.06	2.0	0.9	0.9	0.03	16.0	0.9	0.9	0.06	32.0
0.3	0.9	0.9	0.03	1.0	0.8	0.8	0.06	4.0	0.9	0.9	0.07	8.0	0.7	0.7	0.01	16.0	0.8	0.8	0.06	8.0
0.4	0.9	0.9	0.03	4.0	0.7	0.7	0.03	1.0	0.8	0.8	0.01	2.0	0.9	0.9	0.08	8.0	0.9	0.9	0.02	8.0
0.5	0.9	0.9	0.01	2.0	0.6	0.6	0.05	1.0	0.8	0.8	0.08	2.0	0.7	0.7	0.05	4.0	0.9	0.9	0.01	32.0
0.6	0.9	0.9	0.01	2.0	0.7	0.7	0.06	1.0	0.8	0.8	0.03	4.0	0.9	0.9	0.02	4.0	0.8	0.8	0.03	32.0
0.7	0.9	0.9	0.05	2.0	0.6	0.6	0.05	1.0	0.9	0.9	0.02	2.0	0.9	0.9	0.06	8.0	0.9	0.9	0.08	16.0
0.8	0.8	0.8	0.08	4.0	0.9	0.9	0.02	2.0	0.8	0.8	0.07	4.0	0.8	0.8	0.03	8.0	0.9	0.9	0.01	16.0
0.9	0.1	0.9	0.09	1.0	0.8	0.8	0.05	1.0	0.9	0.9	0.03	2.0	0.8	0.8	0.05	8.0	0.8	0.8	0.02	16.0

(c) Tuning LNS.

Fig. 4. Tuning results.

References

1. Talbi, E. (ed.): Hybrid Metaheuristics. Studies in Computational Intelligence, vol. 434. Springer, Heidelberg (2013)

2. Blum, C., Raidl, G.R.: Hybrid Metaheuristics - Powerful Tools for Optimization. Springer, Heidelberg (2016)
3. Boschetti, M.A., Maniezzo, V., Roffilli, M., Bolufé Röhler, A.: Matheuristics: optimization, simulation and control. In: Blesa, M.J., Blum, C., Gaspero, L., Roli, A., Sampels, M., Schaerf, A. (eds.) HM 2009. LNCS, vol. 5818, pp. 171–177. Springer, Heidelberg (2009). doi:10.1007/978-3-642-04918-7_13
4. Pisinger, D., Ropke, S.: Large neighborhood search. In: Gendreau, M., Potvin, J.-Y. (eds.) Handbook of Metaheuristics. International Series in Operations Research & Management Science, vol. 146, pp. 399–419. Springer, New York (2010)
5. Caserta, M., Voß, S.: A corridor method based hybrid algorithm for redundancy allocation. J. Heuristics **22**(4), 405–429 (2016)
6. Lalla-Ruiz, E., Voß, S.: POPMUSIC as a matheuristic for the berth allocation problem. Ann. Math. Artif. Intell. **76**(1–2), 173–189 (2016)
7. Fischetti, M., Lodi, A.: Local branching. Math. Program. **98**(1), 23–47 (2003)
8. Blum, C., Pinacho, P., López-Ibáñez, M., Lozano, J.A.: Construct, merge, solve & adapt: a new general algorithm for combinatorial optimization. Comput. Oper. Res. **68**, 75–88 (2016)
9. Hansen, P., Mladenović, N.: Variable neighborhood search: principles and applications. Eur. J. Oper. Res. **130**(3), 449–467 (2001)
10. Chu, P.C., Beasley, J.E.: A genetic algorithm for the multidimensional knapsack problem. Discret. Appl. Math. **49**(1), 189–212 (1994)
11. Leung, S., Zhang, D., Zhou, C., Wu, T.: A hybrid simulated annealing metaheuristic algorithm for the two-dimensional knapsack problem. Comput. Oper. Res. **39**(1), 64–73 (2012)
12. Kong, X., Gao, L., Ouyang, H., Li, S.: Solving large-scale multidimensional knapsack problems with a new binary harmony search algorithm. Comput. Oper. Res. **63**, 7–22 (2015)
13. Hanafi, S., Freville, A.: An efficient tabu search approach for the 0–1 multidimensional knapsack problem. Eur. J. Oper. Res. **106**(2–3), 659–675 (1998)
14. López-Ibáñez, M., Dubois-Lacoste, J., Pérez Cáceres, L., Birattari, M., Stützle, T.: The irace package: iterated racing for automatic algorithm configuration. Oper. Res. Perspect. **3**, 43–58 (2016)
15. Blum, C., Blesa, M.J.: Construct, merge, solve and adapt: application to the repetition-free longest common subsequence problem. In: Chicano, F., Hu, B., García-Sánchez, P. (eds.) EvoCOP 2016. LNCS, vol. 9595, pp. 46–57. Springer, Heidelberg (2016). doi:10.1007/978-3-319-30698-8_4
16. Lizárraga, E., Blesa, M.J., Blum, C., Raidl, G.R.: Large neighborhood search for the most strings with few bad columns problem. Soft Comput. (2016, in press)

Decomposing SAT Instances with Pseudo Backbones

Wenxiang Chen$^{(\boxtimes)}$ and Darrell Whitley

Department of Computer Science, Colorado State University,
Fort Collins, CO 80523, USA
{chenwx,whitley}@cs.colostate.edu

Abstract. Two major search paradigms have been proposed for SAT solving: Systematic Search (SS) and Stochastic Local Search (SLS). In SAT competitions, while SLS solvers are effective on uniform random instances, SS solvers dominate SLS solvers on application instances with internal structures. One important structural property is decomposability. SS solvers have long been exploited the decomposability of application instances with success. We conjecture that SLS solvers can be improved by exploiting decomposability of application instances, and propose the first step toward exploiting decomposability with SLS solvers using pseudo backbones. We then propose two SAT-specific optimizations that lead to better decomposition than on general pseudo Boolean optimization problems. Our empirical study suggests that pseudo backbones can vastly simplify SAT instances, which further results in decomposing the instances into thousands of connected components. This decomposition serves as a key stepping stone for applying the powerful recombination operator, partition crossover, to the SAT domain. Moreover, we establish a priori analysis for identifying problem instances with potential decomposability using visualization of MAXSAT instances and treewidth.

Keywords: Satisfiability · Decomposition · Partition crossover · Visualization · Treewidth · Pseudo backbone

1 Introduction

SAT is the first problem proven NP-Complete [1]. Besides its theoretical importance, SAT also finds many practical applications such as bounded model checking [2] and hardware verification [3]. Erdős Discrepancy Conjecture, a longstanding open problem proposed by the famous mathematician Paul Erdős in 1930s, has recently been attacked successfully using a SAT solver [4].

The two major search paradigms for SAT solving are Systematic Search (SS) such as MiniSat [5], and Stochastic Local Search (SLS) such as the ones in the UBCSAT collection [6]. The top SLS solvers in recent SAT competitions [7] can reliably solve uniform random instances with 1 million variables and several million of clauses. The same state-of-art SLS solvers have poor performance on

© Springer International Publishing AG 2017
B. Hu and M. López-Ibáñez (Eds.): EvoCOP 2017, LNCS 10197, pp. 75–90, 2017.
DOI: 10.1007/978-3-319-55453-2_6

hard combinatorial and industrial SAT instances. In this work, we classify both hard combinatorial instances and industrial instances as *Application Instances*.

Application SAT instances often have internal structures, which can be a result of the loosely coupled components originated from the source problem domain [8], and/or the procedure involved in translating a problem from the original domain into a MAXSAT instance [9]. One important structural property, *decomposability*, studies how well the variable interaction of an application instance can be decomposed. The decomposability of an instance has been extensively studied and exploited by SS solvers with success [10–13]. On the contrary, SLS solvers direct the variables to flip only using the objective function, and are completely *oblivious* to the decomposability of application SAT instances. A **Variable Interaction Graph** (VIG) is a useful way to capture information about variable interaction in a MAXSAT instance. A simple version of the variable interaction graph is a graph denoted by G. The set of vertices, denoted by V, are the variables of a MAXSAT instance. If two variables, x_i and x_j appear together in a clause, there is an edge $e_{i,j}$ in the VIG. We conjecture that SLS solvers can be improved by exploiting the decomposability of application instances.

The evolutionary computation community has recently discovered that the decomposability of many combinatorial optimization problems can be exploited using a powerful recombination operator, partition crossover [14–16]. *Partition Crossover* fixes pseudo backbone variables (i.e., variable assignments shared among local optima) to decompose the Variable Interaction Graph (VIG) into q connected components, and then recombines partial solutions to different components in such a way that the best offspring of among all possible 2^q offsprings can be found efficiently. Partition crossover has been shown to be useful in the Traveling Salesperson Problem [17,18] and NK-Landscapes [19], but we are not aware of any work that apply this idea to SAT. The success of partition crossover relies heavily on the decomposition of problem instances. That is, partition crossover can not be applied if a problem instance can not be decomposed. We will study how to decompose application SAT instances, and evaluate the quality of decomposition. This paper will serve as a critical stepping stone for applying partition crossover to exploit the decomposability of application SAT instances. Specifically, the outcome of decomposing VIGs of application instances with pseudo backbones will determine the applicability of partition crossover in SAT.

Decomposing SAT instances with SS solvers is straightforward, since an SS solver assigns one decision variable at a time, and the assigned variables and implied variables through unit propagation [5] can be used to simplify the VIGs, which naturally leads to decomposition. In contrast, SLS solvers search in the space of full tentative assignments to all variables. However, how to "fix" variables so that a VIG can be decomposed in the context of SLS solvers is *nontrivial*. In this paper, we propose a first step toward exploiting decomposability with SLS solvers. Inspired by Zhang's success in guiding SLS solvers with pseudo backbone extracted from good local optima [20] and by the definition that backbone variables are *fixed* variable assignments across all global optima [21], we propose to decompose VIGs with pseudo backbone sampled by SLS solvers. We further propose two SAT specific optimizations that lead to better decomposition than on general pseudo Boolean optimization problems. Moreover, we establish a priori analysis, involving visualization of SAT instances and computation of treewidth, for identifying problem instances with potential decomposability, allowing SAT solving strategies to be chosen accordingly in advance.

2   Preliminaries

Variable interaction is a key structure property influencing the difficulty of SAT instances. We will say that two variables *interact* if they co-occur in some clause. In fact, given a SAT instance that only involves binary clauses (with just pairwise interactions), its satisfiability can be determined in polynomial time [22]. A VIG is typically used to model variable interactions.

Treewidth [23] is an important metric for decomposability of a graph. It has been shown that SAT can be solved in time exponential in treewidth using dynamic programming [24]. Treewidth by definition measures the "tree-likeness" of a graph G and is defined over a Tree Decomposition of G. A *Tree Decomposition* is a mapping from G to a tree T that satisfies:

1. Every vertex of G is in some tree node;
2. any pair of adjacent vertices in G should be in the same tree node;
3. the set of tree nodes containing a vertex v in G forms a connected subtree.

The *width* of a tree decomposition is the maximum size of tree node minus one. The *treewidth* (denoted as tw) of G is the minimal width over all its possible tree decomposition. Figure 1 illustrates an optimal tree decomposition of a graph. It becomes clear that treewidth can measure the decomposability of a graph, since assigning all variables in any tree node guarantees to decompose the graph, i.e., deleting any tree node disconnects a tree. The exact treewidth is NP-Hard to compute [25]. Fortunately, polynomial-time heuristics for approximate treewidth computation [26] are available. The Minimum Degree heuristic is used in our experiments.

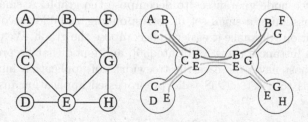

Fig. 1. A graph with eight vertices (left subfigure), and an optimal tree decomposition of it onto a tree with six nodes (right subfigure). $tw = 3 - 1 = 2$.

3 Overview

Our goal in this paper is to demonstrate the feasibility of decomposing VIGs with pseudo backbones. We conjecture that good local optima sampled by an SLS solver will share some common variable assignments, i.e., pseudo backbones are non-empty. Moreover, the pseudo backbone can be used to simplify VIGs of application instances by removing assigned pseudo backbone variables as well as the clauses satisfied by assigned pseudo backbone variables, and possibly decompose the VIGs.

We will start by taking the following three steps. First, we identify a selected set of application instances that demonstrate potential decomposability by examining their treewidths in Sect. 4. Our initial pool of application instances include all 299 satisfiable instances from both hard combinatorial track and industrial track in SAT Competition 2014[1], where the hard combinatorial track contains 150 satisfiable instances and the industrial track contains 149 satisfiable instances[2]. Second, we run one of the best performing SLS solvers on application instances [27], AdaptG^{2}WSAT [28], on the selected instances to collect good local optima (or highly competitive solutions), and then we extract shared variable assignments from the local optima to construct pseudo backbones in Sect. 5. Finally, in Sect. 6 we first propose two SAT-specific optimizations that lead to better decomposition than on general pseudo Boolean optimization problems, and then evaluate how well the pseudo backbones decompose the VIGs, by comparing the original VIGs with the simplified VIGs along two aspects: the quantitative aspect as measured by the number of connected components in the remaining VIG, and the intuitive aspect via the visualization of the remaining VIG.

4 Identifying Application Instances with Potential Decomposability

Recall that treewidth quantifies the size of the most densely connected and least-decomposable subgraph (i.e., size of the largest tree node), and can be used as a metric of the inherent decomposability of a graph [29]. Assigning all variables in any tree node guarantees to decompose the graph. A small treewidth suggests that assigning a small set of corresponding variables according to the tree decomposition will make it easier to decompose the graph. We preprocessed the application instances using SATElite [30], and found that preprocessing can reduce and almost never increase the treewidths of application instances. Preprocessing leads to a set of 258 valid preprocessed application instances. The

---

[1] http://www.satcompetition.org/2014/.
[2] The original benchmark set in the industrial track contains 150 satisfiable instances, but openstacks-sequencedstrips-nonadl-nonnegated-os-sequencedstrips-p30_3.085-SAT.cnf and openstacks-p30_3.085-SAT.cnf are identical, despite their different names.

other application instances are excluded, because they are either solved by the preprocessor or are too large for treewidth computation.

We notice that there is a high variation in the size of preprocessed instances. The number of variables for the 258 preprocessed application instances varies vastly from 55 to 1,471,468 with an average of 47,724 and a standard deviation of 174,429. To accommodate the variation, we use normalized treewidth, i.e., treewidth to number of variables ratio (denoted as $tw.n$), as a metric between zero and one for evaluating inherent decomposability. We therefore intend to select application instances that are known to have small $tw.n$ (i.e., highly decomposable) as the first step toward demonstrating the feasibility of decomposing VIGs with pseudo backbones. If pseudo backbones can not decompose VIGs with small $tw.n$, it would be even less likely to decompose VIGs with larger $tw.n$ with pseudo backbones. This leads to the first research question.

Question 1. Are there many application instances with high decomposability?

Recall that the application instances in our study originate from the hard combinatorial track and the industrial track of the SAT Competition 2014, which has been run for more than a decade. These application instances are representatives of a wide variety of real world SAT applications, and are challenging for the current SAT solvers. To study the overall distribution of $tw.n$ in the preprocessed application instances, we plot the histogram in Fig. 2. The histogram indicates that the highest frequency of $tw.n$ is within the range from 0.00 to 0.01. The first bar has a frequency of 28. This means that there are 28 preprocessed application instances that are guaranteed to be decomposed when assigning all variables in a tree node that contains at most 1% variables. We also observe that $tw.n$ for 199 out of the 258 application instances are less than 0.4. This is encouraging, because many real world application instances from a wide variety of domains are indeed *highly decomposable*.

We decide to select the 27 preprocessed application instances with $tw.n <$ 0.01 as the testbed for exploring the feasibility of decomposition with pseudo backbones. Note that one instance with $tw.n < 0.01$, prime2209-98, is excluded, because it is already separated into over a hundred independent densely connected components and can be solved optimally by SLS solvers. We list the details on the selected instances under "After Preprocessing" of Table 1. Despite the small normalized treewidths, the treewidth in these instances are still much too large to solve using dynamic programming.

We merge data generated across different experiments in one big table. We will focus on the 27 instances from now on in this paper. Merging results across different experiments performed on the same set of instances allows us to synthesize findings and to discover connections among the outcomes of experiments with different emphases. The columns are ordered by the sequence in which the experiments are conducted, and the columns belonging to different experiments are separated using double vertical lines to improve readability.

From *Instance* column of Table 1, we note that the 27 selected instances come from only 5 problem domains: aaai-ipc (planning), AProVE (termination analysis of JAVA program), atco (Air Traffic Controller shift scheduling

Histogram of Normalized Treewidth

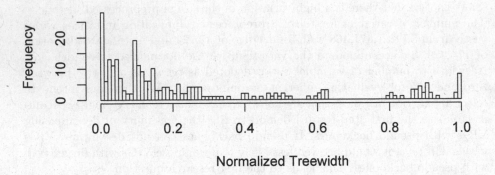

Normalized Treewidth

Fig. 2. Normalized Treewidths of 258 preprocessed application instances. Each bar counts the frequency of the normalized treewidths being within an interval of size 0.01.

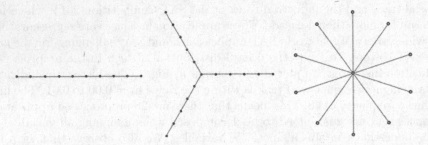

Fig. 3. Three graphs with $tw = 1$, but with different degrees of decomposability.

problem), LABS (Low Autocorrelation Binary Sequence), and SAT-instance (autocorrelation in combinatorial design). The instance with smallest $tw.n$ is LABS_n088_goal008, in which there are 182,015 variables, and yet its treewidth is only 203.

Normalized treewidth as a descriptive measure for decomposability of a graph can be *limited* for two reasons. First, treewidth captures the size of largest tree node in the tree decomposition, while missing the information on the sizes of the rest of tree nodes. Second, the topology of the tree decomposition can also play a role in determining the decomposability of a graph. Consider the three graphs with $tw = 1$ in Fig. 3, all three of them are essentially trees. However, they have very different topologies as well different degrees of decomposability. Removing the central node decomposes them into 2, 3, and 10 connected components, respectively. This concern lead to Question 2.

Question 2. What exactly is the variable interaction topologies of application instances that result in the small normalized treewidths?

To illustrate how the variable interaction topologies of the 27 selected instances lead to the small $tw.n$, we generate their VIGs using a force-directed graph layout algorithm[3]. Sinz [31] suggests VIGs of SAT instances to be laid out using force-directed graph drawing algorithm by Fruchterman and Reingold [32] that is known to reflect graph clustering and symmetry. Fruchterman and Reingold's graph layout algorithm runs in polynomial time, $O(V + E)$, the sum of the number of vertices and edges in the VIG. Among the 27 selected instances, the VIGs of instances from the same class are visually similar. Therefore, one representative VIG is picked from each of 5 problem domains. The 5 representative VIGs are presented in Fig. 4, in which a red circle represents a variable and a black line between two red circles represents the two relative variables co-occur in some clause.

The presented VIGs illustrate the decomposability of the variable interaction topologies of the selected application instances. atco_enc3_opt1_13_48 consists of several linear topologies that loosely interleaves. LABS_n088_goal008 exhibits a densely connected "core" at its center, and the connections become progressively sparser as being away from the core. SAT_instance_N=49 appears axisymmetric with two cores on each side and the connections between the two cores are sparser. aaai10-ipc5 forms several layers of "clusters" with looser connections between clusters. The aforementioned 4 VIGs exhibits high visual decomposability. Lastly, AProVE09-06 seems uniformly and densely connected at its large core. It might be difficult to exploit its decomposability, despite its small normalized treewidth suggests otherwise.

Through the visual analysis on the VIGs, we find that (normalized) treewidth, while reveals the potential decomposability of an instance, does not always tell the full story. Visualization of VIGs can help to gain more intuitive insights. We will further learn how well can a VIG be decomposed in practice with pseudo backbone in Sect. 6.

5   Computing Pseudo Backbone from Good Local Optima

In contrast to the backbone that is extracted from all global optima, *pseudo backbones* are approximated from local optima with low evaluations (aka "Good Local Optima") [20]. We employ AdaptG^{2}WSAT [28] from the UBCSAT implementation [6] to find good local optima, because both our preliminary experiments and a previous study by Kroc *et al.* [27] indicate that AdaptG^{2}WSAT is one of the best performing SLS solvers on application instances. We run AdaptG^{2}WSAT for up to 1000 s, record the best local optima found during the run, and repeat the run for ten times with different random seeds for each of the 27 preprocessed application instances. As a result, 10 good local optima are collected for each instance. The goodness of approximating the true backbone with pseudo backbones depends on the quality of local optima used to generate them.

---

[3] We use the Python package graph-tool (https://graph-tool.skewed.de/). Graph-tool is known for its efficiency on large graphs, due to its C++ core implementation.

Fig. 4. Variable interaction graphs of 5 representative preprocessed application instances.

It is also critical to realize the limit of the current SLS solvers on applications. These concerns leads to Question 3.

Question 3. On application instances with small normalized treewidths, how good are the best local optima found by AdaptG^{2}WSAT within 1000 s?

Table 1 (under *Evaluation*) presents statistics on the quality of the best local optima collected in each of the 10 runs. Except for the 5 atco instances, AdaptG^{2}WSAT is able to find local optima with at most hundreds of unsatisfied clauses, on instances with up to about 3 million clauses. Although these are somewhat good local optima, we also find considerable room for improving upon AdaptG^{2}WSAT on application instances that exhibit high potential decomposability.

Notice that we employ the pseudo backbone concept differently from Zhang's approach [20]. In Zhang's work [20], all local optima are used to compute an empirical probability distribution, *pseudo backbone frequency*, which estimates the likelihood of a variable assigned to true in the real backbone. In our case, we instead prefer to maximize the pseudo backbone size for the purpose of decomposition, i.e., the number of fixed variables across local optima. We expect that the more distinct local optima taken into account, the less variables are assigned consistently. We present an analysis of the sizes of pseudo backbones generated using pairs of local optima (under *Between Pairs of LocOpt*) as well as the size of the (less-)pseudo backbone constructed by taking the intersection of all 10 local optima (under *ALL.LO*) in Table 1.

We observe that, considering the pairs of local optima instead of the 10 local optima all together indeed yields larger pseudo backbone. We also notice that the variation in the pseudo backbone sizes across all $\frac{10 \times (10-1)}{2} = 45$ pairs of local optima is small, as indicated by the small standard deviations and the small difference between min and max. Meanwhile, the difference between mean and ALL.LO is much more pronounced. On the atco instances, average pseudo backbone size is more than 100× larger than that of all 10 local optima (as indicated by ALL.LO). For the purpose of fixing as many pseudo backbone variables as possible to increase the chance of decomposing a given VIG, we choose the pseudo backbones generated from the pairs of local optima for simplifying and decomposing VIGs.

Normalizing the maximum pseudo backbone size constructed from pairs of local optima (max under *Between Pairs of LocOpt*) by the total numbers of variables (n under *After Preprocessing*), we find that the maximum pseudo backbones fix 70.3% of total variables on average, with a standard deviation of 16.4%. On aaai10-ipc5 and LABS instances, maximum pseudo backbones consistently fix as many as around 90% of variables. Combining the large pseudo backbone constructed from pairs of local optima and the low treewidths that are inherent to the selected application instances, the idea of decomposing VIGs with Pseudo backbones appears very promising. In Sect. 6, we will evaluate the practical decomposability of VIGs with pseudo backbones.

Table 1. *After Preprocessing*: number of variables (n), number of clauses (m), treewidth (tw) and normalized treewidth (tw.n) are reported. *Evaluation*: the mean and standard deviation (std) of the best local optima found by AdaptG^{2}WSAT over 10 runs. *Between Pairs of LocOpt*: statistics of the sizes of pseudo backbones constructed from each pair of the 10 local optima. *ALL.LO*: the size of pseudo backbone constructed from all 10 local optima. *#Components*: the number of connected components after decomposition.

Instance	After Preprocessing				Evaluation		Between Pairs of LocOpt				ALL.LO	#Components		
	n	m	tw	tw.n	mean	std	min	mean	std	max		min	median	max
aaai10-ipc5	308480	2910796	2254	0.0073	21.5	1.1	271741	272964	585	274056	231715	7	20	38
AProVE09-06	37726	192754	206	0.0055	472.9	14.5	20676	21407	335	22026	2669	11	1373	1620
atco_enc3_opt1_03_53	991419	3964576	2369	0.0024	494523.1	449.4	495853	497323	630	498900	2768	937	1020	1090
atco_enc3_opt1_04_50	1320073	5313526	3167	0.0024	775484.8	922.2	660175	661204	588	662712	3570	1023	1087	1164
atco_enc3_opt1_13_48	1471468	5921783	3575	0.0024	906206.9	529.2	736017	736956	574	738506	3910	1193	1287	1365
atco_enc3_opt2_05_21	1293612	5193145	3090	0.0024	755748.1	449.2	646598	648166	622	649557	4061	0	0	37
atco_enc3_opt1_13_48	1067657	4305314	2494	0.0023	555417.4	906.2	533617	535064	598	536835	3162	0	0	21
LABS_n044_goal003	20259	70752	100	0.0049	4.8	1.5	16626	17130	185	17661	11915	1	52	374
LABS_n045_goal003	22095	76929	102	0.0046	6.3	1.6	18141	18702	253	19412	13163	1	52	419
LABS_n051_goal004	32522	112980	116	0.0036	5.6	1.3	27368	27956	250	28465	20389	1	69	544
LABS_n052_goal004	34674	120036	115	0.0033	6.3	1.3	29194	29748	265	30397	21513	1	77	548
LABS_n066_goal005	71336	245401	146	0.0020	10.8	1.1	61959	63083	446	64135	49308	100	133	1012
LABS_n078_goal006	118601	405972	179	0.0015	11.5	1.8	104272	105496	569	106950	83702	156	179	1566
LABS_n081_goal007	138803	479962	186	0.0013	75.2	7.3	121057	122232	774	124133	95655	146	291	394
LABS_n082_goal007	147213	509579	192	0.0013	167.8	10.7	127530	129641	1466	132499	102097	168	251	1435
LABS_n088_goal008	182015	638771	203	0.0011	654.1	67.1	134971	152445	7438	160520	103838	231	371	2084
SAT_instance_N=111	72001	287702	269	0.0037	617.9	18.3	41011	42402	684	44759	5986	34	55	1218
SAT_instance_N=49	13766	54806	125	0.0091	117.9	9.3	7500	7971	210	8356	1017	0	26	156
SAT_instance_N=55	17272	68977	145	0.0084	141.3	10.0	9543	10189	333	10852	1513	0	39	186
SAT_instance_N=63	22780	90906	153	0.0067	193.5	16.0	12717	13286	308	14071	1875	21	32	392
SAT_instance_N=69	27286	109087	165	0.0060	239.1	6.7	15294	15825	333	16663	2008	28	51	519
SAT_instance_N=72	29289	117013	184	0.0063	264.1	9.1	16407	16985	336	17879	2196	12	40	370
SAT_instance_N=75	32356	129290	187	0.0058	288.0	11.7	17807	18750	418	19487	2224	25	41	345
SAT_instance_N=77	34129	136225	189	0.0055	303.6	11.9	18924	19685	411	20473	2300	0	41	719
SAT_instance_N=85	41562	166062	203	0.0049	358.8	10.0	22810	24243	552	25229	3232	27	48	707
SAT_instance_N=93	49745	199060	221	0.0044	442.7	10.2	27198	28694	705	30307	3050	16	46	408
SAT_instance_N=99	57407	229153	239	0.0042	467.0	44.6	33073	34138	611	35594	5534	36	54	1075

6   Improving Decomposition on SAT Instances

Selecting the pair of local optima that leads to the largest pseudo backbone does not necessarily decompose a given VIG. Two additional confounding factors also come into play. First, the distribution of pseudo backbone variables over the VIG is critical. The pseudo backbone variables might scatter over many tree nodes in the tree decomposition, which impairs the decomposition. Taking Fig. 1 for an example, assigning variables B and C decomposes the graph into two independent components, whereas assigning variables C and G does not. Second, assigning some variables can lead to considerable propagations, which can further simplify the VIG and facilitate the decomposition. In this section, we first introduce two SAT-specific optimizations that lead to better decomposition than on general pseudo Boolean optimization problems, and then conduct an empirical study to evaluate the practical decomposability of VIGs with pseudo backbones we collected in the previous section.

In general pseudo Boolean optimization, fixing the assignment to a variable triggers the removal of that variable and all edges incident to it [15]. Due to the nature of MAXSAT problems, we introduce two optimizations that promote the decomposition of VIGs.

First, we remove clauses satisfied by the assigned variables. Assigning a pseudo backbone variable v can possibly satisfy a clause C, which leads to direct removal of the entire clause C from the formula. Note that each clause forms a clique in the VIG, since every pair of variables in a clause interacts. Suppose C contains k variables and v only appears in C, assigning v removes up to $\frac{k \times (k-1)}{2}$ edges (i.e., the number of edges in a clique of size k) from the VIG. Now consider a similar pseudo Boolean optimization problem, in which a variable v only appears in a subfunction C that contains k variables. Assigning v in the general pseudo Boolean optimization problem, however, only removes v from C, which leads to the deletion of only the $k - 1$ edges incident to v.

Second, we apply unit propagation after assigning pseudo backbone variables. Unit propagation can also imply the assignments of extra variables besides the pseudo backbone variables. The implied variables again can satisfy some clauses, simplifying and possibly decomposing the VIG even more. In this sense, unit propagation *reinforces* the first optimization. Given the same VIGs, SAT instances clearly have better theoretical potential of being decomposed than general pseudo Boolean optimization instances.

6.1   Empirical Results

Recall that every pair of 10 local optima generates a pseudo backbone. On each application instance, 10 local optima result in $\frac{10 \times (10-1)}{2} = 45$ pairs, which are further used to generate 45 pseudo backbones. We then use the pseudo backbones to simplify each of the application instances, which leads to 45 simplified instances for every application instance. For each application instance, the statistics (min, median and max) on the number of connected components of the instance simplified using 45 different pseudo backbones are presented under *#Components* in Table 1.

We present the median instead of the mean, because we notice a large variance in the number of connected components on many instances. Considering LABS_n052_goal004, the maximum number of components is 548× the minimum number of components. In fact, we argue that the high variance on the number of connected components is an empirical indicator for partition crossover to be *more powerful on SAT* than on other pseudo Boolean optimization problems like NK-landscapes.

The maximum of number of connected components across all 27 non-empty simplified instances varies from 21 to 2084. This means there is always an opportunity for applying partition crossover. In some case, one application of partition crossover can return the best offspring among 2^{2084} ones. As for the typical case scenarios, the median number of connected components varies from 0 to 1373, with an average of 249.

We notice that some instances are drastically simplified. The minimum and median number of components for two *acto* instances are zero, meaning that the two simplified instances are completely empty. Note that pseudo backbones on average only consists of roughly 50% of the variables (648166/1293612 = 0.501 for atco_enc3_opt2_05_21 and 535064/1067657 = 0.501 for atco_enc3_opt1_13_48). Therefore, the other half of the variables are all implied by unit propagation. This suggests that maximizing pseudo backbone size does not always translate into the maximum number of connected components. There is a *trade-off* between simplifying a graph so that it can be decomposed, and avoiding over-simplifying a graph that leads to fewer or none connected components left. Fortunately, the maximum number of connected components for atco_enc3_opt2_05_21 and atco_enc3_opt1_13_48 is 37 and 21, meaning that there are still chances for partition crossover to be useful.

In Fig. 5, we present the decomposed VIGs that yields the median number of connected components. There are several interesting patterns. First, three instances, atco_enc3_opt1_13_48, LABS_n088_goal008 and SAT_instance_N=49, are decomposed into mostly linearly connected components. Second, even though the pseudo backbone for aaai10-ipc5 contains 88% of the variables, the instance still has a large connected component that contains the majority of the remaining variables. Notice that there are several weak links in the largest components that could have been removed using the pseudo backbone, leading to more connected components. This indicates that using the pseudo backbone can miss some "low-hanging fruits" for further decomposing a VIG. Third, recall that the VIG of AProVE09-06 before simplification (see Fig. 4) appears difficult to decompose. Indeed, although the simplified VIG shows that AProVE09-06 is decomposed into many connected components, some of the connected components are non-trivially large and complex, indicating its limited potential of being decomposed further. Visualization can indeed complement treewidth in identifying instances with potential decomposability.

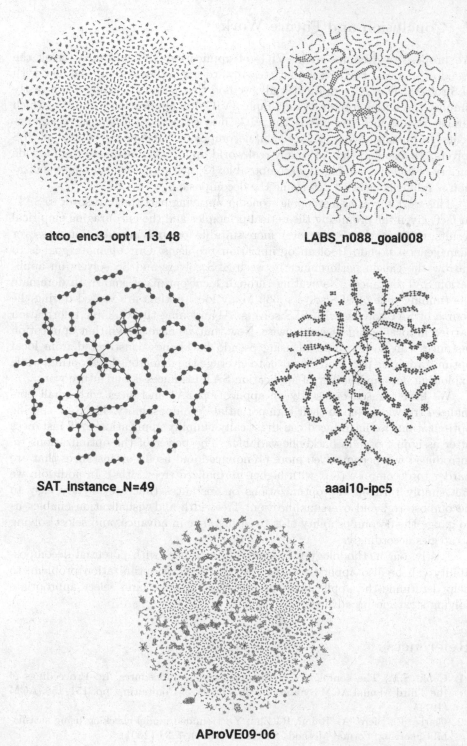

Fig. 5. Decomposed variable interaction graphs of representative application instances that yields the median number of connected components.

7   Conclusion and Future Work

We propose to decompose the VIGs of application SAT instances using the pseudo backbone, taking the first step toward exploiting decomposability with SLS solvers. Through empirical study, we find that pseudo backbones can vastly simplify the Variable Interaction Graphs (VIGs) of application instances, which further results in decomposing the VIGs into thousands of connected components. In pursuit of the primary goal, decomposing VIGs in the context of SLS solvers, we also find that (1) many real world application instances from a wide variety of domains are highly decomposable; (2) neither treewidth nor visualization is fully sufficient to determine the decomposability of VIG.

This work serves as a stepping stone for applying partition crossover to SAT. In fact, given the promising theoretical principles and the encouraging empirical results, we argue that SAT is even more suitable for applying partition crossover than general pseudo Boolean optimization problems. Our ultimate goal is to narrow the gap in performance between SLS solvers and SS solvers on application SAT instances. Navigating through local optima is known to dominate the running time of SLS solvers [33]. Many local optima are visited during the course of SAT solving with SLS solvers. Abandoning the valuable information carried in local optima seems unwise. Now that we have shown that application instances can be decomposed using pseudo backbones constructed from local optima, applying partition crossover to leverage the numerous local optima while exploiting decomposability of application SAT instances is our future work.

We have focused our study on application SAT instances with small normalized treewidth due to their high potential decomposability. The SAT-specific optimizations we introduced can drastically simplify application SAT instances after assigning pseudo backbone variables. The power of the optimizations we introduced will become even more pronounced and useful on instances that are harder to decompose (i.e., with higher normalized treewidths). In addition, we can simply turn off the optimizations on instances that are already easy to decompose to avoid oversimplification. Treewidth and visualization enables us to gauge the decomposability of a given instance in advance, and select solving strategies accordingly.

Lastly, our methodology for identifying instances with potential decomposability can be also applied to other pseudo Boolean optimization problem, to help determine the applicability of partition crossover and select appropriate solving strategies in advance.

References

1. Cook, S.A.: The complexity of theorem-proving procedures. In: Proceedings of the Third Annual ACM Symposium on Theory of Computing, pp. 151–158. ACM (1971)
2. Clarke, E., Biere, A., Raimi, R., Zhu, Y.: Bounded model checking using satisfiability solving. Formal Methods Syst. Des. 19(1), 7–34 (2001)

3. Velev, M.N., Bryant, R.E.: Effective use of Boolean satisfiability procedures in the formal verification of superscalar and VLIW microprocessors. J. Symb. Comput. **35**(2), 73–106 (2003)
4. Konev, B., Lisitsa, A.: Computer-aided proof of Erdős discrepancy properties. Artif. Intell. **224**, 103–118 (2015)
5. Eén, N., Sörensson, N.: An extensible SAT-solver. In: Giunchiglia, E., Tacchella, A. (eds.) SAT 2003. LNCS, vol. 2919, pp. 502–518. Springer, Heidelberg (2004). doi:10.1007/978-3-540-24605-3_37
6. Tompkins, D.A.D., Hoos, H.H.: UBCSAT: an implementation and experimentation environment for SLS algorithms for SAT and MAX-SAT. In: Hoos, H.H., Mitchell, D.G. (eds.) SAT 2004. LNCS, vol. 3542, pp. 306–320. Springer, Heidelberg (2005). doi:10.1007/11527695_24
7. The International SAT Competitions Webpage. http://www.satcompetition.org/
8. Amir, E., McIlraith, S.: Partition-based logical reasoning for first-order and propositional theories. Artif. Intell. **162**(1), 49–88 (2005)
9. Prestwich, S.D.: CNF encodings. Handb. Satisfiability **185**, 75–97 (2009)
10. Ansótegui, C., Giráldez-Cru, J., Levy, J., Simon, L.: Using community structure to detect relevant learnt clauses. In: Heule, M., Weaver, S. (eds.) SAT 2015. LNCS, vol. 9340, pp. 238–254. Springer, Heidelberg (2015). doi:10.1007/978-3-319-24318-4_18
11. Bjesse, P., Kukula, J., Damiano, R., Stanion, T., Zhu, Y.: Guiding SAT diagnosis with tree decompositions. In: Giunchiglia, E., Tacchella, A. (eds.) SAT 2003. LNCS, vol. 2919, pp. 315–329. Springer, Heidelberg (2004). doi:10.1007/978-3-540-24605-3_24
12. Huang, J., Darwiche, A.: A structure-based variable ordering heuristic for SAT. In: IJCAI, vol. 3, pp. 1167–1172 (2003)
13. Monnet, A., Villemaire, R.: Scalable formula decomposition for propositional satisfiability. In: Proceedings of the Third C* Conference on Computer Science and Software Engineering, pp. 43–52. ACM (2010)
14. Whitley, D., Hains, D., Howe, A.: Tunneling between optima: partition crossover for the traveling salesman problem. In: Proceedings of the 11th Annual Conference on Genetic and Evolutionary Computation, pp. 915–922. ACM (2009)
15. Tinós, R., Whitley, D., Chicano, F.: Partition crossover for pseudo-boolean optimization. In: Proceedings of the 2015 ACM Conference on Foundations of Genetic Algorithms XIII, pp. 137–149. ACM (2015)
16. Ochoa, G., Chicano, F., Tinós, R., Whitley, D.: Tunnelling crossover networks. In: Proceedings of the 2015 Annual Conference on Genetic and Evolutionary Computation, GECCO 2015, pp. 449–456. ACM, New York (2015)
17. Lin, S.: Computer solutions of the traveling salesman problem. Bell Syst. Tech. J. **44**(10), 2245–2269 (1965)
18. Tinós, R., Whitley, D., Ochoa, G.: Generalized asymmetric partition crossover (GAPX) for the asymmetric TSP. In: Proceedings of the 2014 Annual Conference on Genetic and Evolutionary Computation, pp. 501–508. ACM (2014)
19. Kauffman, S., Levin, S.: Towards a general theory of adaptive walks on rugged landscapes. J. Theor. Biol. **128**(1), 11–45 (1987)
20. Zhang, W.: Configuration landscape analysis and backbone guided local search: part i: satisfiability and maximum satisfiability. Artif. Intell. **158**, 1–26 (2004)
21. Monasson, R., Zecchina, R., Kirkpatrick, S., Selman, B., Troyansky, L.: Determining computational complexity from characteristic 'phase transitions'. Nature **400**(6740), 133–137 (1999)
22. Aspvall, B., Plass, M.F., Tarjan, R.E.: A linear-time algorithm for testing the truth of certain quantified Boolean formulas. Inf. Process. Lett. **8**(3), 121–123 (1979)

23. Robertson, N., Seymour, P.D.: Graph minors. II. Algorithmic aspects of tree-width. J. Algorithms **7**(3), 309–322 (1986)
24. Bodlaender, H.L.: Dynamic programming on graphs with bounded treewidth. In: Lepistö, T., Salomaa, A. (eds.) ICALP 1988. LNCS, vol. 317, pp. 105–118. Springer, Heidelberg (1988). doi:10.1007/3-540-19488-6_110
25. Arnborg, S., Corneil, D.G., Proskurowski, A.: Complexity of finding embeddings in a k-tree. SIAM J. Algebraic Discret. Methods **8**(2), 277–284 (1987)
26. Bodlaender, H.L.: Discovering treewidth. In: Vojtáš, P., Bieliková, M., Charron-Bost, B., Sýkora, O. (eds.) SOFSEM 2005. LNCS, vol. 3381, pp. 1–16. Springer, Heidelberg (2005). doi:10.1007/978-3-540-30577-4_1
27. Kroc, L., Sabharwal, A., Gomes, C.P., Selman, B.: Integrating systematic and local search paradigms: a new strategy for MaxSAT. In: Proceedings of the 21st International Joint Conference on Artificial Intelligence, IJCAI 2009, pp. 544–551. Morgan Kaufmann Publishers Inc., San Francisco (2009)
28. Li, C.M., Wei, W., Zhang, H.: Combining adaptive noise and look-ahead in local search for SAT. In: Marques-Silva, J., Sakallah, K.A. (eds.) SAT 2007. LNCS, vol. 4501, pp. 121–133. Springer, Heidelberg (2007). doi:10.1007/978-3-540-72788-0_15
29. Gaspers, S., Szeider, S.: Strong backdoors to bounded treewidth SAT. In: 2013 IEEE 54th Annual Symposium on Foundations of Computer Science (FOCS), pp. 489–498. IEEE (2013)
30. Eén, N., Biere, A.: Effective preprocessing in SAT through variable and clause elimination. In: Bacchus, F., Walsh, T. (eds.) SAT 2005. LNCS, vol. 3569, pp. 61–75. Springer, Heidelberg (2005). doi:10.1007/11499107_5
31. Sinz, C.: Visualizing SAT instances and runs of the DPLL algorithm. J. Autom. Reason. **39**(2), 219–243 (2007)
32. Fruchterman, T.M., Reingold, E.M.: Graph drawing by force-directed placement. Softw.: Pract. Exp. **21**(11), 129–1164 (1991)
33. Frank, J.D., Cheeseman, P., Stutz, J.: When gravity fails: local search topology. J. Artif. Intell. Res. **7**, 249–281 (1997)

Efficient Consideration of Soft Time Windows in a Large Neighborhood Search for the Districting and Routing Problem for Security Control

Bong-Min Kim[1,2], Christian Kloimüllner[1(✉)], and Günther R. Raidl[1]

[1] Institute of Computer Graphics and Algorithms, TU Wien,
Favoritenstraße 9–11/1861, 1040 Vienna, Austria
{kloimuellner,raidl}@ac.tuwien.ac.at
[2] Research Industrial Systems Engineering, Concorde Business Park F,
2320 Schwechat, Austria
bong-min.kim@rise-world.com

Abstract. For many companies it is important to protect their physical and intellectual property in an efficient and economically viable manner. Thus, specialized security companies are delegated to guard private and public property. These companies have to control a typically large number of buildings, which is usually done by teams of security guards patrolling different sets of buildings. Each building has to be visited several times within given time windows and tours to patrol these buildings are planned over a certain number of periods (days). This problem is regarded as the *Districting and Routing Problem for Security Control*. Investigations have shown that small time window violations do not really matter much in practice but can drastically improve solution quality. When softening time windows of the original problem, a new subproblem arises where the minimum time window penalty for a given set of districts has to be found for each considered candidate route: What are optimal times for the individual visits of objects that minimize the overall penalty for time window violations? We call this *Optimal Arrival Time Problem*. In this paper, we investigate this subproblem in particular and first give an exact solution approach based on *linear programming*. As this method is quite time-consuming we further propose a heuristic approach based on greedy methods in combination with dynamic programming. The whole mechanism is embedded in a large neighborhood search (LNS) to seek for solutions having minimum time window violations. Results show that using the proposed heuristic method for determining almost optimal starting times is much faster, allowing substantially more LNS iterations yielding in the end better overall solutions.

Keywords: Districting and routing problem for security control · Vehicle routing problem · Soft time windows · Dynamic programming · Linear programming

We thank Günter Kiechle and Fritz Payr from CAPLAS GmbH for the collaboration on this topic. This work is supported by the Austrian Research Promotion Agency (FFG) under contracts 856215 and 849028.

B. Hu and M. López-Ibáñez (Eds.): EvoCOP 2017, LNCS 10197, pp. 91–107, 2017.
DOI: 10.1007/978-3-319-55453-2_7

1   Introduction

Theft and vandalism are a big and growing problem for many private and public companies. Thus, companies need to surveil their property, although permanent surveillance typically is not possible due to limited financial resources. Security companies, which are specialized experts, are therefore frequently engaged with observing the properties of these companies.

To minimize financial expenditures, objects are irregularly visited multiple times per day by security guards instead of dedicating a single security guard to a single object. Security guards have the duty of observing a particular set of objects. The number and times when these objects have to be visited may differ in each considered period. The problem of planning the districting and individual routes for performing the visits has been introduced by Prischink et al. [8] and is called *Districting and Routing Problem for Security Control* (DRPSC).

The previously proposed approach considers time windows in a strict sense. In practice, however, small time window violations typically do not matter much, and a larger flexibility in respect to them often allows substantially better solutions. In this paper, we consider the *Districting and Routing Problem for Security Control with Soft Time Windows* (DRPSC-STW). Soft time windows may be violated to some degree, and their violation is considered in the objective function by penalty terms. In this context, the subproblem of determining optimal visiting times for a given candidate tour so that the total time window penalty is minimized arises. We call this problem *Optimal Arrival Time Problem* (OATP).

To classify the DRPSC-STW in context of the vehicle routing literature, one can see it as a *periodic vehicle routing problem with soft time windows* with additional constraints concerning separation time and maximum tour duration, where objects may have to be visited multiple times in each period. Separation-time constraints are a minimum time difference between two consecutive visits of the same object in a tour. Moreover, each tour for every district and period must not exceed a given maximum tour duration.

In this work, we primarily focus on the OATP and how it can be effectively solved. To this end we propose an approach based on *linear programming* (LP) and a faster heuristic approach using greedy techniques and dynamic programming. These mechanisms are embedded in a large neighborhood search (LNS) [6].

The paper is structured as follows. Related work is given in Sect. 2 and the formal problem definition is stated in Sect. 3. Subsequently, we describe the OATP in Sect. 4 where we also introduce the LP approach. Then, in Sect. 5, the efficient hybrid heuristic for solving the OATP is introduced and in Sect. 6 the LNS metaheuristic for approaching the DRPSC-STW is proposed. Experiments are performed in Sect. 7 and, finally, a conclusion as well as an outlook for future work is given in Sect. 8.

2   Related Work

Prischink et al. [8] introduce the DRPSC and propose two construction heuristics as well as a sophisticated *district elimination algorithm* for the districting part of the problem. In the district elimination algorithm they iteratively eliminate a district, put the objects of these districts in a so called *ejection pool* and then try to insert the objects of this ejection pool again into the set of available districts. We adopt this idea/mechanism for developing a *destroy and recreate* neighborhood inside our LNS. We, thus, remove the objects of two, uniformly at random selected, districts and put them into a so called *ejection pool* but do not delete the districts from which we removed the objects. Then, we execute a single run/step of the proposed district elimination algorithm which tries to reassign the objects in the ejection pool to the available districts, and let the algorithm terminate if the ejection pool is empty at the end of this single iteration.

As the focus of our current work lies on the extension to soft time windows, we also put our attention here on previous work dealing with them. Although much more work is done on problem types with hard time windows, there already exists a significant number of works which introduce efficient methods to effectively handle soft time window constraints.

Ibaraki et al. [4] proposed a dynamic programming (DP) based approach to determine optimal starting times in conjunction with soft time windows which is applicable to a wider range of routing and scheduling applications. The total penalty incurred by time window violations is minimized. Compared to our approach they consider more general piecewise-linear penalty functions. Unfortunately, their approach is not directly applicable in our context as we have to additionally consider minimum separation times between visits of the same objects (i.e., objects can only be visited again if a minimum separation time between two consecutive visits is considered) and a maximum tour length. However, we will show later how this efficient DP method can nevertheless be utilized to some degree in our case.

Hashimoto et al. [3] extended the work of Ibaraki et al. to also consider flexible traveling times, which are also penalized if violated. They show, however, that the problem becomes NP-hard in case.

Taillard et al. [9] solve the vehicle routing problem with soft time windows by using tabu search. They do not consider any penalties for arriving too early but introduce a "lateness penalty" into the objective function. This penalty value is weighted by a given factor and the problem can be transformed into the vehicle routing problem with hard time windows by setting the weight factor to ∞.

Another work which shows the efficiency of applying DP for solving problems with soft time windows is by Ioachim et al. [5]. They solve the shortest path problem with time windows and linear node costs, where the linear node costs correspond to the modeling of soft time windows.

Fagerholt [2] published an approach for *ship scheduling with soft time windows*. He argues that by considering soft time windows, solution quality can be drastically improved and in practice small time window violations do not really matter. As in our work, a maximum allowed time window violation is used and earlier and later service is penalized. The approach can handle also non-convex penalty functions whereas in the literature most often only convex penalty functions are considered. The proposed solution approach uses a discretized time network in which nodes are duplicated according to possible start/arrival times. On the obtained shortest path network problem DP is applied for obtaining optimal arrival times.

To summarize related work, DP can frequently be an effective tool to determine optimal arrival/service times when considering soft time windows. Certain specificities of problems like maximal total tour duration and other constraints, however, frequently become an obstacle and prohibit the direct application of an efficient DP as the subproblem of determining optimal arrival times becomes NP hard. Nevertheless, DP may still be an important ingredient to deal with such situations in practice.

3   Problem Definition

In the DRPSC-STW we are given a set of objects $I = \{1, \ldots, n\}$ and a depot 0, which is the start and end of each route. Travel times among the objects and the depot are given by $t_{i,i'}^{\text{travel}} > 0$ for $i, i' \in I \cup \{0\}$. We assume the triangle inequality to hold among these travel times. For every object $i \in I$ we are given a (small) number of visits $S_i = \{i_1, \ldots, i_{|S_i|}\}$, and we are given a set of periods $P = \{1, \ldots, p\}$. As not all visits have to take place in every period, subsets $W_{i,j} \subset S$ contain the visits of object i requested in period j for all $i \in I$, $j \in P$. The depot is visited two times, namely at the start of the tour and at the end of the tour. To ease modeling we define 0_0 to be the departure from the depot at the beginning and 0_1 to be the arrival at the depot at the end.

Each visit $i_k \in S_i, i \in I$ is associated with a visit time $t_{i_k}^{\text{visit}}$ and a particular time window $T_{i_k} = [T_{i_k}^{e}, T_{i_k}^{l}]$ in which the whole visit should preferably take place, already including its visit time. Visits $i_k \in S_i$ of an object $i \in I$ have to be visited in the given order, i.e., visit k has to be performed before visit k' iff $k < k'$.

Time windows of the visits are now softened such that an earlier start or later finishing of the service at an object is allowed. The maximum allowed earliness and lateness are, however, restricted by Δ, yielding the hard time windows $T_{i_k}^{h} = [T_{i_k}^{he}, T_{i_k}^{hl}] = [T_{i_k}^{e} - \Delta, T_{i_k}^{l} + \Delta]$, which must not be violated in any feasible solution.

An additional important requirement in the context of our security application is that any two successive visits $i_k, i_{k+1} \in W_{i,j}$ of the same object $i \in I$ must be separated by a *minimum separation time* t^{sep}. Obviously, visiting an object twice without a significant time inbetween would not make much sense. The maximum duration of any tour is given by t^{max}.

Solutions to the DRPSC-STW are given by a tuple (D, τ, a) where $D = \{D_1, \ldots, D_m\}$ is the partitioning of objects into districts, $\tau = (\tau_{r,j})_{r=1,\ldots,m, \ j\in P}$ are the routes for each district and period, and a denotes the respective arrival times. Each tour $\tau_{r,j} = (\tau_{r,j,0}, \ldots, \tau_{r,j,l_{r,j}+1})$ with $l_{r,j} = \sum_{i \in D_r} |W_{i,j}|$ starts and ends at the depot, i.e., $\tau_{r,j,0} = 0_0$ and $\tau_{r,j,l_{r,j}+1} = 0_1$, $\forall r = 1, \ldots, m, j \in P$, and performs each visit in the respective ordering of the sequence. Each visit of a tour $\tau_{r,j,u}$ has to be associated with a specific arrival time $a_{r,j,u}$ and thus, $a = (a_{r,j,u})_{r=1,\ldots,m, \ j=1,\ldots,p, \ u=1,\ldots l_{r,j}+1}$. An object always is immediately serviced after arrival but waiting is possible before leaving the object. A tour is feasible, if all visit, travel, and separation times are considered, each visit is performed at least within its hard time window and the total tour duration does not exceed $t^{\max}$.

While in our previous work [7] the primary objective was to minimize the number of districts (m), we consider this number now as pre-specified. For example, it can be obtained in a first optimization round by our previous method based on the hard time windows only. Now, in the DRPSC-STW, our objective is to minimize the total penalty incurred by all time window violations, which is

$$\min \sum_{r=1}^{m} \sum_{j \in P} \sum_{u=1}^{l_{r,j}} \omega_{r,j,u} \tag{1}$$

with

$$\omega_{r,j,u} = \begin{cases} T_{i_k}^{\mathrm{e}} - a_{r,j,u} & \text{if } a_{r,j,u} < T_{i_k}^{\mathrm{e}} \\ a_{r,j,u} + t_{i_k}^{\mathrm{visit}} - T_{i_k}^{\mathrm{l}} & \text{if } a_{r,j,u} + t_{i_k}^{\mathrm{visit}} > T_{i_k}^{\mathrm{l}} \\ 0 & \text{otherwise} \end{cases} \tag{2}$$

4   Optimal Arrival Time Problem

When approaching the DRPSC-STW with an LNS in Sect. 6, we will have to solve for each tour in each period of each candidate solution the following sub-problem: Given a candidate tour $\tau_{r,j} = (\tau_{r,j,0}, \ldots, \tau_{r,j,l_{r,j}+1})$ for some district $r = 1, \ldots, m$ and period $j = 1, \ldots, p$, what are feasible arrival times $a_{r,j,u}$ for the visits $u = 1, \ldots, l_{r,j} + 1$ minimizing $\sum_{u=1}^{l_{r,j}} \omega_{r,j,u}$? Remember that the solution must obey the minimum separation time t^{sep} between any two successive visits of the same object and the maximum tour duration $t^{\max}$. We call this subproblem *Optimal Arrival Time Problem* (OATP).

As we consider in the OATP always only one specific tour $\tau_{r,j}$, i.e., r and j are known and constant, we omit these indices in the following for simplicity wherever this is unambiguous. In particular, we write τ for the current tour, l for the tour's length, τ_h for the h-th visit, a_h for the respective arrival time, and ω_h for the respective time window penalty. Moreover, we introduce some further notations and definitions used in the next sections. Let us more generally define the time window penalty function $\rho_h(t)$ for visit $\tau_h = i_k$ when arriving at time t as the following piecewise linear function, see also Fig. 1:

$$\rho_h(t) = \begin{cases} \infty & \text{if } t < T_{i_k}^{\mathrm{e}} - \Delta \\ T_{i_k}^{\mathrm{e}} - t & \text{if } T_{i_k}^{\mathrm{e}} - \Delta \leq t < T_{i_k}^{\mathrm{e}} \\ t + t_{i_k}^{\mathrm{visit}} - T_{i_k}^{\mathrm{l}} & \text{if } T_{i_k}^{\mathrm{l}} < t + t_{i_k}^{\mathrm{visit}} \leq T_{i_k}^{\mathrm{l}} + \Delta \\ \infty & \text{if } t > T_{i_k}^{\mathrm{l}} + \Delta \\ 0 & \text{otherwise.} \end{cases}$$

Let $V = \{i_k \mid i \in D_r, \ i_k \in W_{i,j}\}$ be the set of all object visits in the current tour. We define the auxiliary function $\kappa : V \mapsto D_r$ mapping visit $i_k \in V$ to its

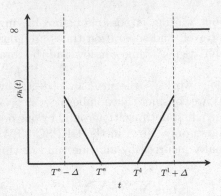

Fig. 1. The time window penalty function $\rho_h(t)$.

corresponding object $i \in D_r$, and function $\sigma(h)$ which finds the nearest successor index h' of the visit $\tau_{h'}$ with $h < h' \le l$ and $\kappa(\tau_h) = \kappa(\tau_{h'})$ if such a successive visit of the same object exists; otherwise $\sigma(h)$ returns -1. Correspondingly, function $\sigma^{-1}(h)$ returns the nearest predecessor index h' of the visit $\tau_{h'}$ with $1 \le h' < h$ and $\kappa(\tau_h) = \kappa(\tau_{h'})$ if such a predecessor exists and -1 otherwise. For convenience, we also define $\zeta_h = t_{\tau_{r,j,h}}^{\text{visit}} + t_{\kappa(\tau_{r,j,h}),\kappa(\tau_{r,j,h+1})}^{\text{travel}}$ as the sum of the visiting time of the hth visit and the travel time from the hth visit to the $(h+1)$st visit.

4.1 Lower and Upper Bounds for Arrival Times

We compute lower and upper bounds for each visit's arrival time by determining routes in which we perform each visit as early as possible and as late as possible. For determining the earliest arrival time at the first visit we have to consider the maximum of the travel time from the depot to the first visit and the earliest possible time of the first visit's hard time window. The earliest possible arrival time for all other visits $h = 1, \ldots, l$ can be computed recursively by considering the dependency on the previous visit $h - 1$, i.e., the visit time and travel time to the current visit h, the beginning of the hard time window $T_{\tau_h}^{\text{e}} - \Delta$ of the current visit h, and the separation time from a possibly existing previous visit of the same object $\sigma^{-1}(h)$ in the tour. This yields:

$$a_h^{\text{earliest}} = \begin{cases} -\infty & \text{if } h < 0 \\ T_{0_0}^{\text{e}} & \text{if } h = 0 \\ \max\left\{ a_{h-1}^{\text{earliest}} + t_{\tau_{h-1}}^{\text{visit}} + t_{\tau_{h-1},\tau_h}^{\text{travel}}, T_{\tau_h}^{\text{e}} - \Delta, a_{\sigma^{-1}(h)}^{\text{earliest}} + t^{\text{sep}} \right\} & \text{if } h > 0 \end{cases}$$

When scheduling a latest tour the last visit of the tour has to be scheduled before arriving at the depot where also the travel time to the depot has to be considered, but on the other hand we have to also consider the end of the hard time window $T_{\tau_l}^{\text{l}} + \Delta$ of the last visit. For all other visits we can compute their latest possible arrival time by considering the next visit's arrival time, the travel time to the next visit, and the visit time at the current visit, the end of the hard time window of the current visit, i.e., $T^l_{\tau_l} + \Delta$, and the separation time by considering a possibly existing successive visit $\sigma(h)$ of the same object where $\kappa(\tau_h) = \kappa(\tau_{h'})$ with $h < h'$:

$$a^{latest}_h = \begin{cases} \infty & \text{if } h < 0 \\ T^l_{0_1} & \text{if } h = l+1 \\ \min\left\{ a^{latest}_{h+1} - t^{visit}_{\tau_h} - t^{travel}_{\tau_h, \tau_{h+1}}, T^l_{\tau_h} + \Delta, a^{latest}_{\sigma(h)} - t^{sep} \right\} & \text{if } 0 \le h \le l \end{cases}$$

If for some h, $a^{earliest}_h > a^{latest}_h$, we immediately terminate as this OATP instance, i.e., underlying route, cannot have a feasible solution.

4.2   Linear Programming Model

The OATP is not an NP-hard optimization problem. We can solve it exactly by means of linear programming (LP) as we show in the following.

Variables a_{i_k} represent the arrival time of the k-th visit of object i, variables $p^e_{i_k}$ are used to compute the penalty when starting the service of visit i_k too early, and variables $p^l_{i_k}$ are used for the penalty when finishing the service of visit i_k too late. The LP is defined as follows:

$$\min \quad \sum_{i_k \in V} p^e_{i_k} + p^l_{i_k} \tag{3}$$

$$\text{s.t.} \quad t^{travel}_{0,\kappa(a_{\tau_1})} + a_{\tau_1} + t^{visit}_{\tau_l} + t^{travel}_{\kappa(\tau_l),0} - a_{\tau_1} + \le t^{max} \tag{4}$$

$$a_{\tau_1} \ge t^{travel}_{0,a_{\tau_1}} + T^e_{0_0} \tag{5}$$

$$a_{\tau_l} + t^{visit}_{\tau_{a_l}} + t^{travel}_{\kappa(\tau_l),0} \le T^l_{0_1} \tag{6}$$

$$a_{\tau_i} \ge a_{\tau_{i-1}} + t^{visit}_{a_{\tau_{i-1}}} + t^{travel}_{\kappa(\tau_{i-1}),\kappa(\tau_i)} \qquad \forall \tau_i \in \tau,\; i = 2,\dots,l \tag{7}$$

$$a_{i_k} \ge a_{i_{k'}} + t^{visit}_{i_{k'}} + t^{sep} \qquad \forall i_k, i_{k'} \in V,\; k > k' \tag{8}$$

$$T^e_{i_k} - \Delta \le a_{i_k} \le T^l_{i_k} + \Delta - t^{visit}_{i_k} \qquad \forall i_k \in V \tag{9}$$

$$p^e_{i_k} \ge T^e_{i_k} - a_{i_k} \qquad \forall i_k \in V \tag{10}$$

$$p^l_{i_k} \ge a_{i_k} + t^{visit}_{i_k} - T^l_{i_k} \qquad \forall i_k \in V \tag{11}$$

$$a_{i_k}, p^e_{i_k}, p^l_{i_k} \ge 0 \qquad \forall i_k \in V \tag{12}$$

Objective function (3) minimizes the total penalty incurred by too late or too early arrival times of visits. Inequality (4) ensures that the makespan of the tour does not exceed the maximum allowed duration t^{max}. Otherwise, the given visit order would be infeasible. Inequality (5) models the travel time from the depot to the first visit of the given order, i.e., the first visit can only be started after traveling from the depot to this visit. Inequality (6) specifies that the tour has to end latest at the end of the time window of the second visit of the depot.

Algorithm 1. Hybrid Heuristic for OATP

1: **Input:** Tour τ
2: **if not** Feasible(τ) **then**
3:     **return** ∞
4: **end if**
5: **if** GreedyHeuristic(τ) = 0 **then**
6:     **return** 0
7: **end if**
8: **return** DPBasedHeuristic(τ)

Inequalities (7) ensure that all travel times between consecutive object visits and visit times are respected. Inequalities (8) guarantee the minimum separation time between two consecutive visits of the same object. Inequalities (9) ensure consideration of the hard time windows. The penalty values are computed by inequalities (10) and (11). If a visit is scheduled too early, then $T_{i_k}^e - a_{i_k} > 0$ and an early penalty is incurred. Obviously, if the earliness penalty $p_{i_k}^e > 0$, then $a_{i_k} + t_{i_k}^{\text{visit}} - T_{i_k}^l < 0$ and thus, $p_{i_k}^l = 0$. This holds vice versa if the lateness penalty $p_{i_k}^l > 0$. If a visit is scheduled within its time window $[T_{i_k}^e, T_{i_k}^l]$, then $p_{i_k}^e = p_{i_k}^l = 0$ as $T_{i_k}^e - a_{i_k} \leq 0$ and $a_{i_k} + t_{i_k}^{\text{visit}} - T_{i_k}^l \leq 0$ and $p_{i_k}^e, p_{i_k}^l \geq 0$, $\forall i_k \in V$ according to Eq. (12).

5   Hybrid Heuristic for the OATP

While the above LP model can be solved in polynomial time, doing this many thousands of times within a metaheuristic for the DRPSC-STW for evaluating any new tour in any period of any district in any candidate solution is still a substantial bottleneck. We therefore consider a typically much faster heuristic approach in the following, which, as our experiments will show, still yields almost optimal solutions. We call this approach *Hybrid Heuristic* (HH) for the OATP as it is, in fact, a sequential combination of different individual components.

The overall approach is shown in Algorithm 1, and the individual components are described in detail in the subsequent sections. First, we show how to efficiently check the feasibility of a given instance (line 2), then, we apply a fast greedy heuristic which tries to solve the problem without penalties (line 5) using an earliest possible start time strategy. Finally, we apply an efficient DP-based heuristic to obtain a solution for the OATP.

5.1   Feasibility Check

The feasibility of a given tour, i.e., existence of feasible arrival times, can be efficiently checked by calculating the minimum tour duration and comparing it to $t^{\max}$. The minimum tour duration can be determined by fixing the arrival time at the depot to $a_{l+1}^{\text{earliest}}$ and calculating the latest arrival times recursively backwards:

$$a_h^{\text{ms}} = \begin{cases} \infty & \text{if } h < 0 \\ a_h^{\text{earliest}} & \text{if } h = l+1 \\ \min\left\{a_{h+1}^{\text{ms}} - t_{\tau_h}^{\text{visit}} - t_{\tau_h,\tau_{h+1}}^{\text{travel}}, T_{\tau_h}^{\text{l}} + \Delta, a_{\sigma(h)}^{\text{ms}} - t^{\text{sep}}\right\} & \text{if } 0 \leq h \leq l \end{cases}$$

The tour is feasible iff $a_{l+1}^{\text{ms}} - a_0^{\text{ms}} \leq t^{\text{max}}$ holds.

5.2 Greedy Heuristic

A fast heuristic for solving the OATP is a greedy strategy that starts each visit as early as possible without violating any soft time window. If this heuristic is successful, no penalty occurs and the obtained solution is optimal. We can formulate this approach as follows:

$$a_h^{\text{greedy}} = \begin{cases} -\infty & \text{if } h < 0 \\ \max\left\{T_{0_0}^{\text{e}} + t_{0,\kappa(\tau_1)}^{\text{travel}}, T_{\tau_h}^{\text{e}}\right\} & \text{if } h = 1 \\ \max\left\{a_{h-1}^{\text{greedy}} + t_{\tau_{h-1}}^{\text{visit}} + t_{\tau_{h-1},\tau_h}^{\text{travel}}, T_{\tau_h}^{\text{e}}, a_{\sigma^{-1}(h)}^{\text{greedy}} + t^{\text{sep}}\right\} & \text{if } h > 1 \end{cases}$$

If for some h, $a_h^{\text{greedy}} > a_h^{\text{latest}}$, then the greedy heuristic cannot solve this problem instance and terminates.

5.3 Efficiently Solving a Relaxation by Dynamic Programming

The greedy strategy is fast, works reasonably well, and frequently yields an optimal solution for easy instances. When the constraints become tighter, however, it often fails. Therefore, we finally use a second, more sophisticated heuristic based on the following considerations.

The required minimum separation times for visits of same objects make the OATP, in contrast to other problems aiming at finding arrival times introducing a minimum penalty, e.g. [4], inaccessible for an efficient exact DP approach. One would need to somehow consider also all objects' last visits when storing and reusing subproblem solutions in the DP recursion.

However, in a heuristic approach we can exploit an efficient DP for the relaxed variant of the OATP in which we remove the separation time constraints. We denote this relaxed OATP by OATP^{rel}. As will be shown in Sect. 5.4, we will modify our instance data before applying this DP in order to obtain a heuristic solution that is feasible for our original OATP.

To solve OATP^{rel} we apply DP inspired by Ibaraki et al. [4]. In contrast to this former work, however, we consider a maximum tour duration.

Let $g_h(t, t_0)$ be the minimum sum of the penalty values for visits $\tau_0, \ldots, \tau_h$ under the condition that all of them are started no later than at time t and the depot is left earliest at time t_0 with $t - t_0 \leq t^{\max}$. Here we assume that $T^e_{0_0} \leq t_0$. Then, $g_h(t, t_0)$ can be expressed recursively by:

$$
g_0(t, t_0) = \begin{cases} \infty & \text{if } t < t_0 \\ 0 & \text{otherwise} \end{cases}
$$

$$
g_h(t, t_0) = \min_{a^{\text{earliest}}_h \leq t' \leq \min\{t, t_0 + t^{\max}\}} g_{h-1}(t' - \zeta_{h-1}, t_0) + \rho_h(t') \quad \text{if } h > 0
$$

Here, we assume the minimum of an empty set or interval to be ∞. The overall minimum time penalty of the tour τ is then $\min_{a^{\text{earliest}}_0 \leq t_0 \leq a^{\text{latest}}_0} g_{l+1}(T^1_{0_1}, t_0)$. Thus, solving OATP^{rel} corresponds to finding a departure time t_0 from the depot which minimizes function $f^{\text{rel}} = g_{l+1}(T^1_{0_1}, t_0)$.

Let t_0 be the value for which $f^{\text{rel}} = g_h(T^1_{0_1}, t_0)$ yields a minimum penalty. Optimal arrival times for the visits and the arrival time back at the depot can now be expressed by:

$$
\begin{aligned}
a^{\text{rel}}_{l+1} &= \underset{T^e_{0_0} \leq t \leq T^1_{0_1}}{\arg\min} \ g_{l+1}(t, t_0) \\
a^{\text{rel}}_h &= \underset{T^e_{0_0} \leq t \leq a^{\text{rel}}_{h+1} - \zeta_h}{\arg\min} \ g_h(t, t_0) \quad \text{if } 0 \leq h \leq l
\end{aligned}
\tag{13}
$$

Now, let us consider the task of efficiently computing $g_h(t, t_0)$ in more detail. Recall that our time window penalty function $\rho_h(t)$ is piecewise linear for all visits $\tau_1, \ldots, \tau_l$ and they have all the same shape as shown in Fig. 1. Therefore, $g_h(t, t_0)$ is also piecewise linear. We store these piecewise linear functions of each recursion step of the DP algorithm in linked lists, whose components represent the intervals and the associated linear functions.

An upper bound for the total number of pieces in the penalty functions for all the visits $\tau_0, \ldots, \tau_{l+1}$ is $5l + 2 = O(l)$. The computation of $g_{h-1}(t - \zeta_{h-1}, t_0) + \rho_h(t)$ and $g_h(t, t_0)$ from $g_{h-1}(t, t_0)$ and $\rho_h(t)$ can be achieved in $O(h)$ time, since the total number of pieces in $g_{h-1}(t, t_0)$ and $\rho_h(t)$ is $O(h)$. In order to calculate the function $g_{l+1}(T^1_{0_1}, t_0)$ for a given tour, we compute $g_h(t, t_0)$ for all $1 \leq h \leq l + 1$. This can be done in $O(l^2)$ time.

Now that we know how to efficiently calculate the minimum time window penalty value for a given departure time from the depot t_0, we draw our attention to the problem of finding a best departure time such that the overall penalty value for a given tour is minimized. Formally, we want to minimize function $g'(t_0) = g_{l+1}(T^1_{0_1}, t_0)$ on interval $t_0 \in [a^{\text{earliest}}_0, a^{\text{latest}}_0]$. Enumerating all possible t_0 values is obviously not a reasonable way to tackle this problem. Fortunately, there is a useful property of function $g'(t_0)$ which enables us to search more efficiently for its minimum.

Proposition 1. *Earliest optimal arrival times can only be delayed further when the depot departure time increases. More formally, let a^0_h for $h = 0, \ldots, l + 1$ be earliest optimal arrival times calculated by $g'(t_0)$ for some t_0 and a^1_h for $h = 0, \ldots, l + 1$ be the earliest optimal arrival times calculated by $g'(t'_0)$ for some t'_0. Then $t_0 \leq t'_0 \implies a^0_h \leq a^1_h$ for $h = 0, \ldots, l + 1$.*

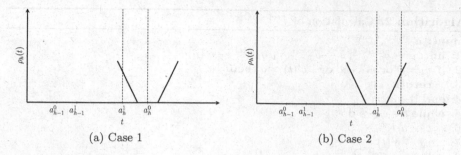

Fig. 2. Visualization of the two case distinctions used in the proof of Proposition 1.

Proof. We show this with a proof by contradiction. Without loss of generality, suppose there is a visit h with $a_{h-1}^0 \leq a_{h-1}^1$ and $a_h^1 < a_h^0$. Let us consider two relevant cases in detail. Other cases can be refuted using similar arguments.

Case 1: Assume a_h^1 is scheduled earlier than a_h^0 and $a_h^1 < T_{\tau_h}^e$, see Fig. 2a. a_h^1 could only have been scheduled earlier than a_h^0 falling below $T_{\tau_h}^e$ threshold if and only if one of its subsequent visits $\tau_{h+1}, \ldots, \tau_{l+1}$ was forced to start earlier. This can only happen if the arrival time constraint, where we have to be back at the depot, is more tightened. But this clearly cannot be the case here, since $t_0 + t^{\max} \leq t_0' + t^{\max}$. In other words, delaying the departure time at the depot also delays the arrival time constraint, when we have to be back at the depot.

Case 2: Assume a_h^1 is scheduled earlier than a_h^0 and $a_h^0 > T_{\tau_h}^l - t_{\tau_h}^{\text{visit}}$, see Fig. 2b. Since $a_h^0 - a_{h-1}^0 > a_h^1 - a_{h-1}^1$, it is easy to see that a_h^0 can be moved further to the left without introducing more penalty. Therefore, a_h^0 cannot be the optimal start time for the visit h, since the T_h^l constraint violation caused by a_h^0 can be reduced further. $\square$

Proposition 2. $\forall t_0', t_0'' \mid g'(t_0') < g'(t_0''), t_0' < t_0'' \implies \forall t_0 \geq t_0'' : g'(t_0'') < g'(t_0)$.

Proof. Let $a_h^{\text{earliest}'}$ for $h = 0, \ldots, l+1$ be the earliest possible arrival times when fixing t_0' as the departure time from the depot and $a_h^{\text{earliest}''}$ for $h = 0, \ldots, l+1$ the earliest possible arrival times when fixing t_0'' as the departure time from the depot. Furthermore, we define a_h'' for $h = 0, \ldots, l+1$ to be earliest optimal arrival times calculated by $g'(t_0'')$.

We have shown that the earliest optimal arrival times can only be delayed further when postponing the departure time from the depot. Thus, the only way the overall penalty value can be increased is when pushing t_0 to the future causes more T^l threshold violations than what you can save by reducing T^e threshold violations.

More formally, if we have $g'(t_0') < g'(t_0'')$ with $t_0' < t_0''$, then there must exist $a_k^{\text{earliest}''} > T_{\tau_k}^l - t_{\tau_k}^{\text{visit}}$ with $a_k^{\text{earliest}'} < a_k^{\text{earliest}''}$ and $a_k^{\text{earliest}''} = a_k''$ for some

---

Algorithm 2. Calculation of f^{rel}

Input: a_0^{earliest}, a_0^{latest}

1: **init:** $a \leftarrow a_0^{\text{earliest}}$, $b \leftarrow a_0^{\text{latest}}$, $v_1 \leftarrow f^{\text{rel}} \leftarrow g'(a)$
2: **if** $v_1 = 0$ **or** $v_1 = \infty$ **or** $\nabla g'(t) > 0$ **then**
3:    **return** v_1
4: **end if**
5: **while** $b - a > \varepsilon$ **do**
6:    $t \leftarrow a + \frac{b-a}{2}$
7:    $v_2 \leftarrow g'(t)$
8:    **if** $v_2 < f^{\text{rel}}$ **then**
9:       $f^{\text{rel}} \leftarrow v_2$
10:    **end if**
11:    **if** $f^{\text{rel}} = 0$ **or** $v_1 = v_2$ **then**
12:       **break**
13:    **end if**
14:    **if** $v_2 = \infty$ **or** $\nabla g'(t) \leq 0$ **then**
15:       $a \leftarrow t$
16:    **else**
17:       $b \leftarrow t$
18:    **end if**
19:    $v_1 \leftarrow v_2$
20: **end while**
21: **return** f^{rel}

$k \in \{0, \ldots, l+1\}$. In other words, if the overall penalty value increases, then there are visits whose earliest possible arrival times are pushed furhter to the future exceeding T^{l} thresholds by t_0'' and their optimal arrival times are equal to earliest possible arrival times.

It is easy to see that once the earliest possible start time a_h^{earliest} starts to increase, it continues to increase strictly monotonically with an increasing departure time from the depot. Therefore, the overall penalties will increase strictly monotonically from t_0'' on with an increasing departure time from the depot until the solution becomes infeasible.   □

These properties show that $g'(t_0)$ is in general a "U-shaped" function when disregarding all infeasible solutions yielding ∞, and we can use a bisection method to search efficiently for a minimum. The calculation of f^{rel} in this way is shown in Algorithm 2.

At each iteration step the middle point t of current search interval is sampled and we calculate an approximate subgradient $\nabla g'(t)$ of g' at t by $\nabla g'(t) = g'(t + \delta) - g'(t)$ where δ is a small constant value. If the subgradient $\nabla g'(t) > 0$, we know that t is in the strictly monotonically rising piece of g' and we continue our search in the left half. Otherwise the search continues in the right half. The bisection method proceeds until the search interval becomes smaller than some predetermined value ε.

5.4 DP-Based Heuristic for OATP

Obviously, OATP^{rel} corresponds to the original OATP if there are no objects that are visited multiple times or $\sum_{i=h}^{\sigma(h)-1} \zeta_i \geq t^{\text{sep}}$ for $h = 1, \dots, l$ with $\sigma(h) \neq -1$. The main idea of our second heuristic is to increase the ζ_i values as necessary so that $\sum_{i=h}^{\sigma(h)-1} \zeta_i \geq t^{\text{sep}}$ holds for all $h = 1, \dots, l$ with $\sigma(h) \neq -1$. Then, when applying the DP, its solution will obviously fulfill the separation-time constraint.

Let visits τ_k and $\tau_{k'}$ with $k < k'$ and $\sum_{i=k}^{k'-1} \zeta_i < t^{\text{sep}}$ be two visits which take place at the same object. Then, one or more $\zeta_i \in \{\zeta_k, \dots, \zeta_{k'-1}\}$ must be extended so that $\sum_{i=k}^{l-1} \zeta_i = t^{\text{sep}}$. In order to decide which ζ_i we want to extend, we first calculate waiting times for all visits with earliest possible arrival times.

The waiting time at the visit τ_h is the amount of time we are forced to wait at the visit τ_{h-1} before we can travel to visit τ_h. Recall that we are forced to wait at visit τ_{h-1} if $a_{h-1} + \zeta_{h-1} < T^{\text{e}}_{\tau_h}$. Thus, the waiting times with earliest possible arrival times can be expressed as $w_h^{\text{earliest}} = \max\{0, a_h^{\text{earliest}} - a_{h-1}^{\text{earliest}} - \zeta_{h-1}\}$, $h = 1, \dots, l$. Using these waiting times as guidance, we extend the ζ_i value at the visit τ_i with the maximum waiting time $w_i^{\text{earliest}} = \max\{w_k^{\text{earliest}}, \dots, w_{l-1}^{\text{earliest}}\}$ where ties are broken randomly. The rationale behind this idea is that large w_h^{earliest} values often indicate the visits in an optimal solution, where extra waiting time actually is introduced to satisfy the separation-time constraints.

Utilizing waiting times computed by earliest possible arrival times works well for the majority of instances but for some instances the ζ_h values are altered unfavorably so that the instances become infeasible. To counteract this problem, we propose alternative waiting times which are calculated using arrival times with minimum tour duration: $w_h^{\text{ms}} = \max\{0, a_h^{\text{ms}} - a_{h-1}^{\text{ms}} - \zeta_{h-1}\}$, $\forall h = 1, \dots, l$. Visits with waiting times larger than 0 indicate visits in the tour with minimum tour duration for which additional waiting time had to be introduced in order to satisfy separation-time constraints. Using w_h^{ms} waiting times we can effectively complement situations where the approach utilizing w_h^{earliest} values yields infeasible or low-quality solutions. Therefore, we solve the DP-based heuristic twice, using both w_h^{earliest} and w_h^{ms} and take the best solution.

Even if the solution of this DP-based heuristic does not guarantee optimality in general, it works well in practice, producing near optimal solutions in significantly shorter computation times than the exact LP approach.

6 Large Neighborhood Search for the DRPSC-STW

Our overall approach for solving the DRPSC-STW follows the classical large neighborhood search metaheuristic [6] with an embedded variable neighborhood descent (VND) for local improvement.

We define our *destroy* and *repair* methods as follows. In order to destroy a current solution candidate, we select two out of m districts uniformly at random and remove all objects from these districts. The removed objects are copied to a so called *ejection pool*. Then, we apply the repair phase of the *district elimination algorithm* proposed by Prischink et al. [8]. The algorithm continues until all objects in the ejection pool are reassigned to the available districts. Using this *destroy* and *repair* methods, we guarantee that the solution stays feasible with the same number of districts. At each LNS iteration a VND is applied to locally improve the incumbent solution.

6.1 Variable Neighborhood Descent

We use three common neighborhood structures from the literature and search in a best improvement fashion. We apply these neighborhoods in the given order since we could not identify any significant advantages using different orderings. Infeasible solutions are discarded.

2-opt: Classical 2-opt neighborhood where all edge exchanges are considered.
swap: Exchanges all pairs of distinct visits within a route.
or-opt: Moves sequences of one to three consecutive visits at another place in the route.

The proposed VND is performed separately for each route of every district. Our local improvement component could also be very well parallelized since different routes can be optimized independently of each other, however this is not in the scope of this work. Since routes having no penalties are already optimal, they are excluded from local improvement.

7   Computational Results

For the computational results, we used the instances which have been created by Prischink et al. [8]. In a first optimization round, we solve the districting part of the DRPSC-STW by means of the district elimination algorithm proposed by Prischink et al., based on the hard time windows only, generating input[1] for the subsequent time window penalty minimization round with the LNS algorithm. As global parameters we have chosen $t^{\max}$ to be $12\,\mathrm{h}$ and the maximum allowed penalty $\Delta = 60\,\mathrm{min}$, which represent typical values used in practical settings. Furthermore, we set HH (Algorithm 1) specific parameters $\delta = 1$ and $\varepsilon = 30$, which have been determined empirically. For our test instances, they give good balance between computational speed and accuracy. Every instance was given a maximum allowed time limit of $900\,\mathrm{s}$ for the execution of the LNS and we have performed 20 runs for every instance. All tests have been executed as single threads on an Intel Xeon E5540 $2.53\,\mathrm{GHz}$ Quad Core processor. The algorithms have been written in C++ and have been compiled with gcc-4.8 and for solving the LP we used Gurobi 7.0.

In Table 1 the results of the LNS-LP and LNS-HH can be found. In the instance column, we specify the instance parameters. Sequentially, the name of the used TSPlib instance (refer to Prischink et al. [8] for a more detailed description), the number of runs performed, the number of objects $|I|$, the total number

---

[1] https://www.ac.tuwien.ac.at/files/resources/instances/drpsc/evoc17.tgz.

Table 1. Results of the LNS with embedded LP and HH as solution evaluation function

Instance							LNS-LP				LNS-HH			
Name	Runs	$|I|$	$|V|$	α	β	v	#best	$\overline{obj}$ [s]	$\bar{t}$ [s]	#eval	#best	$\overline{obj}$ [s]	$\bar{t}$ [s]	#eval
berlin52_1	20	51	133	0.0	0.7	4	13	476.9	640.2	499,624.3	20	29.9	298.0	757,181.5
berlin52_2	20	51	130	0.0	0.7	4	13	235.1	662.5	525,862.1	19	7.9	239.0	586,096.8
berlin52_3	20	51	140	0.0	0.7	4	5	3,230.9	900.0	595,039.3	15	1,169.7	900.0	938,946.7
ch150_1	20	149	360	0.2	0.5	4	4	88,910.1	900.0	663,320.8	16	55,234.8	900.0	1,308,850.8
ch150_2	20	149	402	0.2	0.5	4	0	164,399.8	900.0	662,869.3	20	80,989.3	900.0	1,257,789.1
ch150_3	20	149	357	0.2	0.5	4	3	78,979.9	900.0	748,549.8	17	36,370.5	900.0	1,620,281.2
ft70_1	20	69	167	0.1	0.5	4	2	5,035.8	900.0	953,374.5	18	1,196.9	900.0	2,815,488.6
ft70_2	20	69	180	0.1	0.5	4	1	3,087.0	900.0	975,529.6	20	464.2	878.3	2,719,290.8
ft70_3	20	69	144	0.1	0.5	4	3	5,602.2	900.0	918,004.6	17	1,509.0	900.0	2,496,826.1
gr48_1	20	47	120	0.2	0.7	4	8	92,669.4	900.0	284,934.1	12	70,082.1	900.0	443,507.7
gr48_2	20	47	115	0.2	0.7	4	3	9,445.5	900.0	851,306.0	17	4,233.1	900.0	2,296,485.5
gr48_3	20	47	125	0.2	0.7	4	5	28,606.2	900.0	352,935.5	15	24,982.1	900.0	615,803.7
rd100_1	20	99	152	0.1	0.5	2	3	25,824.7	900.0	876,710.0	17	11,050.4	900.0	2,289,110.3
rd100_2	20	99	160	0.1	0.5	2	4	22,367.5	900.0	828,483.6	16	7,618.6	900.0	2,123,393.7
rd100_3	20	99	152	0.1	0.5	2	4	12,132.5	900.0	826,517.7	17	2,136.9	900.0	2,036,959.2
st70_1	20	69	105	0.1	0.7	2	5	15,052.1	900.0	755,594.8	16	4,380.1	900.0	1,761,210.0
st70_2	20	69	91	0.1	0.7	2	4	18,622.9	900.0	806,985.2	16	8,228.9	900.0	2,126,834.2
st70_3	20	69	106	0.1	0.7	2	3	7,022.6	900.0	696,001.3	20	380.5	673.1	1,140,718.4
tsp225_1	20	224	334	0.2	0.7	2	0	272,118.2	900.0	969,287.0	20	183,974.1	900.0	1,904,494.3
tsp225_2	20	224	341	0.2	0.7	2	0	340,426.5	900.0	692,597.6	20	293,867.6	900.0	1,375,567.5
tsp225_3	20	224	332	0.2	0.7	2	0	161,586.6	900.0	730,581.4	20	141,153.5	900.0	1,471,799.2
Average						4.0	4.0	64,563.4	884.5	727,338.5	17.5	44,240.9	828.0	1,623,173.1

of visits $|V|$, the percentage of large time windows (α), the percentage of mid-sized time windows (β) and the maximum number of allowed visits per object v is given. For the LNS-LP and LNS-HH the number of times the corresponding approach yields the best result, the average objective value over all runs of the instance, the average runtime, and the average number of objective function evaluations are given. Results show clearly that LNS-HH yields better objective values than LNS-LP since it is able to perform much more iterations within the given time limit due to fast objective function evaluations. It is also obvious that by increasing the instance size, the advantage of the efficient HH evaluation function is getting more pronounced. Moreover, a Wilcoxon signed-rank test shows that all observed differences on the overall number of best solutions among the LNS-LP and the LNS-HH are statistically significant with an error level of less than 1%.

We can conclude that LNS-HH is superior compared to LNS-LP due to significant performance advantage in the evaluation function, even though the HH-based evaluation function is only a heuristic method which in general does not yield proven optimal solutions although it can be observed that the optimality gap of HH is in most cases neglectably small.

8 Conclusions and Future Work

In this work we analyze the DRPSC-STW where the already introduced DRPSC is extended by soft time windows. This problem is of high practical relevance as it is possible to significantly improve solution quality by introducing only a negelectable penalty.

As metaheuristic we propose an LNS for approaching the DRPSC-STW. A critical bottleneck of our LNS is the evaluation of solution candidates where one has to find the minimum penalty given a particular visit order. We show that this evaluation function can be efficiently implemented by an LP-based approach, and furthermore we developed a sophisticated hybrid heuristic which was able to drastically outperform the LP-based variant.

We have formulated an efficient method to determine optimal arrival times of a given visit order which can be embedded inside a metaheuristic framework to solve the penalty minimization part of the DRPSC-STW. On the one hand this is not only relevant for the DRPSC-STW, as soft time windows play in general an important role in many practical scenarios.

Future research goals include the extension of the current LNS by incorporating adaptiveness into the destroy and repair moves. Furthermore, the authors want to note that it is also possible to extend the VND local search into a VNS by including a shaking neighborhood like randomized k-swap neighborhood, c.f. [1]. This way, one can combine micro- and macro-diversifications during the search.

References

1. Davidovic, T., Hansen, P., Mladenovic, N.: Variable neighborhood search for multi-processor scheduling problem with communication delays. In: Proceedings of MIC, vol. 4, pp. 737–741 (2001)
2. Fagerholt, K.: Ship scheduling with soft time windows: an optimisation based approach. Eur. J. Oper. Res. **131**(3), 559–571 (2001)
3. Hashimoto, H., Ibaraki, T., Imahori, S., Yagiura, M.: The vehicle routing problem with flexible time windows and traveling times. Discrete Appl. Math. **154**(16), 2271–2290 (2006)
4. Ibaraki, T., Imahori, S., Kubo, M., Masuda, T., Uno, T., Yagiura, M.: Effective local search algorithms for routing and scheduling problems with general time-window constraints. Transp. Sci. **39**(2), 206–232 (2005)
5. Ioachim, I., Gelinas, S., Soumis, F., Desrosiers, J.: A dynamic programming algorithm for the shortest path problem with time windows and linear node costs. Networks **31**(3), 193–204 (1998)
6. Pisinger, D., Ropke, S.: Large neighborhood search. In: Gendreau, M., Potvin, J.Y. (eds.) Handbook of Metaheuristics, chap. 13, pp. 399–419. Springer, Heidelberg (2010)
7. Prischink, M.: Metaheuristics for the districting and routing problem for security control. Master's thesis, TU Wien, Institute of Computer Graphics and Algorithms, May 2016. https://www.ac.tuwien.ac.at/files/pub/prischinkd_16.pdf, supervised by G. Raidl, B. Biesinger, and C. Kloimdllner
8. Prischink, M., Kloimüllner, C., Biesinger, B., Raidl, G.R.: Districting and routing for security control. In: Blesa, M.J., Blum, C., Cangelosi, A., Cutello, V., Nuovo, A.D., Pavone, M., Talbi, E.G. (eds.) HM 2016. LNCS, vol. 9668, pp. 87–103. Springer, Heidleberg (2016). doi:10.1007/978-3-319-39636-1_7
9. Taillard, É., Badeau, P., Gendreau, M., Guertin, F., Potvin, J.Y.: A tabu search heuristic for the vehicle routing problem with soft time windows. Transp. Sci. **31**(2), 170–186 (1997)

Estimation of Distribution Algorithms for the Firefighter Problem

Krzysztof Michalak[✉]

Department of Information Technologies, Institute of Business Informatics,
Wroclaw University of Economics, Wroclaw, Poland
krzysztof.michalak@ue.wroc.pl

Abstract. The firefighter problem is a graph-based optimization problem in which the goal is to effectively prevent the spread of a threat in a graph using a limited supply of resources. Recently, metaheuristic approaches to this problem have been proposed, including ant colony optimization and evolutionary algorithms.

In this paper Estimation of Distribution Algorithms (EDAs) are used to solve the FFP. A new EDA is proposed in this paper, based on a model that represents the relationship between the state of the graph and positions that become defended during the simulation of the fire spreading. Another method that is tested in this paper, named EH-PBIL, uses an edge histogram matrix model with the learning mechanism used in the Population-based Incremental Learning (PBIL) algorithm with some modifications introduced in order to make it work better with the FFP. Apart from these two EDAs the paper presents results obtained using two versions of the Mallows model, which is a probabilistic model often used for permutation-based problems. For comparison, results obtained on the same test instances using an Ant Colony Optimization (ACO) algorithm, an Evolutionary Algorithm (EA) and a Variable Neighbourhood Search (VNS) are presented.

The state-position model proposed in this paper works best for graphs with 1000 vertices and more, outperforming the comparison methods. For smaller graphs (with less than 1000 vertices) the VNS works best.

Keywords: Estimation of distribution algorithms · Graph-based optimization · Firefighter problem

1 Introduction

The Firefighter Problem (FFP) originally formulated by Hartnell in 1995 [10] is a combinatorial optimization problem in which spreading of fire is modelled on a graph and the goal is to protect nodes of the graph from burning. Despite the name of the problem, the same formalism can also be used to describe spreading of other threats, such as floods, diseases in humans as well as in livestock, viruses in a computer network and so on.

© Springer International Publishing AG 2017
B. Hu and M. López-Ibáñez (Eds.): EvoCOP 2017, LNCS 10197, pp. 108–123, 2017.
DOI: 10.1007/978-3-319-55453-2_8

There are three main areas of study concerning the Firefighter Problem. The first area is the theoretical study of the properties of the problem. In the survey [7] many aspects of the FFP are discussed, for example the complexity of the problem, algorithms and special cases, such as the FFP on infinite and finite grids. Other works study the FFP on specific graph topologies, such as grids [6], trees [5], digraphs [13], etc.

Another area of study is the application of classical optimization algorithms. For example, methods such as the linear integer programming have been used for solving the single-objective version of the FFP [6]. Some authors combined linear integer programming methods with simple heuristics, such as "save vertices with highest degrees first" [9].

An area that has just recently emerged concerns attempts to employ meta-heuristic methods. This line of work originated with a work by Blum et al. [2] presented at the EvoCOP conference in 2014. This first attempt was made using the Ant Colony Optimization (ACO) approach and concerned the single-objective FFP. Later the same year another paper has been published [17] in which the multiobjective version of the FFP (MOFFP) has been formulated. The next paper concerning the MOFFP employed a multipopulation evolutionary algorithm with migration [18]. The multipopulation algorithm has later been used in the non-deterministic case [20], combining simulations with evolutionary optimization in an approach known as simheuristics which was proposed in a recent survey [12] as a proper approach to nondeterministic optimization. In another paper the Variable Neighborhood Search (VNS) method has been applied to the single-objective version of the FFP [11].

The papers published so far have been based on such metaheuristic approaches as the Ant Colony Optimization (ACO) and Evolutionary Algorithms (EAs). To the best of the knowledge of the author of this paper no attempts to use EDAs for this problem have previously been made. This paper starts investigation in this direction by proposing a new State-Position (S-P) model for the use in the FFP as well as by studying other models.

The rest of this paper is structured as follows. In Sect. 2 the firefighter problem is defined. In Sect. 3 the EDA approach is described and probabilistic models tested in this paper are presented. Section 4 describes the experimental setup and presents the results. Section 5 concludes the paper.

2   Problem Definition

The Firefighter Problem is defined on an undirected graph $G = \langle V, E \rangle$ with N_v vertices. Each vertex of this graph can be in one of the states from the set $L = \{'B', 'D', 'U'\}$ with the interpretation $'B' = $ burning, $'D' = $ defended and $'U' = $ untouched. Formally, we will use a function $l : V \to L$ to assign labels to the vertices of the graph G.

Spreading of fire is simulated in discrete time steps $t = 0, 1, \ldots$. At $t = 0$, the graph is in the initial state S_0. Most commonly, in the initial state vertices from a non-empty set $\emptyset \neq S \subset V$ are burning ('B') and the remaining ones are untouched ('U') (no vertices are initially defended). In each time instant $t > 0$ we are allowed to assign firefighters to a predefined number N_f of still untouched ('U') nodes of the graph G. These nodes become defended ('D') and are immune to fire for the rest of the simulation. Next, fire spreads along the edges of the graph G from the nodes labelled 'B' to all the adjacent nodes labelled 'U'. The simulation stops, when fire can no longer spread. It can happen either when all nodes adjacent to fire are defended ('D') or when all undefended nodes are burning ('B').

Solutions to the firefighter problem can be represented as permutations of the numbers $1, \ldots, N_v$. During the simulation, in each time step, the first N_f yet untouched nodes ('U') are taken from the permutation π and become defended ('D'). In every time step exactly N_f nodes become defended, except for the final time step in which the number of the untouched nodes may be less than N_f.

The evaluation of a solution (permutation) π is performed by simulating the spread of fire from the initial state until the fire can no longer spread. During the simulation nodes of the graph G become protected in the order determined by the permutation π. In the classical version of the FFP the evaluation of the solution π is equal to the number of nodes not touched by fire (those, that are in one of the states 'D' or 'U') when the simulation stops. In the paper [17] the multiobjective version of the FFP was proposed in which there are m values $c_i(v)$, $i = 1, \ldots, m$ assigned to each node v in the graph. In the context of fire containment multiple criteria could represent, for example, the financial value $c_1(v)$ and the cultural importance $c_2(v)$ of the items stored at the node v. Multiobjective evaluation can also be useful when preventing the spread of epidemics in livestock. In such scenario we would probably be interested in protecting different species to a certain degree. In this paper a single-objective version of the FFP is studied, in order to start the work on probabilistic models for the FFP, which are naturally less complex in the single-objective case. However, to retain similarity to the multiobjective case, costs are assigned to nodes and solutions are evaluated using these costs. To stick to the formalism used for the multiobjective case, these costs will be denoted as $c_1(v)$, even though there are no $c_i(v)$ with $i > 1$ in this paper. Evaluating a solution π requires simulating how fire spreads when firefighter assignment is done according to π. The evaluation of the solution can then be calculated as the sum of the costs assigned to those nodes that are not burnt at the end of the simulation:

$$e(\pi) = \sum_{v \in V : l(v) \neq 'B'} c_1(v) \tag{1}$$

where:
   $c_1(v)$ - the cost assigned to the node v.

3   Estimation of Distribution Algorithms

Estimation of Distribution Algorithms (EDAs) work in a way that bears a certain resemblance to Evolutionary Algorithms (EAs). The algorithm operates in a loop

---

Algorithm 1. A general structure of an EDA algorithm used in this paper.

IN: N_{pop} - The size of the population
N_{sample} - The size of a sample used for probabilistic model update

$P := \text{InitPopulation}(N_{pop})$
$M := \text{InitModel}()$
$B := \emptyset$

while not StoppingCondition() **do**
> // Evaluation
> Evaluate(P)
>
> // Caching of the best specimen
> $B := \text{GetBestSpecimens}(P, 1)$
>
> // Update of the probabilistic model
> $P_{sample} := \text{GetBestSpecimens}(P, N_{sample})$
> $M := \text{UpdateModel}(P_{sample}, M)$
>
> // New population
> $P := \text{CreateNewSpecimens}(M, N_{pop} - 1)$
> $P := P \cup B$

---

in which a population of specimens is used to represent and evaluate solutions of a given optimization problem (see Algorithm 1). The main difference between EDAs and evolutionary algorithms is that in the EAs genetic operators are used to produce the next generation of specimens, while in EDAs a probabilistic model is built from the population and specimens for the new generation are drawn from this model. In the algorithm used in this paper the mechanism of elitism is used, that is, the best specimen is always preserved and is transferred from the previous generation to the next one.

A general structure of EDAs used in this paper is presented in Algorithm 1. This algorithm uses the following procedures:

InitPopulation - Initializes a new population by creating N_{pop} specimens, each with a genotype initialized as a random permutation of N_v elements.

InitModel - Initializes the probabilistic model. The initialization procedure depends on the chosen problem and the EDA algorithm.

StoppingCondition - Checks if the stopping condition has been satisfied. In this paper the total running time T_{max} is used as a stopping criterion.

Evaluate - Evaluates specimens in a given population. The evaluation is performed by simulating the spread of fire and, after fire no longer spreads, calculating the overall value of the nodes in the graph that are not consumed by fire.

GetBestSpecimens - Returns a given number of specimens with the highest evaluation values from a given population. This procedure is used for storing the best specimen used by the elitism mechanism and for getting a sample from the population used for updating the probabilistic model.

UpdateModel - Updates the probabilistic model based on the current population. The exact procedure depends on the chosen model and the representation of solutions.

CreateNewSpecimens - Generates a required number of specimens using the information contained in the probabilistic model.

For an EDA algorithm three elements are necessary: a probabilistic model, a method of updating this model based on a sample of specimens and a procedure for generating new solutions based on the probabilistic model. All probabilistic models used in this paper are defined on a space Π_n of permutations of a fixed length n (with $n = N_v$).

The first two, the Mallows model and the generalized Mallows model describe exponential probability distributions on the space of permutations. Even though they were proposed in 1957 [15] and 1986 [8] respectively, they are nowadays actively researched in various applications such as recommender systems [14, 16]. Recently, both models have been proposed for the usage in EDAs [3, 4]. Because of space limitations, the details of the Mallows models are not given in this paper. A discussion of the learning and sampling processes can be found for example in [3].

Third method studied in this paper (EH-PBIL) uses an edge histogram matrix model used, among others, in the Edge Histogram-Based Sampling Algorithm (EHBSA) [21] and the learning rule known from the PBIL algorithm [1] to update a probability matrix, which in turn is used for generating new solutions. A similar method was used in the paper [19] for the Travelling Salesman Problem (TSP), but the model update procedure used in this paper was modified to fit the specifics of the FFP.

Fourth model, named State-Position model and introduced in this paper, is dedicated for the FFP and models the relation between the state of the graph and positions at which firefighters should be placed.

3.1   The EH-PBIL Method

This method uses an edge histogram model used for example in the Edge Histogram-Based Sampling Algorithm (EHBSA) [21] - a matrix $\mathbb{P}_{[N_v \times N_v]}$ in which each element p_{ij} represents a probability that the number j will appear in good solutions right after the number i. However, the model update mechanism is different from that used in the paper [21]. Also, an additional component is added to the model which is a weight vector W_s of length N_v containing weights that are used to calculate the chance of each number $i \in \{1, \ldots, N_v\}$ being used as the first element in the permutation. This element is added because, contrary to the TSP, in the FFP it is very important which element is the first in the solution.

Learning the Edge Histogram Matrix Model

Learning of this model follows the model update rule known from the PBIL algorithm [1]. From the sample P_{sample} used for updating the EDA models the worst solution (permutation) $\pi^{(-)}$ and the best one $\pi^{(+)}$ are selected according to the evaluation function (1). From these solutions two probability matrices $\mathbb{P}^{(-)}$ and $\mathbb{P}^{(+)}$ are built. In $\mathbb{P}^{(+)}$ all the elements are set to 0 except those elements $p_{ij}^{(+)}$ for which j immediately follows i in the permutation $\pi^{(+)}$. The matrix $\mathbb{P}^{(-)}$ is built in the same way from $\pi^{(-)}$.

The matrices $\mathbb{P}^{(+)}$ and $\mathbb{P}^{(-)}$ are used to update $\mathbb{P}$ by applying the model update mechanism from PBIL for each element p_{ij} in $\mathbb{P}$ separately, except for the elements on the diagonal which are always set to 0. This process uses two learning rate parameters: the positive learning rate η_+ and the negative learning rate η_-. Their sum is denoted $\eta = \eta_+ + \eta_-$. According to the PBIL algorithm, if $p_{ij}^{(-)} = p_{ij}^{(+)}$ the element p_{ij} of $\mathbb{P}$ is set to:

$$p_{ij} = p_{ij} \cdot (1 - \eta_+) + p_{ij}^{(+)} \cdot \eta_+, \tag{2}$$

and if $p_{ij}^{(-)} \neq p_{ij}^{(+)}$ the element p_{ij} is set to:

$$p_{ij} = p_{ij} \cdot (1 - \eta) + p_{ij}^{(+)} \cdot \eta. \tag{3}$$

Finally, each element p_{ij} (except the elements on the diagonal) is mutated with probability P_{mut} by setting:

$$p_{ij} = p_{ij} \cdot (1 - \mu) + \alpha * \mu, \tag{4}$$

where:

α - a 0 or 1 value drawn randomly with equal probabilities $P(0) = P(1) = \frac{1}{2}$,
μ - a mutation-shift parameter controlling the intensity of mutation.

The weight vector W_s is modified by taking the first element k in a genotype of each specimen and increasing $W_s[k]$ by the evaluation of the specimen from which k was taken. Thus, weights for those numbers that are used as first elements in good solutions are increased.

Sampling of the Matrix Model

A new permutation π is generated from the model as follows. The first element in the permutation may be selected in two different ways. The first is to draw a number uniformly from the set $1, \ldots, N_v$. This initialization method was used in the paper [19] for the TSP, because in the TSP the solution does not depend on which element in the tour is considered the first. In the FFP it does matter, however, which node is defended as the first one. Therefore, when the model is updated, a weight vector W_s is constructed which contains weights corresponding to how often a given number was used at the first position in the permutation. The first element $\pi[1]$ is then randomly drawn from the set $\{1, \ldots, N_v\}$ with probabilities proportional to the elements of W_s. Because the second method of initialization cannot select as the first element any of the nodes that did not appear at the first position already, it turned out that it is the most effective to combine both methods of initialization. Consequently, the uniform initialization is performed with a probability P_{unif} and the weight-vector-based initialization with a probability $1 - P_{unif}$.

When generating the element $\pi[i]$, $i = 2, \ldots, N_v$ a sum $\overline{p}$ is calculated:

$$\overline{p} = \sum_{j \notin \{\pi[1], \ldots, \pi[i-1]\}} p_{\pi[i-1]j}. \tag{5}$$

If $\overline{p} > 0$ then a number $j \notin \{\pi[1], \ldots, \pi[i-1]\}$ is randomly selected with probability $\frac{p_{\pi[i-1]j}}{\overline{p}}$. If $\overline{p} = 0$, a number from $j \notin \{\pi[1], \ldots, \pi[i-1]\}$ is randomly selected with uniform probability.

3.2   The State-Position Model

The State-Position (S-P) model proposed in this paper represents the relationship between the graph state S, the position (number of the vertex) at which the firefighter was assigned v, and the mean evaluation e which was finally achieved after using this particular assignment in the graph state S. Graph states are represented as vectors of states of the vertices in the graph, so each graph state is an element of the space L^{N_v}. The model M is thus represented by a list of ordered triples:

$$M = [\langle S_1, v_1, e_1 \rangle, \langle S_2, v_2, e_2 \rangle, \ldots \langle S_n, v_n, e_n \rangle] \tag{6}$$

where $S_i \in L^{N_v}$, $v_i \in V$ and $e_i \in \mathbb{R}$ for $i = 1, \ldots, n$.

Because the same position can be defended in the same graph state, but in various solutions (attaining each the same or a different final evaluation), triples with the same elements S and v (but different e values) or multiple copies of the same triple may be constructed when analyzing solutions found in the sample P_{sample}. To reduce the size of the model, the evaluations obtained for the same state S_i and vertex v_i are averaged and only the mean e_i is stored.

Additionally, for each node v in the graph the model stores a weight $Q[v]$ calculated as the sum of the reciprocals of the positions (counting from 1) of node v in the solutions used to build the model. These weights are used for selecting nodes to defend if no selection can be done based on the M model.

Learning of the State-Position Model

The model M is built using solutions in the sample P_{sample} (see Algorithm 1). From each permutation π in the sample, several triples are generated by simulating the spreading of fire from the initial graph state. Each time a node v is protected, a pair containing the current graph state S and the node v is stored. After the simulation finishes, the final state is evaluated and triples are formed from the stored $\langle S, v \rangle$ pairs and the evaluation e. This procedure is presented in Algorithm 2. Note, that the symbol $\oplus$ used in this algorithm is an operator for adding an item to a list.

Algorithm 2. Learning of the State-Position model

IN: P_{sample} - A sample from the population

        S_0 - The initial state of the graph

        N_f - The number of firefighters assigned in one time step

OUT: M - The State-Position model built from the sample

        Q - The vector of weights assigned to the graph nodes

                when the model is built

$M := \emptyset$

for $\pi \in P_{sample}$ **do**

    // *Fire spreading simulation stores graph states and the defended nodes*

    $R := \emptyset$

    $S := S_0$

    while CanSpread(S) **do**

        $V := $ SelectPositions(S, π, N_f)

        for $v \in V$ **do**

            $S[v] := $ 'D'

            $R := R \cup \{\langle S, v \rangle\}$

        $S := $ SpreadFire(S)

    // *Evaluation of the final graph state*

    $e := $ EvaluateState(S)

    // *Addition of the evaluated state-position pairs to the model*

    for $\langle S, v \rangle \in R$ **do**

        $M := M \oplus \langle S, v, e \rangle$

    // *Calculation of the weights of individual nodes*

    for $i := 1, \dots, N_v$ **do**

        $Q[\pi[i]] := Q[\pi[i]] + \frac{1}{i}$

$M := $ CalculateMeanEvals(M)

Algorithm 2 uses the following procedures:

CanSpread - Returns a logical value indicating if in the graph state S the fire can still spread, that is, if there are untouched nodes adjacent to burning ones.

SelectPositions - Returns a set of N_f numbers that correspond to untouched nodes in the state S and are placed nearest the beginning of the permutation π (i.e. the first N_f nodes in the state 'U' appearing in π).

SpreadFire - Performs one step of the fire spreading by changing to 'B' the state of all untouched ('U') nodes adjacent to the burning ('B') ones.

EvaluateState - Evaluates the final state S by calculating the sum of values assigned to those vertices that are not burning ('B') in the state S.

CalculateMeanEvals - Aggregates the evaluations stored for state-position pairs by calculating, for each unique pair $\langle S', v' \rangle$, an average evaluation from all triples $\langle S, v, e \rangle$ in which $S' = S$ and $v' = v$. After the aggregation the model M contains triples with unique values of S and v (no duplicate pairs $\langle S, v \rangle$ exist).

After the learning of the model is completed the set M contains ordered triples $\langle S, v, e \rangle$. In each such triple a state of the graph S is combined with the position (node number) v which got protected and the average evaluation e that was finally achieved. Using this scheme, assignments that are good at a given graph state are rewarded by high final evaluations.

Sampling of the State-Position Model

Sampling of the State-Position model is performed by generating new specimens one by one using the information stored in the model. Each specimen is generated by performing the simulation of the fire spreading starting from the initial graph state S_0 and selecting positions to defend using the model M. The procedure of generating one specimen is presented in Algorithm 3.

Initially, nodes are added in a simulation loop in which the current state of the graph is compared to the states stored in the model M. The distance between graph states is measured using the Hamming distance $H(S, S')$ (which is, simply, the number of vertices $u \in 1, \ldots, N_v$ for which $S[u] \neq S'[u]$). For each triple $\langle S', v', e' \rangle$ stored in the model M such that $v' = v$ the weight $w[v]$ of the node v is increased proportionally to the evaluation e' and inversely proportionally to a function $f()$ of the Hamming distance between the current state S and the state stored in the model S'. The function $f()$ is used to determine how the weight of the position should change with the distance between graph states. In this paper the following functions were tested:

Linear - $f(x) = 1 + x$. Makes the weight of the position decrease inversely with the distance between graph states. The 1 is added to avoid errors when the states S and S' are equal and H(S, S') = 0.
Square - $f(x) = 1 + x * x$. A function whose inverse decreases faster than that of the linear one.
Sqrt - $f(x) = 1 + \sqrt{x}$. A function whose inverse decreases more slowly than that of the linear one.
Exponential - $f(x) = 3^x$. A function producing a very narrow, exponentially vanishing peak around the given graph state S. The basis of 3 was selected because there are three possible states of each node of the graph, so the value of 3^x represents the number of the graph states with all the possible values at the x positions at which S differs from S'. The exponential function produces a value of $f(x) = 1$ for $x = 0$ as the other functions, so in each case the weight assigned to the current graph state is 1.

When there is at least one node v with a positive weight $w[v]$ the selection of the node to defend is performed using a roulette wheel selection procedure with probabilities proportional to the weights in w. Otherwise, the selection is performed using weights stored in Q which are inversely proportional to the positions at which the nodes appeared in the population (only untouched nodes are considered). Thus, selection using weights from Q gives higher priorities to

Algorithm 3. Sampling of the State-Position model

IN: S_0 - The initial state of the graph
       N_f - The number of firefighters assigned in one time step
       M - The State-Position model built from the sample
       Q - The vector of weights assigned to the graph nodes
         when the model is built

OUT: π - A new solution generated from the model

// *Simulate spreading of fire*
$\pi := \emptyset$
$S := S_0$
while CanSpread(S) **do**
$\quad$ $w := [0, 0, \ldots, 0]$
$\quad$ $W := [0, 0, \ldots, 0]$
$\quad$ **for** $v := 1, \ldots, N_v$ **do**
$\quad\quad$ **if** $S[v] = {}'U'$ **then**
$\quad\quad\quad$ **for** $\langle S', v', e' \rangle \in M$, *s.t.* $v' = v$ **do**
$\quad\quad\quad\quad$ $w[v] := w[v] + e' \frac{1}{f(H(S,S'))}$
$\quad\quad\quad$ $W[v] := Q[v]$
$\quad$ **for** $i := 1, \ldots, N_f$ **do**
$\quad\quad$ **if** $\sum w > 0$ **then**
$\quad\quad\quad$ $v :=$ RouletteWheelSelection(w)
$\quad\quad$ **else**
$\quad\quad\quad$ $v :=$ RouletteWheelSelection(W)
$\quad\quad$ $w[v] := 0$
$\quad\quad$ $W[v] := 0$
$\quad\quad$ $\pi := \pi \oplus v$
$\quad\quad$ $S[v] := {}'D'$
$\quad$ $S :=$ SpreadFire(S)

// *Nodes not used in the simulation are added using Q weights*
$W := [0, 0, \ldots, 0]$
for $v := 1, \ldots, N_v$ **do**
$\quad$ **if** $v \notin \pi$ **then**
$\quad\quad$ $W[v] := Q[v]$

while $\sum W > 0$ **do**
$\quad$ $v :=$ RouletteWheelSelection(W)
$\quad$ $W[v] := 0$
$\quad$ $\pi := \pi \oplus v$

return π

nodes that tend to appear towards the beginning of the solutions. Of course, only some nodes are defended during the simulation. The remaining ones are added to the solution π using the roulette wheel selection with probabilities proportional to weights in Q. This time nodes are used regardless of the state 'B' or 'U' in which they were during the simulation, except for the nodes in state 'D' which are already in the solution π. This last step is used in order to retain some information concerning the precedence of the nodes in the population, even if those nodes are not used for defense.

In addition to CanSpread() and SpreadFire() procedures used in Algorithm 2, Algorithm 3 uses the **RouletteWheelSelection()** procedure that performs the roulette wheel selection procedure using weights in the given vector w. An index of each of the elements in the vector can be returned with the probability proportional to the weight at that index. For example if the weight vector is $w = [12, 4, 1, 3]$ the probability of returning 1 is 0.6, for 2 it is 0.2, for 3 it is 0.05 and for 4 it is 0.15.

4   Experiments and Results

In the experiments the EDA approach was tested with four different probabilistic models (Simple Mallows, Generalized Mallows, EH-PBIL and State-Position model). In the case of the EH-PBIL model, the parameters were set following the original paper on the PBIL algorithm [1] to: $\eta_+ = 0.1$ (learning rate), $\eta_- = 0.075$ (negative learning rate), $P_{mut} = 0.02$ (mutation probability) and $\mu = 0.05$ (mutation shift parameter). The values of $P_{unif} = 0.0, 0.2, 0.4, 0.6,$ 0.8 and 1.0 were tested in order to determine the influence of this parameter introduced in this paper on the working of the algorithm. The State-Position model was used with the four functions mentioned before (Exponential, Linear, Sqrt and Square).

For comparison, tests were performed with the Ant Colony Optimization (ACO) algorithm proposed in [2] and the VNS method proposed in [11]. An Evolutionary Algorithm (EA) was also tested with three crossover operators that performed best in the previous paper [17], that is the CX, OBX and PBX. For the mutation operator the insertion mutation was used because it worked best in the aforementioned paper. Crossover and mutation probabilities were set to $P_{cross} = 0.9$ and $P_{mut} = 0.05$. The population size was set to $N_{pop} = 100$ for all the methods. The sample size for the EDAs was set to 20% of the population size, so $N_{sample} = 20$. The EDAs and EAs employed the elitism mechanism in which one, the best, solution was always promoted to the next generation.

For testing Erdős-Renyi graphs, represented using adjacency matrices, with $N_v = 500, 750, 1000, 1250, 1500, 1750, 2000, 2250, 2500$ and 5000 vertices were used. The probability with which an edge was added between any two different vertices (independently of the others) was $P_{edge} = \frac{3}{N_v}$. This value was selected during preliminary experiments in such a way that the obtained problem instances were not too easy (the entire graph easily protected) nor too difficult (all the nodes except the defended ones always lost). The other parameters of the problem instances were $N_s = 1$ starting point and $N_f = 2$ firefighters allowed per a time step. In order to ensure that the generated instances were difficult enough, only such graphs were used in which the number of edges adjacent to the starting points exceeded the number N_f of firefighters allowed per a time step. This requirement was formulated to eliminate a trivial solution which is to use the N_f firefighters to cut off the starting points from the rest of the graph during the first time step. While such a solution is a very good one (most of the graph is saved in such case) it is also trivial to apply and therefore is not really indicative of the actual problem solving capacity of the tested algorithms. Costs drawn uniformly from the range $[0, 100]$ were assigned to the nodes of the graphs. A set of 50 different graphs was prepared as described above for each graph size N_v and each method was tested on the same 50 graphs.

Comparison of the algorithms was done on the basis of the median calculated over these 50 runs from the evaluations of the best solution found by each of the algorithms. Median values were used because statistical testing could then be performed using the Wilcoxon test without ensuring normality of the distributions. The results were compared at $T_{max} = 300, 600, 900$ and $7200\,$s for $N_v = 500, 750, 1000$; $N_v = 1250, 1500, 1750$; $N_v = 2000, 2250, 2500$; and $N_v = 5000$ respectively (see Tables 1 and 2). In the tables the best value for each test problem size N_v is marked in bold.

Table 1. Median value of the saved nodes obtained in the experiments.

N_v		500	750	1000	1250	1500
T_{max}		300 s			600 s	
ACO		$1.369 \cdot 10^3$	$1.332 \cdot 10^3$	$1.320 \cdot 10^3$	$3.331 \cdot 10^3$	$3.837 \cdot 10^3$
EA	CX	$1.834 \cdot 10^3$	$1.831 \cdot 10^3$	$1.909 \cdot 10^3$	$4.023 \cdot 10^3$	$4.614 \cdot 10^3$
	OBX	$2.039 \cdot 10^3$	$2.035 \cdot 10^3$	$2.027 \cdot 10^3$	$3.993 \cdot 10^3$	$4.504 \cdot 10^3$
	PBX	$2.013 \cdot 10^3$	$2.057 \cdot 10^3$	$2.013 \cdot 10^3$	$3.971 \cdot 10^3$	$4.474 \cdot 10^3$
Mal-	Generalized	$1.421 \cdot 10^3$	$1.337 \cdot 10^3$	$1.388 \cdot 10^3$	$3.757 \cdot 10^3$	$4.305 \cdot 10^3$
lows	Simple	$1.523 \cdot 10^3$	$1.525 \cdot 10^3$	$1.495 \cdot 10^3$	$3.790 \cdot 10^3$	$4.398 \cdot 10^3$
	0.0	$1.470 \cdot 10^3$	$1.442 \cdot 10^3$	$1.465 \cdot 10^3$	$3.697 \cdot 10^3$	$4.249 \cdot 10^3$
	0.2	$1.450 \cdot 10^3$	$1.482 \cdot 10^3$	$1.502 \cdot 10^3$	$3.743 \cdot 10^3$	$4.276 \cdot 10^3$
EH-	0.4	$1.454 \cdot 10^3$	$1.471 \cdot 10^3$	$1.450 \cdot 10^3$	$3.780 \cdot 10^3$	$4.314 \cdot 10^3$
PBIL	0.6	$1.386 \cdot 10^3$	$1.420 \cdot 10^3$	$1.413 \cdot 10^3$	$3.703 \cdot 10^3$	$4.261 \cdot 10^3$
	0.8	$1.364 \cdot 10^3$	$1.350 \cdot 10^3$	$1.361 \cdot 10^3$	$3.634 \cdot 10^3$	$4.174 \cdot 10^3$
	1.0	$1.296 \cdot 10^3$	$1.343 \cdot 10^3$	$1.344 \cdot 10^3$	$3.646 \cdot 10^3$	$4.138 \cdot 10^3$
	Exponential	$1.821 \cdot 10^3$	$1.875 \cdot 10^3$	$1.951 \cdot 10^3$	$4.229 \cdot 10^3$	$4.831 \cdot 10^3$
State-	Linear	$1.942 \cdot 10^3$	$2.057 \cdot 10^3$	$2.195 \cdot 10^3$	$4.464 \cdot 10^3$	$5.025 \cdot 10^3$
Position	Sqrt.	$1.694 \cdot 10^3$	$1.812 \cdot 10^3$	$1.955 \cdot 10^3$	$4.173 \cdot 10^3$	$4.715 \cdot 10^3$
	Square	$2.076 \cdot 10^3$	$2.178 \cdot 10^3$	$\mathbf{2.372 \cdot 10^3}$	$\mathbf{4.643 \cdot 10^3}$	$\mathbf{5.216 \cdot 10^3}$
VNS		$\mathbf{2.832 \cdot 10^3}$	$\mathbf{2.678 \cdot 10^3}$	$2.351 \cdot 10^3$	$4.521 \cdot 10^3$	$5.121 \cdot 10^3$
FWER		$2.09 \cdot 10^{-8}$	$5.53 \cdot 10^{-7}$	$8.43 \cdot 10^{-1}$	$1.04 \cdot 10^{-2}$	$5.20 \cdot 10^{-4}$

Table 2. Median value of the saved nodes obtained in the experiments.

N_v		1750	2000	2250	2500	5000
T_{max}		600 s	900 s			7200 s
ACO		$4.314 \cdot 10^3$	$1.425 \cdot 10^3$	$5.207 \cdot 10^3$	$5.828 \cdot 10^3$	$1.525 \cdot 10^3$
EA	CX	$4.874 \cdot 10^3$	$2.148 \cdot 10^3$	$5.811 \cdot 10^3$	$6.338 \cdot 10^3$	$2.454 \cdot 10^3$
	OBX	$4.796 \cdot 10^3$	$2.170 \cdot 10^3$	$5.863 \cdot 10^3$	$6.339 \cdot 10^3$	$2.476 \cdot 10^3$
	PBX	$4.808 \cdot 10^3$	$2.139 \cdot 10^3$	$5.777 \cdot 10^3$	$6.348 \cdot 10^3$	$2.408 \cdot 10^3$
Mal-	Generalized	$4.674 \cdot 10^3$	$1.606 \cdot 10^3$	$5.494 \cdot 10^3$	$6.205 \cdot 10^3$	$1.746 \cdot 10^3$
lows	Simple	$4.736 \cdot 10^3$	$1.620 \cdot 10^3$	$5.631 \cdot 10^3$	$6.257 \cdot 10^3$	$1.803 \cdot 10^3$
	0.0	$4.678 \cdot 10^3$	$1.546 \cdot 10^3$	$5.626 \cdot 10^3$	$6.238 \cdot 10^3$	$1.749 \cdot 10^3$
	0.2	$4.709 \cdot 10^3$	$1.652 \cdot 10^3$	$5.617 \cdot 10^3$	$6.238 \cdot 10^3$	$1.781 \cdot 10^3$
EH-	0.4	$4.712 \cdot 10^3$	$1.610 \cdot 10^3$	$5.505 \cdot 10^3$	$6.239 \cdot 10^3$	$1.768 \cdot 10^3$
PBIL	0.6	$4.732 \cdot 10^3$	$1.566 \cdot 10^3$	$5.479 \cdot 10^3$	$6.218 \cdot 10^3$	$1.712 \cdot 10^3$
	0.8	$4.649 \cdot 10^3$	$1.499 \cdot 10^3$	$5.471 \cdot 10^3$	$6.189 \cdot 10^3$	$1.673 \cdot 10^3$
	1.0	$4.685 \cdot 10^3$	$1.493 \cdot 10^3$	$5.490 \cdot 10^3$	$6.152 \cdot 10^3$	$1.627 \cdot 10^3$
	Exponential	$5.287 \cdot 10^3$	$2.240 \cdot 10^3$	$6.081 \cdot 10^3$	$6.858 \cdot 10^3$	$2.476 \cdot 10^3$
State-	Linear	$5.527 \cdot 10^3$	$2.534 \cdot 10^3$	$6.445 \cdot 10^3$	$7.059 \cdot 10^3$	$2.982 \cdot 10^3$
Position	Sqrt.	$5.200 \cdot 10^3$	$2.224 \cdot 10^3$	$6.161 \cdot 10^3$	$6.703 \cdot 10^3$	$2.640 \cdot 10^3$
	Square	$\mathbf{5.713 \cdot 10^3}$	$\mathbf{2.756 \cdot 10^3}$	$\mathbf{6.591 \cdot 10^3}$	$\mathbf{7.252 \cdot 10^3}$	$\mathbf{3.209 \cdot 10^3}$
VNS		$5.253 \cdot 10^3$	$2.019 \cdot 10^3$	$6.092 \cdot 10^3$	$6.514 \cdot 10^3$	$1.647 \cdot 10^3$
FWER		$2.63 \cdot 10^{-7}$	$1.04 \cdot 10^{-2}$	$1.36 \cdot 10^{-8}$	$1.55 \cdot 10^{-8}$	$2.131 \cdot 10^{-7}$

The comparison with respect to the running time of the algorithms was chosen because of a large variety of methods used for comparison (the same approach was used in the paper on VNS [11]). Two other commonly used comparison criteria, the number of generations and the number of solution evaluations, are not well-suited for the comparisons made in this paper. Comparison by the number of generations overlooks the fact that certain methods may perform additional, costly computations in each generation, which is not uncommon in the EDAs. Comparison by the number of solution evaluations suffers from the same problem and also may not be reliable in the case of different solution representations (e.g. those used by the EA/EDA and the VNS).

From the results presented in the tables it can be seen, that for smaller graphs the VNS is very effective, but for larger graphs the State-Position EDA performs best. The tables contain also the value of the Family-Wise Error Rate for the hypothesis that the median values produced by the best performing method for a given N_v are statistically different than those produced by the other methods. This FWER value was calculated as $1 - \prod_i (1 - v_i)$, where v_i are p-values obtained in pairwise comparisons between the best performing method and each of the other methods. The calculated value is the upper bound for the probability that at least one of the comparison methods attains the same median values as the best-performing one. For graph sizes of $N_v = 1250$ and more the calculated FWER values are at most $1.04 \cdot 10^{-2}$. This shows that the difference between the State-Position EDA with the Square function and the other methods is statistically significant. For $N_v = 1000$ the State-Position EDA still produced the best result, but because of the small difference from the results produced by the VNS the FWER is high, so the statistical significance cannot be confirmed. For $N_v = 500$ and 750 the State-Position EDA was outperformed by the VNS, with the difference statistically significant in both cases.

The results produced by the algorithms for $N_v = 2500$ with respect to the running time are shown in Fig. 1. Note, that in case when multiple variants or parametrizations of one method were tested, only one variant is presented in the figure, the one that produced the best result at $T_{max} = 900$ s. The algorithms presented in the figure are the ACO, the EA using the OBX crossover, the Simple Mallows, the EH-PBIL with $P_{unif} = 0.4$, the State-Position EDA with the Square function and the VNS.

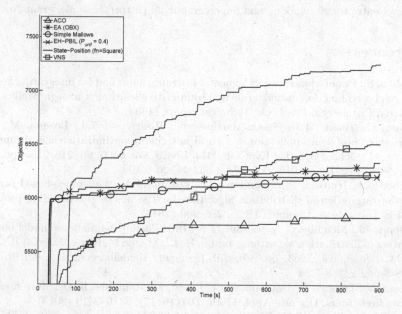

Fig. 1. Results produced by the algorithms for $N_v = 2500$ with respect to the running time. Note, that in case when multiple variants or parametrizations of one method were tested, only one variant is presented in the figure, the one that produced the best result at $T_{max} = 900$ s.

5   Conclusion

In this paper EDA algorithms using several probabilistic models were tested for the Firefighter Problem (FFP). A new model was proposed which represents the relationship between the state of the graph and the positions which are defended by firefighters. The proposed State-Position model outperformed the other EDAs, in particular the Mallows model which is commonly used for permutation-based problems and therefore was used in this paper as one of the comparison models. Compared to all the tested methods, the State-Position EDA produced the best results for graphs with $N_v = 1000$ and more. For $N_v = 500$ and 750 the VNS method shown the best performance.

The results presented in this paper form an interesting starting point for further research. First, the State-Position model builds a representation that is easily interpretable in the domain of the problem. This may constitute a basis for knowledge extraction, for example in the form of rules guiding the placement of firefighters in various graph states. Another possible extension is to apply EDAs to the multiobjective FFP either by developing multiobjective models, or by using the Sim-EDA approach proposed in [19].

Acknowledgment. Calculations have been carried out using resources provided by Wroclaw Centre for Networking and Supercomputing (http://wcss.pl), grant No. 407.

References

1. Baluja, S.: Population-based incremental learning: a method for integrating genetic search based function optimization and competitive learning. Carnegie Mellon University, Pittsburgh, PA, USA, Technical report (1994)
2. Blum, C., Blesa, M.J., García-Martínez, C., Rodríguez, F.J., Lozano, M.: The firefighter problem: application of hybrid ant colony optimization algorithms. In: Blum, C., Ochoa, G. (eds.) EvoCOP 2014. LNCS, vol. 8600, pp. 218–229. Springer, Heidelberg (2014). doi:10.1007/978-3-662-44320-0_19
3. Ceberio, J., Irurozki, E., Mendiburu, A., Lozano, J.A.: A distance-based ranking model estimation of distribution algorithm for the flowshop scheduling problem. IEEE Trans. Evol. Comput. **18**(2), 286–300 (2014)
4. Ceberio, J., Mendiburu, A., Lozano, J.A.: Introducing the mallows model on estimation of distribution algorithms. In: Lu, B.-L., Zhang, L., Kwok, J. (eds.) ICONIP 2011. LNCS, vol. 7063, pp. 461–470. Springer, Heidelberg (2011). doi:10.1007/978-3-642-24958-7_54
5. Costa, V., Dantas, S., Dourado, M.C., Penso, L., Rautenbach, D.: More fires and more firefighters. Discrete Appl. Math. **161**(16–17), 2410–2419 (2013)
6. Develin, M., Hartke, S.G.: Fire containment in grids of dimension three and higher. Discrete Appl. Math. **155**(17), 2257–2268 (2007)
7. Finbow, S., MacGillivray, G.: The firefighter problem: a survey of results, directions and questions. Australas. J. Comb. **43**, 57–77 (2009)
8. Fligner, M.A., Verducci, J.S.: Distance based ranking models. J. R. Stat. Soc. **48**(3), 359–369 (1986)
9. Garca-Martnez, C., et al.: The firefighter problem: empirical results on random graphs. Comput. Oper. Res. **60**, 55–66 (2015)
10. Hartnell, B.: Firefighter! An application of domination. In: 20th Conference on Numerical Mathematics and Computing (1995)
11. Hu, B., Windbichler, A., Raidl, G.R.: A new solution representation for the firefighter problem. In: Ochoa, G., Chicano, F. (eds.) EvoCOP 2015. LNCS, vol. 9026, pp. 25–35. Springer, Heidelberg (2015). doi:10.1007/978-3-319-16468-7_3

12. Juan, A.A., et al.: A review of simheuristics: extending metaheuristics to deal with stochastic combinatorial optimization problems. Oper. Res. Perspect. **2**, 62–72 (2015)
13. Kong, J., Zhang, L., Wang, W.: The surviving rate of digraphs. Discrete Math. **334**, 13–19 (2014)
14. Lu, T., Boutilier, C.: Learning mallows models with pairwise preferences. In: Getoor, L., Scheffer, T. (eds.) Proceedings of the 28th International Conference on Machine Learning (ICML 2011), pp. 145–152. ACM (2011)
15. Mallows, C.L.: Non-null ranking models. Biometrika **44**(1–2), 114–130 (1957)
16. Meila, M., Chen, H.: Dirichlet process mixtures of generalized mallows models. Uncertainty Artif. Intell. **1**(1), 358–367 (2010)
17. Michalak, K.: Auto-adaptation of genetic operators for multi-objective optimization in the firefighter problem. In: Corchado, E., Lozano, J.A., Quintián, H., Yin, H. (eds.) IDEAL 2014. LNCS, vol. 8669, pp. 484–491. Springer, Heidelberg (2014). doi:10.1007/978-3-319-10840-7_58
18. Michalak, K.: The Sim-EA algorithm with operator autoadaptation for the multiobjective firefighter problem. In: Ochoa, G., Chicano, F. (eds.) EvoCOP 2015. LNCS, vol. 9026, pp. 184–196. Springer, Heidelberg (2015). doi:10.1007/978-3-319-16468-7_16
19. Michalak, K.: Sim-EDA: a multipopulation estimation of distribution algorithm based on problem similarity. In: Chicano, F., Hu, B., García-Sánchez, P. (eds.) EvoCOP 2016. LNCS, vol. 9595, pp. 235–250. Springer, Heidelberg (2016). doi:10.1007/978-3-319-30698-8_16
20. Michalak, K., Knowles, J.D.: Simheuristics for the multiobjective nondeterministic firefighter problem in a time-constrained setting. In: Squillero, G., Burelli, P. (eds.) EvoApplications 2016. LNCS, vol. 9598, pp. 248–265. Springer, Heidelberg (2016). doi:10.1007/978-3-319-31153-1_17
21. Tsutsui, S.: Probabilistic model-building genetic algorithms in permutation representation domain using edge histogram. In: Guervós, J.J.M., Adamidis, P., Beyer, H.-G., Schwefel, H.-P., Fernández-Villacañas, J.-L. (eds.) PPSN 2002. LNCS, vol. 2439, pp. 224–233. Springer, Heidelberg (2002). doi:10.1007/3-540-45712-7_22

LCS-Based Selective Route Exchange Crossover for the Pickup and Delivery Problem with Time Windows

Miroslaw Blocho and Jakub Nalepa[✉]

Silesian University of Technology, Gliwice, Poland
jakub.nalepa@polsl.pl

Abstract. The pickup and delivery problem with time windows (PDPTW) is an NP-hard discrete optimization problem of serving transportation requests using a fleet of homogeneous trucks. Its main objective is to minimize the number of vehicles, and the secondary objective is to minimize the distance traveled during the service. In this paper, we propose the longest common subsequence based selective route exchange crossover (LCS-SREX), and apply this operator in the memetic algorithm (MA) for the PDPTW. Also, we suggest the new solution representation which helps handle the crossover efficiently. Extensive experimental study performed on the benchmark set showed that using LCS-SREX leads to very high-quality feasible solutions. The analysis is backed with the statistical tests to verify the importance of the elaborated results. Finally, we report one new world's best routing schedule found using a parallel version of the MA exploiting LCS-SREX.

Keywords: Memetic algorithm · LCS · Crossover · PDPTW

1 Introduction

Solving rich vehicle routing problems (VRPs) is a vital research topic due to their practical applications which include delivery of food, beverages and parcels, bus routing, delivery of cash to ATM terminals, waste collection, and many others. There exist numerous variants of rich VRPs reflecting a wide range of real-life scheduling scenarios [1]—they often combine multiple realistic constraints which are imposed on the desired solutions (a solution is feasible if all of the constraints are satisfied). Although exact algorithms retrieve an optimal routing schedule, they are still very difficult to exploit in practice, because of their unacceptable execution times for massively-large problems. Therefore, approximate algorithms became the main stream of the research activities—they deliver high-quality (not necessarily optimal) schedules in significantly shorter time.

The pickup and delivery problem with time windows (PDPTW) is the NP-hard problem of serving a number of transportation requests using a fleet of trucks. Each request is a pair of the pickup and delivery operations which must be performed in the appropriate order (the precedence constraint). Each travel

© Springer International Publishing AG 2017
B. Hu and M. López-Ibáñez (Eds.): EvoCOP 2017, LNCS 10197, pp. 124–140, 2017.
DOI: 10.1007/978-3-319-55453-2_9

point should be visited within its time window (the time window constraint), the size of vehicles cannot be exceeded (the capacity constraint), and all trucks should start and finish the service in a single depot. The PDPTW is a hierarchical objective discrete problem—the main objective is to minimize the number of vehicles, whereas the secondary one is to optimize the total travel distance.

The exact algorithms for the PDPTW are most often applied to small-scale instances due to their execution times. Hence, a plethora of approximate methods have been exploited for this task—they include various heuristic and metaheuristic techniques (both sequential and parallel). Memetic algorithms (MAs)—the hybrids of evolutionary techniques (utilized to explore the solution space) and refinement procedures (used for the exploitation of solutions already found) have been proven very efficient in solving the PDPTW [2], as well as other VRPs [1].

In this paper, we propose a new crossover operator for the PDPTW—the longest common subsequence based selective route exchange crossover (LCS-SREX). The original SREX operator was proposed in [2], and it was shown very efficient in solving the PDPTW. In SREX, the routes which are to be affected during the crossover are selected based on the number of common travel points in the parent schedules (however, these solutions are very often not similar in terms of the longest common partial routes). This approach could easily lead to large numbers of unserved transportation requests after applying SREX, which had to be re-inserted into the partial offspring solutions. Here, we tackle this issue—in LCS-SREX, the selection of routes is based on the analysis of the longest common subroutes of the parents to minimize the number of unserved customers after the crossover. Also, we suggest the new solution representation to handle the crossover efficiently. Our extensive experimental study performed on the 400-customer Li and Lim's benchmark set of problem instances of various characteristics showed that applying LCS-SREX leads to retrieving very high-quality solutions. We investigated how the number of children generated for each pair of parents during the recombination affects the optimization process. The analysis is coupled with statistical tests to verify the importance of the retrieved results. Finally, we report one new world's best solution to the benchmark instance elaborated using a parallel version of our algorithm.

This paper is structured as follows. In Sect. 2, the PDPTW is formulated. Section 3 presents the state-of-the-art algorithms for the PDPTW. In Sect. 4, we present the MA and discuss in detail the LCS-SREX operator. The extensive experimental study is reported and analyzed in Sect. 5. Section 6 concludes the paper and serves as the outlook to the future work.

2   Problem Formulation

The PDPTW is defined on a directed graph $G = (V, E)$, with a set V of $C + 1$ vertices (v_i, $i \in \{1, \ldots, C\}$). These vertices represent the customers, v_0 is the depot, whereas the set of edges $E = \{(v_i, v_{i+1}) | v_i, v_{i+1} \in V, v_i \neq v_{i+1}\}$ are the travel connections. The travel costs $c_{i,j}$, $i, j \in \{0, 1, \ldots, C\}$, $i \neq j$, are equal to the distance between the corresponding travel points (in the Euclidean metric).

Each request h_i, $i \in \{1, 2, \ldots, N\}$, where $N = C/2$, is a coupled pair of pickup (P) and delivery (D) customers—p_h and d_h, respectively, where $P \cap D = \emptyset$, and $P \cup D = V \setminus \{v_0\}$. For each h_i, the amount of delivered $(q^d(h_i))$ and picked up $(q^p(h_i))$ goods is defined, and $q^d(h_i) = -q^p(h_i)$. Each customer v_i defines its demand (being either delivery or pickup), service time s_i (serving the depot does not take time, hence $s_0 = 0$), and time window $[e_i, l_i]$ within which its service must be started (however, it can be finished after the corresponding time slot has been closed). The fleet of available trucks is homogeneous (let K denote its size)—the capacity of each vehicle is Q. Each route r in the solution σ (being a set of routes), starts and finishes at the depot v_0.

A solution σ (K routes) is feasible if (i) Q is not exceeded for any vehicle (the capacity constraint is fulfilled), (ii) the service of every request starts within its time window (time window constraint), (iii) every customer is served in exactly one route, (iv) every vehicle leaves and returns to v_0 within its time windows, and (v) each pickup is performed before the corresponding delivery (precedence constraint). The primary objective of the PDPTW is to minimize the number of vehicles (K). Afterwards, the total travel distance is optimized. Let σ_A and σ_B be two solutions. σ_A is then of a higher quality compared with σ_B if $(K(\sigma_A) < K(\sigma_B))$ or $(K(\sigma_A) = K(\sigma_B)$ and $T(\sigma_A) < T(\sigma_B))$, where T is the total distance.

3 Related Literature

State-of-the-art techniques for tackling the PDPTW [3] (and numerous other rich VRP variants [4]) are divided into exact and approximate approaches. The former algorithms deliver the optimal solutions, whereas the latter obtain high-quality feasible solutions in acceptable time. The exact algorithms were devised for relatively small problem instances (up to 30 requests [5]) due to their enormous computation time. Although they are being actively developed, this execution time becomes their important bottleneck and it is still difficult to apply them in practice for solving massively large real-life scenarios.

The exact techniques encompass, among others, column generation methods, branch-and-cut, branch-and-price solvers, and dynamic-programming-based techniques [6–8]. The set-partitioning-like integer formulation of the problem was presented by Baldacci et al. [9]—two dual ascent heuristics were coupled with the cut-and-column generation for finding the dual solution of the linear programming relaxation of this formulation.

The heuristic algorithms usually tackle the PDPTW in two steps (due to the hierarchical objective of the PDPTW [2])—the fleet size is minimized in the first stage, and the total distance is optimized afterwards. This approach allows for designing and implementing efficient techniques for both stages independently. Approximate algorithms to minimize the number of trucks include construction and improvement heuristics. The construction (often referred to as the *insertion-based* methods) heuristics create solutions from scratch by inserting requests iteratively into the partial solution, according to certain criteria, e.g., the maximum cost savings, the minimum additional travel distance, or the cost of reducing the time window slacks [10,11]. These insertions should not violate the solution feasibility. If any of the constraints is violated, then the solution is usually backtracked to the last feasible state, and the further attempts of re-inserting the unserved customers are undertaken.

Improvement heuristics modify a low-quality solution by executing refinement procedures in search for better neighboring solutions. The metaheuristics usually embed construction and improvement heuristics. Temporarily infeasible solutions, along with deterioration of the solutions quality during the optimization process are very often allowed in these algorithms [12]. Heuristic algorithms for the PDPTW comprise tabu [5] and neighborhood searches [13], guided ejection searches (GESes), enhanced in our very recent work [14], evolutionary algorithms [15,16], hybrid techniques (e.g., combining simulated annealing [17]), and many more [2,18]. A very interesting formulation of a variant of the PDPTW with simultaneous pickup and delivery operations, along with a new mat-heuristics to tackle this problem have been reported recently [3]. Parallel algorithms were explored for solving rich VRPs [19], including the PDPTW [20,21]. Thorough surveys on approximate approaches for the PDPTW were presented by Parragh et al. [12], and recently by Cherkesly et al. [22].

4   Memetic Algorithm for the PDPTW

In this section, we present in detail the MA for minimizing the distance traveled during the service in the PDPTW. We put a special emphasis on discussing the longest common subsequence based crossover operator which is the main contribution of this paper. Also, we discuss the new representation of solutions which involves assigning a hash function to each arc connecting the neighboring travel points. This representation is pivotal and helps handle the selection of routes for crossover efficiently using dynamic programming.

4.1   Algorithm Outline

In the MA (Algorithm 1), the number of routes is minimized in the first stage (lines 1–5), and the total travel distance is optimized afterwards (lines 7–27). In this work, we utilize the enhanced guided ejection search (GES) to minimize the number of trucks (line 1), and to generate the initial population of solutions—each individual contains $K(\sigma_1)$ routes (line 2). In GES, each request is served in a separate route at first, and the attempts to decrease the number of routes are undertaken. A random route is removed, and the requests are put into the *ejection pool* (containing unserved requests). Then, these requests are re-inserted to the partial solution (either feasibly or infeasibly, with additional local moves to restore the feasibility). The optimization is terminated if the maximum execution time has been exceeded. More details on GES can be found in [14].

This algorithm can be easily replaced by another (perhaps more efficient) technique without affecting the MA. The time limit for minimizing the number of trucks is τ_{RM}, whereas the limit for generating the initial population of individuals is denoted as τ_{PopGen}. If the number of solutions is less than the desired population size N_{pop} (because of the time limit), the remaining solutions are constructed by copying and perturbing the already-found individuals. Even if all required solutions are found, all of them are perturbed at the end of this initial stage (lines 3–5) to diversify the population. This perturb operation involves executing local search moves (pair-relocate and pair-exchange), which do not violate the constraints (but can potentially decrease the solution quality).

Algorithm 1. Memetic algorithm for minimizing distance in the PDPTW.

1: $\sigma_1 \leftarrow$ **ROUTE-MINIMIZATION**();
2: Generate the population of N_{pop} solutions $\{\sigma_1, \sigma_2, \ldots, \sigma_{N_{pop}}\}$;
3: **for each** solution σ_i in population **do**
4:     $\sigma_i \leftarrow$ **PERTURB**(σ_i);
5: **end for**
6: $done \leftarrow$ **false**;
7: **while not** $done$ **do**
8:     Determine N_{pop} random pairs (σ_A^p, σ_B^p);   ▷ Selection
9:     **for all** pairs (σ_A^p, σ_B^p) **do**
10:         **if** $\sigma_A^p = \sigma_B^p$ **then**
11:             $\sigma_A^p \leftarrow$ **PERTURB**(σ_A^p);
12:         **end if**
13:         $\sigma_{best}^c \leftarrow \sigma_A^p$;
14:         Generate N_{ch} children $\{\sigma_1^c, \sigma_2^c, \ldots, \sigma_{N_{ch}}^c\}$ for (σ_A^p, σ_B^p);
15:         **for** $i \leftarrow 1$ **to** N_{ch} **do**
16:             $\sigma_i^c \leftarrow$ **LOCAL-SEARCH**(σ_i^c);
17:             **if** $T(\sigma_i^c) < T(\sigma_{best}^c)$ **then**
18:                 $\sigma_{best}^c \leftarrow \sigma_i^c$;
19:             **end if**
20:         **end for**
21:         $\sigma_A^p \leftarrow \sigma_{best}^c$;   ▷ Replace σ_A^p with the best child in the next generation
22:     **end for**
23:     **if** (termination condition is met) **then**
24:         $done \leftarrow$ **true**;
25:     **end if**
26: **end while**
27: **return** best solution σ_{best} in the entire population;

The population of feasible solutions undergoes the memetic evolution to minimize the total travel distance (lines 7–26). Each iteration starts with determining N_{pop} random pairs of parents σ_A^p and σ_B^p for crossover (line 8). Then, N_{ch} children are retrieved for each pair of parents (note that if the structures of the individuals selected as parents appear the same, then one solution is perturbed, line 11). The parent σ_A^p becomes the best initial child σ_{best}^c (line 13). The N_{ch} offspring solutions are constructed (line 14). Each child σ_i^c is enhanced by applying a number of feasible local search moves, and the best offspring solution σ^c_{best}—with the minimum T—is selected (line 18)—for details, see Sect. 4.2.

The local search is visualized in Algorithm 2—it involves executing local moves in search of higher-quality neighboring solutions (i.e., with the lower T's). It is worth noting that we propose to exploit not only the pair-relocation moves (as shown in [2]), but also the pair-exchange moves, which are much harder to implement efficiently in the case of the PDPTW. The pair-exchange moves are significantly more time-consuming, and they are performed only if the pair-relocation neighborhood was found to be empty (hence, the pair-relocation moves are analyzed at first). The pair-relocation move involves ejecting the pickup and the corresponding delivery customer from one route, and inserting them feasibly in all possible ways into other routes. On the other hand, the pair-exchange move encompasses ejecting the pickup and delivery customers from two different routes and re-inserting them in all possible ways into other routes.

Algorithm 2. Improving a solution with local search moves.

1: **function** LOCALSEARCH(σ)
2:     $\sigma_b \leftarrow \sigma$;
3:     *improvement* ← **true**;
4:     **while** *improvement* = **true do**
5:         Find σ' through feasible local search moves on σ_b;
6:         **if** $T(\sigma') < T(\sigma_b)$ **then**
7:             $\sigma_b \leftarrow \sigma'$;
8:         **else**
9:             *improvement* ← **false**;
10:        **end if**
11:    **end while**
12:    **return** σ_b;
13: **end function**

In order to create the pair-relocation and pair-exchange neighborhoods efficiently, we exploit several pruning strategies to identify "branches" of the search tree, where any other insertion/removal of customers would lead to the feasible and better solutions in terms of the lower travel distance (as shown for the VRP with time windows, VRPTW, in the our previous works [1]). These strategies are applied together with the forward/backward penalty slacks for analyzing the capacity and time window violations [2]. These techniques allow for verifying if the solution will remain in the feasible state in $O(1)$ time.

After applying local search moves to all children, the best offspring replaces σ^p_A (Algorithm 1, line 21), and survives to the next generation. The MA may be terminated if: (i) the maximum execution time has been exceeded, (ii) the routing schedule of desired quality has been retrieved, (iii) the maximum number of generations have been processed, or (iv) the best individual in the population could not be further improved for a given number of consecutive generations.

4.2   Longest Common Subsequence Based SREX

The selective route exchange crossover (SREX) was successfully used for generating child solutions from two parent PDPTW schedules σ_A^p and σ_B^p [2]. In this work, we enhance the entire procedure of constructing offspring solutions (our new operator is referred to as the longest common subsequence based SREX, LCS-SREX)—it is presented in Algorithm 3. The main idea behind SREX is to create children by replacing some of the routes from the first parent σ_A^p with other routes selected from the second parent σ_B^p.

Algorithm 3. Generating child solutions using LCS-SREX.

1: **function** GENERATECHILDSOLUTION(σ_A^p, σ_B^p)
2:      **for** $i \leftarrow 1$ **to** N_{total} **do**
3:          $\{S_{A_i}, S_{B_i}\} \leftarrow$ random initial subsets of routes from σ_A^p and σ_B^p;
4:          $\{X_{A_i}, X_{B_i}\} \leftarrow \{S_{A_i}, S_{B_i}\}$ represented as arc sequences;
5:          $\text{LCS}_i \leftarrow \text{LCS}(X_{A_i}, X_{B_i})$;   ▷ Longest common subsequence
6:          $improvement \leftarrow$ **true**;
7:          **while** $improvement =$ **true do**
8:              $\{S'_{A_i}, S'_{B_i}\} \leftarrow$ MakeNeighborhood(S_{A_i}, S_{B_i});
9:              $\{X'_{A_i}, X'_{B_i}\} \leftarrow \{S'_{A_i}, S'_{B_i}\}$ represented as arc sequences;
10:             $\text{LCS}'_i \leftarrow \text{LCS}(X'_{A_i}, X'_{B_i})$;
11:             **if** $\text{LCS}'_i > \text{LCS}_i$ **then**
12:                 $\{S_{A_i}, S_{B_i}\} \leftarrow \{S'_{A_i}, S'_{B_i}\}$;
13:                 $\text{LCS}_i \leftarrow \text{LCS}'_i$;
14:             **else**
15:                 $improvement \leftarrow$ **false**;
16:             **end if**
17:          **end while**
18:      **end for**
19:      Sort N_{total} items in the descending order by LCS_i;
20:      Eliminate $\{S_{A_i}, S_{B_i}\}$ duplicates;
21:      Select N_{cross} best subsets $\{S_{A_i}, S_{B_i}\}$;
22:      **for** $i \leftarrow 1$ **to** N_{cross} **do**
23:          $\sigma'_o \leftarrow \sigma_A^p$;
24:          $\sigma''_o \leftarrow \sigma_A^p$;
25:          Remove routes S_{A_i} from σ'_o;
26:          Eject from σ'_o nodes $\in \sigma_B^p$ and $\notin \sigma_A^p$;
27:          Insert routes S_{B_i} to σ'_o;
28:          Remove routes S_{A_i} from σ''_o;
29:          Insert to σ''_o the routes S_{B_i} having nodes which were ejected from σ'_o
30:          Insert unserved requests to both σ'_o and σ''_o feasibly;
31:          $\sigma_{o_i} \leftarrow$ better from σ'_o and σ''_o;
32:      **end for**
33:      **return** best child solution from $\{\sigma_{o_1}, \sigma_{o_2}, ..., \sigma_{o_{N_{cross}}}\}$;
34: **end function**

In order to explain the rest of the algorithm, the following notations are defined. The set of routes in σ_A^p and σ_B^p are defined as R_A and R_B. $S_A (\subseteq R_A)$ and $S_B (\subseteq R_B)$ are the *set of replaced routes* in the parent σ_A^p (during the recombination), and the *set of inserted routes* taken from σ_B^p, respectively. In this paper, we propose: (i) the route representation being the sequence of arcs instead of the sequence of customer nodes, and (ii) the new process for selecting subsets of routes for LCS-SREX, which considers the longest common subsequence (LCS) values of arcs. For this reason, we introduce the additional notations of X_A and X_B—the route arc sequences generated from S_A and S_B, respectively. $LCS(X_A, X_B)$ is the LCS value of arcs from X_A and X_B.

First, N_{total} pairs of S_A and S_B subsets are constructed (lines 2–18). The random subsets of routes from σ_A^p and σ_B^p are selected as the initial pair $\{S_{A_i}, S_{B_i}\}$— both S_{A_i} and S_{B_i} must encompass the same number of routes. In the next step, the pair of arc sequences $\{X_{A_i}, X_{B_i}\}$ is built from S_{A_i} and S_{B_i}. The arc sequence represents the subsequent arcs in the given route including arcs to and from the depot. As an example, consider the route $r_t = \langle 0, 3, 6, 2, 4, 0 \rangle$, which is converted to the following arc sequence: $X_t = \{\{0, 3\}, \{3, 6\}, \{6, 2\}, \{2, 4\}, \{4, 0\}\}$. Each arc $\{a, b\}$ is assigned the unique hash value using the following hash function h [23]:

$$h(a, b) = \begin{cases} a^2 + a + b, & a \geq b \\ a + b^2, & a < b \end{cases} \tag{1}$$

where a and b correspond to the start and to the finish customer identifiers of the arc $\{a, b\}$. Finally, $LCS(X_{A_i}, X_{B_i})$ for two arc sequences is calculated. We employ the dynamic programming algorithm to find the LCS for the arc sequences represented as the arrays of hash values (in $O(x_l \cdot y_l)$ time, where x_l and y_l are the lengths of two sequences). Thus, we take advantage of the fact that the LCS values for the subproblems are memorized and reused when needed. Importantly, we exploit the new solution representations—the arrays of hashes representing arcs in two subsets (e.g., $X[0 \ldots x_l - 1]$ for the first subset, and $Y[0 \ldots y_l - 1]$ for the second one), and the array L for storing the LCS values elaborated for the subproblems. Therefore, we have:

$$L[i][j] = \begin{cases} 0, & i = 0 \text{ or } j = 0 \\ L[i-1][j-1] + 1, & i > 0, j > 0, X[i] = Y[j], \\ \max(L[i-1][j], L[i][j-1]), & i > 0, j > 0, X[i] \neq Y[j] \end{cases} \tag{2}$$

and the final LCS value is stored in $L[x_l - 1][y_l - 1]$. Once the initial LCS_i is calculated, the subsequent attempts of finding the neighboring subsets having larger LCS values are carried out (lines 7–17). The attempts are terminated if this value cannot be further improved (line 15). The neighbor subsets of routes $\{S'_{A_i}, S'_{B_i}\}$ are constructed by adding a random route to both S_{A_i} and S_{B_i}. Then, the pair of arc sequences $\{X'_{A_i}, X'_{B_i}\}$ is elaborated from the subsets S'_{A_i} and S'_{B_i} (line 9). Finally, LCS'_i for X'_{A_i} and X'_{B_i} are calculated, and if the resulting LCS value is larger than current one (line 11), the subsets of routes $\{S_{A_i}, S_{B_i}\}$, along with the LCS_i value are updated (lines 12–13).

In the next stage, all N_{total} subsets of routes are sorted in the descending order according to their LCS values (line 19). Subsequently, the possible duplicates of $\{S_{A_i}, S_{B_i}\}$ are removed (line 20), and the N_{cross} best subsets of routes are selected for further processing (line 21). The proposed selection of $\{S_{A_i}, S_{B_i}\}$ based on the analysis of the LCS values proved to be crucial for the effective generation of offspring solutions. In SREX [2], the subsets S_A and S_B were selected for crossover if they had the lowest number of customer nodes existing in S_A, and not existing in S_B. This approach favored solutions having common customer nodes, but usually not similar in terms of common partial routes. It could easily lead to large numbers of unserved requests which had to be re-inserted back into the partial child solution. This issue is tackled in LCS-SREX, in which the probability of having numerous unserved requests is reduced.

The child solutions are retrieved for all N_{cross} best subsets of routes $\{S_{A_i}, S_{B_i}\}$ (lines 22–32). Each iteration starts with assigning the parent solution σ_A^p to the first (σ_o') and second (σ_o'') offspring schedules (lines 23–24). Then, the routes identified by S_{A_i} are removed from σ_o'. The customer nodes existing in σ_B^p and not belonging to σ_A^p are ejected from σ_o'. Next, the routes identified by S_{B_i} are appended to σ_o'. The routes S_{A_i} are removed from the second offspring σ_o'' in the first step. The routes S_{B_i} that contain customer nodes ejected from σ_o' are inserted into σ_o''. Eventually, multiple attempts to insert all unserved requests to both σ_o' and σ_o'' are performed. If the attempts are successful, then the best offspring is selected and assigned to σ_{o_i} (line 31).

Once all N_{cross} child solutions are found, the best offspring (with the shortest travel distance) is selected from $\{\sigma_{o_1}, \sigma_{o_2}, \ldots, \sigma_{o_{N_{cross}}}\}$ (line 33).

5   Experimental Results

The MA for minimizing the travel distance in the PDPTW was implemented in the C++ programming language, and run on an Intel Xeon 3.2 GHz computer (16 GB RAM). Its maximum execution time was $\tau_M = 2$ min. To verify how applying the new crossover, along with the number of children generated for each pair of selected parents affect the final scores, we analyzed six MA variants summarized in Table 1. Each variant was executed 5 times for each instance (hence, each variant was run 300 times), and the best, average, and worst (with the minimum, average, and the maximum final T values) results were logged.

Table 1. Investigated variants of the MA.

Setting ↓	(a)	(b)	(c)	(d)	(e)	(f)
Crossover	SREX	**LCS-SREX**	SREX	**LCS-SREX**	SREX	**LCS-SREX**
N_{cross}	1	1	20	20	40	40

We focus on 400-customer Li and Lim's tests[1]. Six test classes (C1, C2, R1, R2, RC1 and RC2) reflect various real-life scheduling factors: C1 and C2 encompass *clustered* customers, in R1 and R2 the customers are *randomly* scattered around the map, whereas RC1 and RC2 instances contain a *mix of random and clustered* customers. The classes C1, R1 and RC1 have *smaller* capacities and *shorter* time windows compared with the second-class instances (the smaller K should be necessary to serve customers in the second-class tests). Tests have unique names: lα_β_γ, where α is the class, β relates to the number of customers ($\beta = 4$ for 400 customers), and γ is the identifier ($\gamma = 1, 2, \ldots, 10$).

As mentioned in Sect. 4, the number of trucks (K) is minimized at first, and the population of feasible solutions (which evolves in time in order to decrease the total travel distance) encompasses routing schedules of the same fleet size. Here, we employed the enhanced guided ejection search [14] to optimize K, and to generate the initial populations of size $N_{pop} = 10$ (the population size is relatively small in order to decrease the processing time of a single generation). The numbers of routes retrieved for each 400-customer problem instance have been gathered in the supplementary material[2]. This stage (i.e., minimizing K and creating the initial populations) was extremely fast and took less than 60 s for all instances ($\tau_{RM} + \tau_{PopGen} < 60$ s).

5.1   Analysis and Discussion

The results retrieved using all algorithm variants are summarized in Table 2—the T values are averaged for all problem instances. We present the minimum (best), average, and maximum (worst) T's. In order to investigate the results elaborated using the MA with and without our new LCS-SREX crossover applied, we compare the algorithms in the pairwise manner: the (a) variant is compared with the (b) variant (the better result is boldfaced), (c) with (d), and (e) with (f). Additionally, the background of the cells containing the best T's across all investigated algorithms (for each Li and Lim's class) is grayed. The results show that applying the LCS-SREX operator allows for retrieving the feasible solutions with smaller final T's for most cases. Although there exist instances for which the baseline SREX crossover operator performed better (see e.g., RC1 for the (c) vs. (d) variants), it is the MA with LCS-SREX utilized which outperforms other variants on average for all numbers of generated children. It is interesting to note that we have applied LCS-SREX in the initial implementation of a parallel MA for minimizing distance in the PDPTW—it allowed for retrieving a new world's best solution to the lR1_4_2 benchmark test (the parallel MA was run on 32 processors, and $\tau_M = 120$ min.). The details of this solution are available in the supplementary material (T was decreased to 9968.19 compared with $T = 9985.28$ in the previous world's best schedule for 31 routes). Since the work on the parallel MA for the PDPTW is very preliminary (we incorporated it into our parallel VRPTW framework [19]), we do not focus on that in this paper.

---

[1] See: http://www.sintef.no/projectweb/top/pdptw/li--lim-benchmark/.

[2] The supplementary material is available at: http://sun.aei.polsl.pl/~jnalepa/ EvoCOP2017/evocop2017_supplement.zip.

Table 2. The travel distances averaged across all problem instances retrieved using all investigated algorithm variants. The best T's are boldfaced (in the pairwise comparison of the algorithms), and the background of the best T across all variants are grayed.

		(a)	(b)	(c)	(d)	(e)	(f)
	Min.	10112.81	9952.98	7538.87	7534.25	7420.58	7413.93
C1	Avg.	10426.51	10389.97	7622.76	7591.94	7473.37	7466.80
	Max.	10811.26	10791.47	7721.85	7658.64	7541.44	7532.60
	Min.	4243.42	4238.47	4559.34	4172.00	4204.02	4203.80
C2	Avg.	4331.13	4324.95	5193.29	4204.64	4236.60	4236.14
	Max.	4421.84	4420.69	5259.15	4249.43	4294.00	4279.88
	Min.	10085.33	10045.53	8587.91	8502.07	8500.29	8513.16
R1	Avg.	10366.73	10453.14	8704.50	8446.74	8597.51	8583.70
	Max.	10616.15	10474.84	8849.27	8547.80	8730.22	8668.75
	Min.	6844.40	6820.61	7079.56	6932.99	7282.74	7294.35
R2	Avg.	6944.84	6914.41	7139.39	7081.19	7484.71	7473.26
	Max.	7122.64	7027.84	7371.67	7248.36	7699.88	7697.50
	Min.	8978.88	8858.41	7545.67	7801.82	7754.69	7762.18
RC1	Avg.	9094.38	9056.19	7369.95	7844.18	7783.62	7781.49
	Max.	9255.68	9253.75	7434.46	7895.33	7817.35	7816.46
	Min.	5709.53	5693.31	5746.73	5632.73	5685.17	5659.26
RC2	Avg.	5794.75	5700.04	5717.20	5710.12	5760.31	5758.15
	Max.	5909.28	5903.58	5831.43	5822.50	5861.51	5861.42
	Min.	7662.40	7601.55	6843.01	6762.64	6807.92	6807.78
Avg.	Avg.	7826.39	7806.45	6957.85	6813.13	6889.35	6883.26
	Max.	8022.81	7978.70	7077.97	6903.67	6990.73	6976.10

The number of children created during the reproduction process significantly affects the final scores. Intuitively, the larger N_{cross} should lead to the higher-quality solutions faster (each pair of the selected parents is intensively exploited in search of well-fitted child schedules). On the other hand, crossing over similar parents (i.e., routing schedules of a similar structure) multiple times may be unnecessary since the offspring solutions will most likely be of very similar quality. The experimental results indicate that the MA with $N_{cross} = 20$ (which is $N_{cross} = 2 \cdot N_{pop}$) gave the best T values—see the (d) variant in Table 2. Hence, too small number of children ($N_{cross} = 1$) was insufficient to exploit the population, whereas too large value ($N_{cross} = 40$) appeared not necessary (retrieving more offsprings could not drastically improve the final score). Albeit this initial investigation suggests that the number of children should be roughly twice as large as the population size, it requires further investigation (also, the initial efforts show that the adaptive change of N_{cross} may be very beneficial, as presented for another challenging variant of the VRP [1]).

The impact of the N_{cross} is also highlighted in Fig. 1 (we used the full algorithm variant descriptions instead of abbreviations for clarity). Here, we

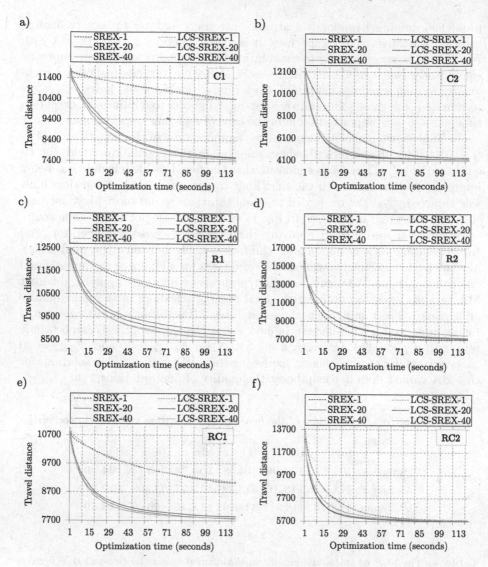

Fig. 1. Average best travel distance averaged across the independent runs obtained using the investigated algorithm variants for all Li and Lim's classes.

average the T values (of the best individual) across all problem instances (for all classes and for all investigated variants) retrieved during the evolution. The initial populations are of similar quality in all cases, and the decrease in the total travel distance obtained using the MA with $N_{cross} = 20$ and $N_{cross} = 40$ is much larger compared with $N_{cross} = 1$. Interestingly, for the R2 class, generating a single offspring solution is enough to converge to high-quality schedules very fast. It means that creating more children is not necessary in the case of problems containing randomly scattered customers that are to be served using vehicles of relatively large capacities. It is worth mentioning that the MA with $N_{cross} > 1$ converges quickly to the solutions whose quality is not later improved significantly (see e.g., C2 and RC2—most algorithm variants elaborated high-quality routing schedules in approx. 45 s). Hence, the MA could have been safely terminated once these solutions were obtained, in order to minimize its execution time—it is a very important issue in the case of real-time applications.

In Table 3, we gather the average convergence time of the investigated MAs (i.e., the time after which the best solution in the population could not be enhanced *at all*). These values indicate that the best feasible schedule was being improved continuously until the time limit was reached (thus, if the time limit was enlarged, then the quality of the final solutions would most likely increase slightly). However, as rendered in Fig. 1, the changes are not very significant.

In order to verify if (i) applying the LCS-SREX crossover operator, and (ii) generating various numbers of children notably influence the final schedules, we performed the two-tailed Wilcoxon tests for all pairs of the investigated algorithm variants. We verify the null hypothesis saying that "applying different MA variants leads to obtaining the final routing schedules of the same quality on average". The results are gathered in Table 4 (the differences that are statistically important are boldfaced). In most cases, the null hypothesis can be safely rejected (at $p < 0.05$). However, using SREX for $N_{cross} = 20$ and $N_{cross} = 40$ gave very similar results—hence, increasing the number of offspring solutions for this MA variant does not help boost the quality of final solutions.

Table 3. The convergence time of the investigated algorithm variants (in seconds).

	(a)	(b)	(c)	(d)	(e)	(f)
Minimum	105.25	**104.99**	104.58	**103.73**	105.75	**104.89**
Average	113.33	**113.10**	113.22	**112.27**	113.09	**112.87**
Maximum	**118.75**	118.93	118.96	**118.46**	120.00	**118.59**

Table 4. The level of statistical significance obtained using the two-tailed Wilcoxon tests. The differences which are statistically important (at $p < 0.05$) are boldfaced.

	(b)	(c)	(d)	(e)	(f)
(a)	**0.0008**	**<0.0001**	**<0.0001**	**<0.0001**	**<0.0001**
(b)	—	**<0.0001**	**<0.0001**	**<0.0001**	**<0.0001**
(c)		—	**<0.0001**	0.3222	**0.0097**
(d)			—	**0.0091**	**0.0083**
(e)				—	**<0.0001**

Comparing different algorithm variants for a number of different datasets is a challenging and difficult task (e.g., the Algorithm A can outperform the Algorithm B for the first dataset, and can be significantly worse for another dataset). Therefore, we executed the non-parametric Friedman test [24] (which has been exploited in numerous works from the field of machine learning where the comparison of the multiple algorithms over multiple datasets is an important concern [25, 26]), in order to rank the investigated MA variants, and to check if these rank differences are statistically important for all datasets (i.e., C1, C2, R1, and so forth). The results of this test are gathered in Table 5 (for the minimum, average, and maximum total travel distance T)—the best (lowest) ranking is boldfaced. Also, we highlight the methods which are significantly different for each MA variant (at $p < 0.05$). The results confirm that the MA with the new LCS-SREX crossover and $N_{cross} = 20$ allows for retrieving the best routing schedules across all classes of problems. As for the pairwise comparison (with and without our new crossover operator), applying LCS-SREX is beneficial and leads to the higher-quality solutions in the majority of cases (only for $N_{cross} = 40$ and the minimum T, the rankings of the memetic methods with and without LCS-SREX appear the same). Therefore, the new LCS-SREX operator should become the default operator choice in the MA for tackling the PDPTW.

Table 5. The results of the Friedman test (at $p < 0.05$) for the minimum, average, and maximum distance T. The best ranking is boldfaced (the lower ranking, the better).

	(a)	(b)	(c)	(d)	(e)	(f)
Minimum T						
Ranking	5.14	4.14	4.14	**2.14**	2.71	2.71
Different from	d, e, f	d	d	a, b, c	a	
Average T						
Ranking	5.14	3.86	3.43	**2.14**	3.71	2.71
Different from	c, d, f	d	a	a, b, e	d	a
Maximum T						
Ranking	5.29	4.29	3.29	**2.00**	3.57	2.57
Different from	c, d, e, f	d, f	a	a, b, e	a, d	a, b

6   Conclusions and Future Work

In this paper, we proposed a new crossover operator (LCS-SREX) for the PDPTW. It extends the original SREX in order to minimize the number of unserved customers which must be re-inserted into the partial child solution after the recombination process. In LCS-SREX, the longest common subroutes of the parent solutions are analyzed to select the appropriate routes for the crossover operation. The suggested solution representation allows for finding the LCS values of the parents quickly using dynamic programming. The experimental study performed on the widely-used Li and Lim's benchmark set (we focused on the 400-customer tests) showed that the proposed algorithmic solutions significantly affect the final solutions, and allow for retrieving higher-quality schedules compared with SREX (the two-tailed Wilcoxon tests showed that the differences are statistically important). We investigated the impact of the number of children generated for each pair of parents in the MA (exploiting both SREX and LCS-SREX). Also, we executed the non-parametric Friedman statistical tests to verify the rankings of the investigated algorithms across all classes of benchmark problems. This investigation showed that the MA with the LCS-SREX crossover applied outperforms the baseline SREX-based technique. Finally, we report one new world's best solution (for the lR1_4_2 test) retrieved using a parallel version of the MA with the LCS-SREX crossover operator applied.

Our ongoing research encompasses the work on the adaptive schemes to dynamically select the appropriate number of child solutions generated for each pair of parents, and comparing the MA with LCS-SREX with other state-of-the-art techniques. Also, we work on a parallel framework for solving the PDPTW (and other challenging optimization problems), which will initially involve the LCS-SREX-based MA for the total distance minimization. Finally, we aim at verifying the impact of other MA parameters on the quality of final solutions (including the population size and the mutation rate). We plan to apply the proposed algorithmic solutions to tackle large-scale real-life dynamic scheduling scenarios (in which the dynamic changes may happen, e.g., the road network could be updated due to the traffic congestion).

Acknowledgments. This research was supported by the National Science Centre under research Grant No. DEC-2013/09/N/ST6/03461, and performed using the Intel CPU and Xeon Phi platforms provided by the MICLAB project No. POIG.02.03.00.24-093/13.

References

1. Nalepa, J., Blocho, M.: Adaptive memetic algorithm for minimizing distance in the vehicle routing problem with time windows. Soft Comput. **20**(6), 2309–2327 (2016)
2. Nagata, Y., Kobayashi, S.: Guided ejection search for the pickup and delivery problem with time windows. In: Cowling, P., Merz, P. (eds.) EvoCOP 2010. LNCS, vol. 6022, pp. 202–213. Springer, Heidelberg (2010). doi:10.1007/978-3-642-12139-5_18
3. Kececi, B., Altiparmak, F., Kara, I.: A hybrid constructive mat-heuristic algorithm for the heterogeneous vehicle routing problem with simultaneous pick-up and delivery. In: Chicano, F., Hu, B., García-Sánchez, P. (eds.) EvoCOP 2016. LNCS, vol. 9595, pp. 1–17. Springer, Heidelberg (2016). doi:10.1007/978-3-319-30698-8_1
4. Grandinetti, L., Guerriero, F., Pezzella, F., Pisacane, O.: The multi-objective multi-vehicle pickup and delivery problem with time windows. Soc. Behav. Sci. **111**, 203–212 (2014)
5. Nanry, W.P., Barnes, J.W.: Solving the pickup and delivery problem with time windows using reactive tabu search. Transp. Res. **34**(2), 107–121 (2000)

6. Cordeau, J.F., Laporte, G., Ropke, S.: Recent models and algorithms for one-to-one pickup and delivery problems. In: Golden, B., Raghavan, S., Wasil, E. (eds.) The Vehicle Routing Problem: Latest Advances and New Challenges, pp. 327–357. Springer, Boston (2008)

7. Baldacci, R., Mingozzi, A., Roberti, R.: Recent exact algorithms for solving the vehicle routing problem under capacity and time window constraints. Eur. J. Oper. Res. **218**(1), 1–6 (2012)

8. Bettinelli, A., Ceselli, A., Righini, G.: A branch-and-price algorithm for the multi-depot heterogeneous-fleet pickup and delivery problem with soft time windows. Math. Program. Comput. **6**(2), 171–197 (2014)

9. Baldacci, R., Bartolini, E., Mingozzi, A.: An exact algorithm for the pickup and delivery problem with time windows. Oper. Res. **59**(2), 414–426 (2011)

10. Lu, Q., Dessouky, M.M.: A new insertion-based construction heuristic for solving the pickup and delivery problem with time windows. Eur. J. Oper. Res. **175**(2), 672–687 (2006)

11. Zhou, C., Tan, Y., Liao, L., Liu, Y.: Solving multi-vehicle pickup and delivery with time widows by new construction heuristic. In: Procedings of CISDA, pp. 1035–1042 (2006)

12. Parragh, S.N., Doerner, K.F., Hartl, R.F.: A survey on pickup and delivery problems. J. fur Betriebswirtschaft **58**(1), 21–51 (2008)

13. Ropke, S., Pisinger, D.: An adaptive large neighborhood search heuristic for the pickup and delivery problem with time windows. Transp. Sci. **40**(4), 455–472 (2006)

14. Nalepa, J., Blocho, M.: Enhanced guided ejection search for the pickup and delivery problem with time windows. In: Nguyen, N.T., Trawiński, B., Fujita, H., Hong, T.-P. (eds.) ACIIDS 2016. LNCS (LNAI), vol. 9621, pp. 388–398. Springer, Heidelberg (2016). doi:10.1007/978-3-662-49381-6_37

15. Pankratz, G.: A grouping genetic algorithm for the pickup and delivery problem with time windows. OR Spectr. **27**(1), 21–41 (2005)

16. Nagata, Y., Kobayashi, S.: A memetic algorithm for the pickup and delivery problem with time windows using selective route exchange crossover. In: Schaefer, R., Cotta, C., Kołodziej, J., Rudolph, G. (eds.) PPSN 2010. LNCS, vol. 6238, pp. 536–545. Springer, Heidelberg (2010). doi:10.1007/978-3-642-15844-5_54

17. Bent, R., Hentenryck, P.V.: A two-stage hybrid algorithm for pickup and delivery routing problems with time windows. Comput. Oper. Res. **33**(4), 875–893 (2006)

18. Kalina, P., Vokrinek, J.: Parallel solver for vehicle routing and pickup and delivery problems with time windows based on agent negotiation. In: Proceedings of IEEE SMC, pp. 1558–1563 (2012)

19. Nalepa, J., Blocho, M.: Co-operation in the parallel memetic algorithm. Int. J. Parallel Program. **43**(5), 812–839 (2014)

20. Blocho, M., Nalepa, J.: A parallel algorithm for minimizing the fleet size in the pickup and delivery problem with time windows. In: Proceedings of 22nd European MPI Users' Group Meeting, EuroMPI 2015, pp. 15:1–15:2. ACM, New York (2015)

21. Nalepa, J., Blocho, M.: A parallel algorithm with the search space partition for the pickup and delivery with time windows. In: Proceedings of 3PGCIC, pp. 92–99 (2015)

22. Cherkesly, M., Desaulniers, G., Laporte, G.: A population-based metaheuristic for the pickup and delivery problem with time windows and LIFO loading. Comput. Oper. Res. **62**, 23–35 (2015)

23. Szudzik, M.: An Elegant Pairing Function. Wolfram Research, Champaign (2006). pp. 1–12

24. Demšar, J.: Statistical comparisons of classifiers over multiple data sets. J. Mach. Learn. Res. **7**, 1–30 (2006)
25. Krawczyk, B., Woźniak, M., Herrera, F.: On the usefulness of one-class classifier ensembles for decomposition of multi-class problems. Pattern Recogn. **48**(12), 3969–3982 (2015)
26. Trajdos, P., Kurzynski, M.: A dynamic model of classifier competence based on the local fuzzy confusion matrix and the random reference classifier. Appl. Math. Comput. Sci. **26**(1), 175 (2016)

Multi-rendezvous Spacecraft Trajectory Optimization with Beam P-ACO

Luís F. Simões[1]([✉]), Dario Izzo[2], Evert Haasdijk[1], and A.E. Eiben[1]

[1] Vrije Universiteit Amsterdam, Amsterdam, The Netherlands
{luis.simoes,e.haasdijk,a.e.eiben}@vu.nl
[2] European Space Agency, ESTEC, Noordwijk, The Netherlands
dario.izzo@esa.int

Abstract. The design of spacecraft trajectories for missions visiting multiple celestial bodies is here framed as a multi-objective bilevel optimization problem. A comparative study is performed to assess the performance of different Beam Search algorithms at tackling the combinatorial problem of finding the ideal sequence of bodies. Special focus is placed on the development of a new hybridization between Beam Search and the Population-based Ant Colony Optimization algorithm. An experimental evaluation shows all algorithms achieving exceptional performance on a hard benchmark problem. It is found that a properly tuned deterministic Beam Search always outperforms the remaining variants. Beam P-ACO, however, demonstrates lower parameter sensitivity, while offering superior worst-case performance. Being an anytime algorithm, it is then found to be the preferable choice for certain practical applications.

Keywords: Beam Search · Ant Colony Optimization · P-ACO · Bilevel optimization · Multi-objective optimization · Spacecraft trajectories

1 Introduction

The design of multi-rendezvous spacecraft trajectories poses a considerable challenge to aerospace engineers. This is due, in part, to the combinatorial nature of the problem that emerges with the increase in number of bodies to visit along a mission (e.g., planets, moons, asteroids). The complexity of the task stems from an interplay of multiple factors under optimization, including: the decision of which of the bodies of interest to visit, the order in which they are to be visited, and the design of the actual trajectory arcs to take the spacecraft between them. Maximization of such a mission's scientific return may demand for as many bodies to be visited as possible, in the shortest possible amount of time, while consuming the lowest possible amount of propellant mass. The underlying optimization problem can be seen as a variant of the well known Traveling Salesman Problem (TSP), with nodes corresponding to the celestial bodies under consideration, and edge weights a function of the costs (time and

---

Code available at https://github.com/lfsimoes/beam_paco__gtoc5.

© Springer International Publishing AG 2017
B. Hu and M. López-Ibáñez (Eds.): EvoCOP 2017, LNCS 10197, pp. 141–156, 2017.
DOI: 10.1007/978-3-319-55453-2_10

mass) to propel the spacecraft between them. These costs vary as bodies move in space along their trajectories, but also as a function of the spacecraft's state: a lighter spacecraft that has already shed some of its mass is simpler to maneuver.

As evidence into the complexity, and relevance, of the above described problems, consider the following: since 2005, the aerospace engineering community has periodically organized GTOC, the Global Trajectory Optimization Competition [1]. In it, different groups have taken turns at creating "nearly-impossible" problems of interplanetary trajectory design to pose to the community. Of the 8 competitions organized to date, 4 were multiple-asteroid rendezvous problems of the kind considered here, and 3 others were multiple fly-by problems posing similar combinatorial optimization challenges.

In [19], a multi-objective Beam Search algorithm is described, and applied to a low-thrust model of the GTOC7 problem. In this research, we propose a series of extensions to it. First, we provide an improved orbital phasing indicator, and a procedure to create from it a probability distribution over candidate bodies to extend missions with. Second, we hybridize the Beam Search procedure with the well known Ant Colony Optimization algorithm [7]. We conclude by evaluating the resulting algorithms on a Lambert model of the GTOC5 problem [11]. Two main research questions are investigated in this paper:

1. Can the randomization of Beam Search, via probabilistic branching choices, improve performance?
2. Does the pheromone-based positive reinforcement of sequences lead to improved performance?

This paper is organized as follows: in Sect. 2 we list related work in the combinatorial optimization of spacecraft trajectories. Section 3 describes the Beam Search algorithm, and the proposed randomized variants. In Sect. 4 we describe the GTOC5 problem used in our experiments, and in Sect. 5 the interfacing between it and the search algorithms. Section 6 reports on an experimental evaluation, and Sect. 7 discusses its results. Conclusions are drawn in Sect. 8.

2   Related Work

Beam Search [2,28] has emerged as the *de facto* standard approach to tackle the combinatorial optimization sub-problems present in most GTOC competitions. Though at times called by other names, it is common to find the general architecture of a tree search that has its computational cost bounded via the selection of a limited number of nodes to branch at each depth-level (non-selected nodes at that depth being immediately discarded). We can find examples of such algorithms in the winning solutions to GTOC4 [10], GTOC5 [24], and in the second ranked solution to GTOC7 [19], which the present research builds on. The Lazy Race Tree Search described in [20], which at the time presented the best known solution to the GTOC6 problem, can also be seen as a Beam Search variant. In it, the "beam" is composed of all nodes, possibly originating from different tree depths, that fall within a given mission time window. The most promising nodes in that sliding window are branched, and the remaining ones discarded.

Evolutionary Algorithms have been explored as an alternative to solve combinatorial problems in mission analysis. In the GTOC5 problem considered here, for instance, [9,21] used Genetic Algorithms with "hidden genes", to evolve chromosomes encoding asteroid sequences. These approaches were however outperformed in the GTOC5 competition by tree-based approaches. In [18] an evolutionary approach is described for designing debris removal missions. In this highly dynamic trajectory problem, the Inver-over Genetic Algorithm was found to provide competitive solutions to those constructed by different approaches.

Ant Colony Optimization (ACO) [7] was used by some teams in the GTOC competitions over the years. However, to our knowledge, no scientific publication has been produced to date with the details of such deployment. The most successful use of ACO in a GTOC competition was possibly that by the NASA/JPL team, winners of GTOC7. According to the GTOC portal [1], a "very competitive solution was found by JPL using an Ant Colony Optimization approach. Eventually a different solution turned out to be better and was thus submitted"[1]. Other applications of ACO algorithms to optimize the sequences of bodies to visit along a spacecraft's trajectory can be found in [4,5,26]. In them, the test problems over which algorithms are evaluated have few bodies to select from ($\approx$10), and the found sequences visit $\approx$5 bodies. In contrast, sequences of up to 17 asteroids are assembled here, from a database of 7075 available asteroids.

A hybridization of Beam Search and ACO was previously presented in [3]. A different hybridization is introduced here, "Beam P-ACO", that differs mainly in the ACO variant under use, and in being a multi-objective algorithm.

3   Beam Search

Beam Search [2,28] is a tree search algorithm where computational cost is bounded by employing heuristics that allow for non-promising solutions under construction to be discarded. It can be executed as a variant of depth-first or breadth-first search. When operating as a variant of breadth-first search, as is done here, Beam Search traverses the tree one depth-level at a time. From all the solutions generated at one level, only a limited subset (the so called "beam") will be selected for carrying over to the next level. An evaluation of solutions' quality determines whether they are included in the beam, or instead permanently discarded from the search. The size of the beam is designated as the "beam width", and is here represented as bw. The extension of partial solutions in the beam can be performed towards all possible successor nodes, or instead towards only a limited number, enabling further control over the search's computational cost. A "branching factor" parameter, here represented as bf, indicates that each solution in the beam will be extended only towards bf successor nodes, chosen as a function of how good the solutions they lead to are estimated to be.

---

[1] The NASA/JPL team's GTOC7 submission report and workshop slides, containing details of their ACO deployment, can be found in the GTOC portal [1].

In summary, at one level of the tree, each of the beam's bw solutions will be expanded towards bf new nodes, resulting in $bw * bf$ new partial solutions. These are then evaluated, and the bw best of those will become the beam that is carried over to the next tree level. The search process is therefore driven by two heuristics: (1) h_s, which evaluates partial solutions (a path down from the tree's root node), and (2) h_e, which evaluates candidate successor nodes for extending solutions with. The more accurately these heuristics point to the complete optimal solution, the more successful the search will be. Beam Search being an "incomplete" search algorithm however, the identification of the globally optimal solution is not guaranteed [28].

3.1 Multi-objective Heuristics

The first extension we introduce to the conventional Beam Search framework is the addition of multi-objective heuristics. That is, either h_s, or h_e (or both) will evaluate partial solutions or candidate extensions, respectively, according to multiple objectives. In such a scenario, it is then necessary to employ techniques that will allow for the ranking of alternatives, or probabilistic selection among them, that take into account the multiple evaluations.

In our current implementation, h_s evaluates partial solutions according to multiple objectives, while the evaluation of candidate extensions by h_e remains single-objective. Beam Search must then be able to select at each tree level the best bw solutions from among the newly generated extensions. To that end, we employ a Pareto dominance approach. Specifically, we apply non-dominated sorting [6] to the pool of extensions, and include in the beam as many of the best Pareto fronts as needed to reach the beam size bw. That process will most likely lead to a final Pareto front whose full inclusion in the beam would exceed bw. A tie-breaker criterion must then be defined to determine which of those (equally good) solutions to keep, and which to discard (see Sect. 5.3).

3.2 Probabilistic Branching

In the "Stochastic Beam" algorithm, the beam's construction remains deterministic, but the branching stage will now be subjected to probabilistic decisions. Given a solution in the beam, with a probability q_0 it will be extended towards the bf nodes with best h_e evaluation. With a probability of $1 - q_0$, the choice of bf nodes will instead be a biased sampling without replacement, proportional to h_e. Note that through a parameter setting of $q_0 = 1$ the algorithm reverts to the previously described (deterministic) Beam Search.

Beam Search will always converge to the same solution, given the same root node. Stochastic Beam however, can be executed multiple times, and possibly converge to different solutions in each run. These solutions may outperform those found by Beam Search, by including links to nodes that Beam Search incorrectly prunes, due to ranking above the bf threshold. Non-determinism therefore provides a degree of robustness against imperfections in the h_e heuristic.

3.3   Hybridization with Ant Colony Optimization

A hybridization of Beam Search with Ant Colony Optimization brings two mains changes to the algorithmic framework: (1) multiple tree searches are now performed, in consecutive runs designated as "generations", and (2) positive feedback takes place, in the form of "pheromones" that change the h_e heuristic evaluation of candidate successor nodes, therefore biasing the dynamics of tree searches in subsequent generations.

Of the many ACO variants in existence, we chose to hybridize with the Population-based Ant Colony Optimization algorithm (P-ACO), introduced in [12–15]. A detailed analysis in [23] found it to be "competitive to the state-of-the-art ACO algorithms with the advantage of finding good solution quality in a shorter computation time". Also, a recent thorough benchmarking of a high number of approaches to solve the Traveling Salesman Problem found P-ACO to be the best of the tested global optimization algorithms, as well as the best overall algorithm when seeded and hybridized with local search [27].

In "Beam P-ACO", a partial solution at node i, with available successors S evaluates the quality of extending towards node j by

$$h'_e(i,j) = \frac{\tau(i,j)^\alpha h_e(i,j)^\beta}{\sum_{s \in S} \tau(i,s)^\alpha h_e(i,s)^\beta}$$

where τ is the pheromone concentration along an edge, and h_e (known as η in the common ACO notation) is the problem-specific heuristic. The weighting factors α and β determine the relative contributions of pheromone and heuristic values to the branching decision. As in the previously described Stochastic Beam algorithm, with a probability q_0 the solution is extended towards the bf nodes with best h'_e evaluation, while with a probability of $1 - q_0$, that choice is instead a biased sampling without replacement proportional to h'_e. A setting of $\alpha = 0$ would result in all pheromone information being ignored, and Beam P-ACO would then revert to the Stochastic Beam algorithm.

The pheromone concentration τ along an edge takes in P-ACO discrete values in a given range $[\tau_{init}, \tau_{max}]$. We define these parameters as $\tau_{init} = 1/(n - 1)$ and $\tau_{max} = 1$, where n is the total number of nodes. This implements the convention from [14] of having the row/column sum of initial pheromone values be 1 (assuming a problem where revisits are disallowed, and the diagonal of the pheromone matrix is therefore 0). Given the pheromone range, and k, the maximum size of a population of top solutions, we can define $\tau_\Delta = (\tau_{max} - \tau_{init})/k$ as the pheromone increment deposited in an edge by a solution in the population that follows it along its path. If l solutions in the population include the edge (i,j), its pheromone concentration will then be $\tau(i,j) = \tau_{init} + l\tau_\Delta$.

3.4   Pareto Elitism

To complete the definition of Beam P-ACO, a population update model must be defined, a model that handles multi-objective evaluations produced by the h_s

heuristic. We propose here a variation of the method defined in [15, Sect. 3.1]. An archive will collect the set of non-dominated solutions found so far. At the end of each tree search, the best found solutions are merged into it. After updating the archive, the population that defines the pheromone matrix is reset.

In standard P-ACO [13], the population is implemented as a FIFO-queue (each generation's best solution enters the population, possibly displacing the oldest solution it contains so as not to exceed the population size k). We propose here instead to have one FIFO-queue per each of the problem's n nodes, each with a size limit of k. When resetting the population, all FIFO-queues are emptied. Then, solutions in the archive are shuffled, and one by one, they are added to the population (of edges). Each edge (i, j) in those solutions will result in j being added to the ith FIFO-queue. The ith FIFO-queue then directly maps to the ith row of the pheromone matrix, and its contents define which nodes do pheromones most bias a solution at node i to branch towards.

If solutions visit all n nodes, as is the case in many problems (e.g., TSP), this process will result in only the last k solutions of the shuffled archive adding pheromones to the population. A random unbiased sampling of k archive solutions would then be sufficient. However, in problems where solutions include only some of the nodes (such as here in the GTOC5 problem, where $n = 7075$, but a solution will visit <20 nodes), only small amounts of pheromone would be deposited, and plenty of information in the archive would be ignored. By following in such problems the process described above, the pheromone matrix may now receive contributions from more than k solutions, while still only receiving $\leq k$ contributions at the level of each individual node. This way, in the limit, all solutions in the archive may end up depositing pheromones.

4   The GTOC5 Trajectory Design Problem

The trajectory design problem posed in the 5th edition of the Global Trajectory Optimization Competition (GTOC5) is used as benchmark in the current research. The full problem specification can be found in [11]. The problem dataset, along with additional information related to this edition of the competition, can be found in the GTOC portal [1]. Our current work makes use of the problem model developed by the competition's 4th ranked team [21].

In the GTOC5 problem, a spacecraft leaves the Earth at some point along an 11-year time-window, to embark on a 15 year (max.) mission of asteroid exploration. The spacecraft starts with a mass of 4000 kg, of which 3500 are reserved for propellant mass, and the scientific equipment used at asteroids. A total of 7075 asteroids are available as possible targets to visit along the mission. The exploration of a single asteroid is carried out in two stages. First, the spacecraft must rendezvous (match position and velocity) with the asteroid, and leave there a 40 kg scientific payload. Later in the mission, the spacecraft performs a fly-by of the asteroid, and sends a 1 kg "penetrator" towards it. Upon impact, this penetrator would release a cloud of debris, that would be investigated by the payload left there. Partial scores are given for the rendezvous and fly-by maneuvers. An asteroid on which both are performed contributes 1 point to the full score. In the problem models developed by most teams, including the one used here, the problem is simplified by having an asteroid's fly-by maneuver performed immediately after its rendezvous: the spacecraft departs the asteroid, moves some distance away, and then accelerates back towards it. Under this simplification, a trajectory that manages to complete all pairs of maneuvers on a given sequence of m asteroids will score m points.

GTOC5 was won by a team from NASA/JPL with a score 18 trajectory [24]. In contrast, the model being used here, with the initial trajectory conditions listed in Sect. 6.1, has only ever allowed the discovery of trajectories with at most a score 16. Nevertheless, for the goal of evaluating the performance of algorithms that solve the combinatorial part of the problem (finding the asteroid sequence that enables the greatest possible score), the used model is perfectly suitable.

5   Bilevel Optimization of GTOC5 Trajectories

The design of GTOC5 trajectories is here tackled as a multi-objective bilevel optimization problem [25]. At an upper level, optimization seeks to identify a good subset of asteroids, and the order in which they are to be visited. At a lower level, each chosen pair of asteroids triggers an optimization of the trajectory leg that takes the spacecraft between them. This section details how the different Beam Search variants are employed for solving the upper-level combinatorial problem, and how the lower-level process optimizes transfer legs.

5.1   Orbital Phasing Indicators as Heuristic Estimators

The problem of assembling an efficient heuristic able to help select possible asteroid targets is crucial and very difficult at the same time. The ground truth (i.e. the optimal ΔV cost obtained optimizing the transfer leg) is far too expensive to be be computed for all possible asteroid targets and for all states encountered along the search. A solution to this problem was recently proposed via the development of so-called orbital phasing indicators [19]. These essentially allow to introduce for each epoch t_s a metric over the set of all possible asteroids, a metric that can, in turn, be used to detect asteroid neighborhoods efficiently. It was shown in [19] how the orbital indicator, defined as $d_o(\mathcal{A}_1, \mathcal{A}_2, \Delta T) = |\mathbf{x}_t - \mathbf{x}_s|$, where $\mathbf{x} = \left[\frac{1}{\Delta T}\mathbf{r}(t_s) + \mathbf{v}(t_s), \frac{1}{\Delta T}\mathbf{r}(t_s) \right]$ and $\mathbf{r}(t_s)$ and $\mathbf{v}(t_s)$ correspond to the asteroid ephemeris, positively correlates to some extent to the ground truth (i.e. the asteroids that are actually easy to reach via an orbital transfer). It essentially considers a snapshot at t_s of the asteroid population and, using a zero order approximation for the dynamics, predicts what asteroids are the closest in terms of transfer ΔV. As such, it is bound to neglect the known state of the asteroid population at the arrival time t_t, which seems as a loss of available information. A simple modification to the orbital indicator, though, allows to account for the final asteroid geometry and thus to improve the overall correlation to the ground truth. In essence, one can consider the orbital indicator backward in time, starting from the arrival asteroid, to get a new indicator (note that asteroid velocities will have to have their sign inverted). The average between the two, i.e. the orbital indicator and the backward orbital indicator, is what we use here and call improved orbital indicator, defined as $d_{o'}(\mathcal{A}_1, \mathcal{A}_2, \Delta T) = |\mathbf{x}_t - \mathbf{x}_s|$, where

$$\mathbf{x} = \left[\frac{1}{\Delta T}\mathbf{r}(t_s) + \mathbf{v}(t_s), \frac{1}{\Delta T}\mathbf{r}(t_s), \frac{1}{\Delta T}\mathbf{r}(t_t + \Delta T) - \mathbf{v}(t_t + \Delta T), \frac{1}{\Delta T}\mathbf{r}(t_t + \Delta T) \right]$$

For every node i being branched during the search, to which corresponds a trajectory presently at asteroid $\mathcal{A}_i$, we compute $d_{o'}(\mathcal{A}_i, \mathcal{A}_j, \Delta T)$ for all of the problem's n nodes, assuming a reference transfer time of $\Delta T = 125$ days. From it we build a probability distribution over successor nodes as $h_e(i, j) = (1 - p(i, j)/n)^\gamma$, where $p(i, j)$ is the rank in $\{0, \ldots, n - 1\}$ of $d_{o'}(\mathcal{A}_i, \mathcal{A}_j, \Delta T)$ among all estimated costs. This results in a selection probability that decays exponentially with increasing rank, at a rate tuned through γ. A setting of $\gamma = 50$ is used in this problem. There is then a $\approx 30\%$ chance of branching towards an asteroid ranked among the 50 best, and $\approx 84\%$ among the 250 best. Finally, the problem disallows revisits to asteroids. So, for any node i, $h_e(i, j) = 0$ if j was already visited at any previous point.

5.2   Optimization of Transfer Legs

During the tree search, branching a partial solution into a given node triggers an optimization process. Its end result will be the definition of the transfer leg that allows the spacecraft to rendezvous with the corresponding asteroid. Adding the new leg extends the mission's trajectory, which can then be reevaluated by the h_s heuristic. In the approach followed here, only the trajectory's rendezvous legs need to be optimized. The cost estimates for self-fly-by legs are instead approximated by a linear acceleration model [21, Sect. 3].

Each transfer will have a duration in a set time window of $[60, 500]$ days. A grid of 50 evenly spaced values defines the candidate transfer times, ΔT (≈ 9 days separation between grid points). For each ΔT, we employ PyKEP's [16][2] multiple revolution Lambert solver [17,19] to design a trajectory arc having that exact duration. If multiple revolution solutions exist for a given ΔT, the one with lowest ΔV is chosen. Two constraints are imposed: (1) ΔT should exceed the parabolic time of flight given by Barker's equation[3], and (2) the leg's maximum acceleration should be $<90\%$ of the maximum supported by the spacecraft. These constraints lead many transfers to have no feasible solution, for any ΔT (the targeted asteroid is simply unreachable). From all ΔT points in the grid that do have a feasible solution, the one with lowest ΔV is chosen to define the new rendezvous leg. In this approach, the optimization of one leg then equates to finding the solutions to at most 50 Lambert's problems. Note

---

[2] http://esa.github.io/pykep/.

[3] Legs failing this check are immediately discarded, saving computation time that would otherwise be spent generating Lambert arcs with excessive ΔV.

that this final ΔV cost is the ground-truth to the $d_{o'}$ indicator described in the previous section, from which h_e is defined. Note also that we take here a greedy choice of ΔT, imposing in that way a transfer leg upon the combinatorial search problem that might in the long term be sub-optimal with respect to its own goals.

5.3 Trajectory Evaluation, Ranking, and Selection

Trajectories are evaluated by h_s with respect to three criteria: (1) the mission's score, (2) the total mass required for propellant, and scientific equipment left at asteroids, and (3) the total time of flight. The ideal mission will have the greatest possible score, while requiring the lowest possible amounts of mass and time.

As mentioned in Sect. 3.1, a Pareto dominance approach [6] is used for handling the multiple objectives. However, in this problem mass and time evaluations are only fairly comparable among trajectories that share the same score. As such, ranking a set of trajectories involves (1) binning trajectories according to their score, and (2) applying non-dominated sorting [6] over the mass and time costs of trajectories within each bin. Identifying the top trajectories in a given set takes place by iterating through bins in descending order of score, gradually extracting their Pareto fronts. If only a subset of a Pareto front's trajectories is required, those with lowest mass cost are favored.

In a tree search, this process is applied at each depth-level to construct the beam with the best bw of the newly extended solutions. Before that, however, h_s is used for pruning nodes corresponding to missions that require >3500 kg, or >15 years. Should that result in an empty pool of extended solutions, or the pool otherwise be empty because no feasible transfer legs were found, the tree search has then reached its final level and is terminated. The beam of solutions carried over from the previous tree level will then be its final output. In Beam P-ACO, this signals the end of a generation. The contents of that final beam are then merged with the archive of non-dominated solutions found so far. By applying the previously described ranking process, the archive will always be a Pareto front of trajectories that all share the maximum score reached to date. Beam P-ACO will at this point refresh pheromones. Note that the combinatorial problem is asymmetric: an edge (i, j) present in a good solution is not predictive of an edge (j, i) being likely to lead to good solutions. As such, an (i, j) edge in an archived solution only results in pheromones along the (i, j) direction.

6 Experimental Evaluation

An experimental evaluation was carried out to assess the performance of the different Beam Search variants, using the GTOC5 problem as a test case. Performance is here evaluated in terms of multiple criteria. Primarily, we care about search algorithms that enable the consistent discovery of asteroid sequences having the greatest possible length (score). Among equally scored missions, we care for the best possible coverage of the Pareto front of mass and time costs required

to achieve such score. Finally, these considerations must be traded-off against
the computational cost to obtain such solutions.

6.1   Setup

We adopt as measure of computational cost the number of trajectory legs optimized throughout the search. This is in practice the main performance bottleneck, especially if instead of a Lambert model one were to use a low-thrust one.
A threshold of 100000 optimized legs was used as stopping criterion.

The performance of deterministic Beam Search was evaluated over a dense
grid of settings for the beam width (bw) and branching factor (bf) parameters.
In total, 118 different setups were evaluated, all having upfront an estimated
cost of $\leq$100000 optimized legs required in order to complete execution at tree
depth 16, where missions reach and fully score the 16th asteroid (see Fig. 1).
Being a deterministic algorithm, only a single run was executed per setup.

Stochastic Beam and Beam P-ACO were evaluated under the 5 different configurations of bf and bw highlighted in red in Fig. 1. Being stochastic algorithms,
100 independent runs were performed per setup. Deterministic branching decisions were taken with a probability of $q_0 = 0.5$ ($q_0 = 1$ in Beam Search). In Beam
P-ACO, pheromone and heuristic values contributed equally to the h_e heuristic:
$\alpha = \beta = 1.0$ (in the other Beam Search variants, implicitly $\alpha = 0.0$). Pheromone
concentrations were limited to at most $k = 3$ contributions to each node.

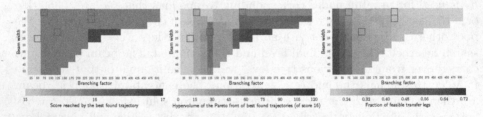

Fig. 1. Beam Search results from the parameter sweep over beam width (bw) and
branching factor (bf) configurations (118 setups in total, with costs below the threshold
of 100000 optimized trajectory legs). Darker is better. Highlighted in red: configurations
used in the Stochastic Beam and Beam P-ACO experiments. (Color figure online)

All tree searches reported here had the same root node. Its initial conditions
were originally obtained during the GTOC5 competition through a time-optimal
low-thrust optimization of the launch leg, as described in [21, Sect. 2] and exem-
plified in [16, Sect. 6.1]. After applying the linear model to define the self-fly-by
leg, the initial state is then a trajectory that has already scored 1 point at asteroid *2001 GP2* (id: 1712), and is ready to depart from it at epoch 59325.360 MJD,
having already expended 253.518 kg and 198.155 days from the total budgets.

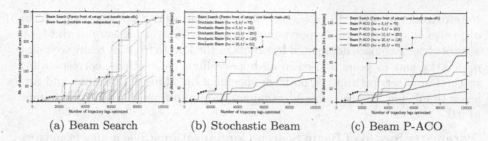

Fig. 2. Quantity of top solutions: cumulative number of *distinct* trajectories of score 16 or 17 found over time. Note the change of scale in the vertical axis.

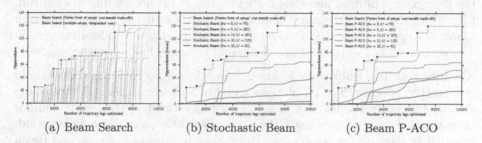

Fig. 3. Quality of top solutions: growth of the dominated area of objective space over time (hypervolume of the Pareto front of score 16 trajectories found so far).

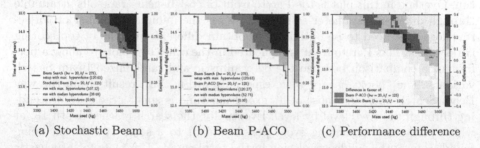

Fig. 4. Empirical Attainment Functions: probabilities of objective space vectors being dominated (or matched) in an algorithm's run. The darker a region is, the likelier it is that in a run a solution will be found that dominates it. Considers the Pareto fronts of score 16 trajectories found up to the 100000 optimized legs threshold.

6.2   Results

The results obtained in the experimental evaluation are shown in Figs. 1, 2, 3 and 4. Analyses into the quality of solutions found in a run consider their h_s evaluations, in particular, the extent to which they minimize mass and time costs. Though missions of score 17 were found in these experiments, they were only rarely found, and furthermore only a single distinct mission was ever found with that score. Therefore, analyses into the quality of solutions found consider exclusively the score 16 missions found in runs. Specifically, we evaluate the

Pareto fronts of score 16 missions found, and measure their coverage of the objective space using the hypervolume indicator [29] – a measure of the total area of objective space dominated by points in the Pareto front. A reference point of 3500 kg and 15 years is used in all hypervolume calculations, corresponding to the limits set forth in the problem specification. Runs that do not find any score 16 mission, and therefore have an empty Pareto front, have a hypervolume of 0.0.

Parameter Sweep of Beam Search Configurations: The results from these experiments are shown in Fig. 1. Figure 1(a) shows the highest score reached among the missions designed in each run. Figure 1(b) shows the hypervolume of the Pareto front of mass and time costs in score 16 missions found in each run. Figure 1(c) shows the prevalence of unfeasible solutions in the continuous search spaces of asteroid transfers. In the setups that reached score 17, for instance, on average only ≈23% of the attempted asteroid transfers had a feasible solution.

Quantity of Top Solutions: Figure 2 shows the results from the evaluation of the quantity of score 16 or 17 solutions found by each algorithm over time. Only distinct trajectories count here to a run's totals – two trajectories are equal if their asteroid sequence is exactly the same. The three plots shown correspond, from left to right, to the results from Beam Search, Stochastic Beam, and Beam P-ACO. Figure 2(a) shows one curve for each of Beam Search's 118 evaluated setups. The algorithm being deterministic, each curve depicts also one single run. Overlaid in this plot is the Pareto front of cost-benefit trade-offs attainable through different parameter settings: it shows the minimum number of trajectory legs that need to be optimized to obtain different amounts of top scoring trajectories. This Pareto front is replicated in the other two plots to allow a performance comparison between the deterministic and randomized Beam Search variants. The "hooks" shape seen, especially in Fig. 2(a), results from the way Beam Search operates: most of its execution time is spent gradually descending through the tree, level by level. Eventually, the search reaches depth 15, at which point each branching event possibly leads to a score 16 solution being found. Hence, the sudden explosion in the total count. The plots for Stochastic Beam and Beam P-ACO also display this effect, but in them values are averaged over 100 runs, and consecutive tree searches (generations) are chained together.

Quality of Top Solutions: Figures 3 and 4 show the results from the evaluation of the quality of score 16 solutions found by each algorithm over time. Figure 3 is structured in the same way as Fig. 2, so the description made above for it also applies here. Figure 3 shows how the total dominated area of objective space (hypervolume) grows over time, as new score 16 trajectories are found and enter the Pareto front. Figure 4 takes a closer look at the setup $bw = 20$, $bf = 125$, over which both randomized Beam Search algorithms are seen in Fig. 3 reaching median performance (out of the five evaluated setups). The plots shown are Empirical Attainment Functions (EAFs) [8], which depict the likelihood of objective space vectors being dominated (or matched) in an algorithm's run. It aggregates into one visualization the final Pareto fronts of score 16 trajectories found across all 100 runs. In Fig. 4(a–b), the boundaries of the shaded areas show the 0.25, 0.5 and 0.75 "attainment surfaces". In other words, the areas having at least 25, 50 or 75% chance of being attained (dominated or matched) by points in a run's final Pareto front. Figure 4(c) takes the difference between the EAFs for Beam P-ACO and Stochastic Beam, and shows how they compare in terms of likelihood to attain different areas of objective space, making clear the extent to which Beam P-ACO outperforms Stochastic Beam. The Pareto front with greatest hypervolume found by deterministic Beam Search is shown for reference.

7   Analysis and Discussion

Conceptually, we showed in this research a formal equivalence between four combinatorial optimization algorithms: Beam Search, Stochastic Beam, Beam P-ACO, and P-ACO. We demonstrated how they can all be implemented in the same algorithm, and made accessible through minor parameter changes.

A surprising result, that validates the introduced phasing indicator, as well as the baseline multi-objective Beam Search algorithm employed here, was the discovery of a score 17 mission (17 asteroids visited, and fully investigated). Searches over this model of the GTOC5 problem, using these same initial conditions, had previously only reached a maximum score of 16 [21]. Furthermore, those searches, employing a Branch & Prune tree search algorithm, took days to complete during the GTOC5 competition. The Beam Search variants under consideration could all find that single score 17 mission in runs lasting 10 to 20 min. Pure P-ACO is the exception here, having never surpassed a score 15 in our experiments (with a $bf = 1$, the high chance of an asteroid transfer's optimization problem not having a feasible solution greatly limits performance).

The first research question in Sect. 1 called for a demonstration of examples where randomized Beam Search variants would outperform the deterministic approach. If we evaluate in terms of mean performance, then, as we can see from Figs. 2 and 3, the current experimental evaluation could not find any such cases. At all computational cost thresholds we can find deterministic setups outperforming both of the randomized Beam Search variants, both in the quantity of score 16 solutions found, and in their quality. The deterministic algorithm was tuned to a considerably greater extent than the randomized ones (118 setups, against only 5), so it is possible that we are presenting a skewed view of each algorithm's capabilities. Alternatively, it may be that the specific problem we consider here is so resource-constrained, that the construction of long asteroid sequences actually demands for greedy branching decisions to be taken at every single step. In such a case, the search will not benefit from the greater tolerance for local sub-optimality present in the randomized Beam Search variants.

The second research question in Sect. 1 considers how the addition of feedback (pheromones) in Beam P-ACO changes performance, by comparison with Stochastic Beam, where search proceeds along multiple independent generations with no feedback between them. Figures 2, 3 and 4 show a clear positive effect of feedback on performance, both in the quantity and quality of top solutions found. This effect is seen to be larger the smaller the branching factor is. In other words, the greater the branching factor, the likelier it is for good solutions to be identified in a single generation, and thus the lower the benefit from feedback.

An identical comparison between randomized Beam Search variants, with and without pheromone updates, can be found in [3, Sect. 6.2] (for a different hybrid algorithm, and different problems). There, pheromones are also found to improve performance, though in small amounts. The current research goes beyond the analysis in [3] by uncovering the inverse relationship between branching factor and benefit from pheromones, and in demonstrating the superior performance of deterministic Beam Search over the randomized variants, given proper tuning.

Overall, an apt description of the behaviours displayed by the randomized algorithms is that they approximate the performance of deterministic Beam Search setups that use larger beam widths and branching factors – probabilistic branching effectively picks a subset of the successor nodes that deterministic Beam Search with larger branching factors would follow. The eventual success or failure of a run will then depend on how well those decisions align with those a better informed algorithm would take. This can be seen on display in Fig. 4. There, we see that in the extreme a randomized Beam Search run can closely approximate the Pareto front of score 16 trajectories found by an "optimally" tuned deterministic Beam Search (with a $bf = 125$, less than half of the $bf = 275$ needed in the deterministic setting). However, it can also fail to find a single score 16 trajectory: 7 of the 100 runs in Fig. 4(a), and 4 in Fig. 4(b) could only reach a score 15, thus ending in this analysis with a min. hypervolume of 0.0. Overall, the median hypervolume of 52.75 reached by Beam P-ACO in Fig. 4(b) is higher than that reached in ≈60% of the 118 deterministic Beam Search setups, and also higher than that reached in 50% of the 32 deterministic setups that had equal or larger beam width and branching factors.

In a real setting, in the preliminary design phase, missions would be constructed not from a single set of initial conditions, as done here, but from a great number of them, possibly numbering in the thousands. Being anytime algorithms, the randomized Beam Search variants investigated here can be employed in a racing approach [22], with multiple tree searches being executed in parallel. Computational effort would then be dynamically allocated across searches, as a function of the growing statistical evidence as to which searches lead to better missions. In such a setting, the randomized Beam search variants would be preferable choices, over the deterministic algorithm.

8   Conclusion

We considered here a hard real-world problem of spacecraft trajectory design, featuring an interplay of combinatorial and (constrained) continuous optimization sub-problems, dealing at different levels with uncertain and multi-objective quality functions. In this challenging domain, we investigated a number of extensions to Beam Search, the traditionally used approach to solve such problems.

We provided an improved orbital phasing indicator, and its transformation into a probability distribution over candidate bodies to extend missions with. We then hybridized the search process with the well known Ant Colony Optimization algorithm, and investigated the behaviours of the resulting randomized Beam Search variants. We found them to have lower sensitivity to the beam width and branching factor parameter settings, while offering in each generation a partially-informed approximation to the behaviours of deterministic setups running at higher computational costs.

Acknowledgment. Luís F. Simões was supported by FCT (Ministério da Ciência e Tecnologia) Fellowship SFRH/BD/84381/2012.

References

1. Global trajectory optimization competition portal. http://sophia.estec.esa.int/gtoc_portal/
2. Bisiani, R.: Beam search. In: Shapiro, S.C. (ed.) Encyclopedia of Artificial Intelligence, vol. 1, pp. 56–58. Wiley, Hoboken (1987)
3. Blum, C.: Beam-ACO-hybridizing ant colony optimization with beam search: an application to open shop scheduling. Comput. Oper. Res. **32**(6), 1565–1591 (2005)
4. Ceriotti, M., Vasile, M.: Automated multigravity assist trajectory planning with a modified ant colony algorithm. J. Aerosp. Comput. Inf. Commun. **7**(9), 261–293 (2010)
5. Ceriotti, M., Vasile, M.: MGA trajectory planning with an ACO-inspired algorithm. Acta Astronaut. **67**(9–10), 1202–1217 (2010)
6. Deb, K.: Multi-objective Optimization Using Evolutionary Algorithms. Wiley, Chichester (2001)
7. Dorigo, M., Stützle, T.: Ant Colony Optimization. A Bradford Book. MIT Press, Cambridge (2004)
8. da Fonseca, V.G., Fonseca, C.M., Hall, A.O.: Inferential performance assessment of stochastic optimisers and the attainment function. In: Zitzler, E., Thiele, L., Deb, K., Coello Coello, C.A., Corne, D. (eds.) EMO 2001. LNCS, vol. 1993, pp. 213–225. Springer, Heidelberg (2001). doi:10.1007/3-540-44719-9_15
9. Gad, A.H.: Space trajectories optimization using variable-chromosome-length genetic algorithms. Ph.D. thesis, Michigan Technological University (2011)
10. Grigoriev, I.S., Zapletin, M.P.: Choosing promising sequences of asteroids. Autom. Remote Control **74**(8), 1284–1296 (2013)
11. Grigoriev, I.S., Zapletin, M.P.: GTOC5: problem statement and notes on solution verification. Acta Futura **8**, 9–19 (2014)
12. Guntsch, M.: Ant algorithms in stochastic and multi-criteria environments. Ph.D. thesis, Karlsruher Institut für Technologie (2004). http://d-nb.info/1013929756
13. Guntsch, M., Middendorf, M.: A population based approach for ACO. In: Cagnoni, S., Gottlieb, J., Hart, E., Middendorf, M., Raidl, G.R. (eds.) EvoWorkshops 2002. LNCS, vol. 2279, pp. 72–81. Springer, Heidelberg (2002). doi:10.1007/3-540-46004-7_8
14. Guntsch, M., Middendorf, M.: Applying population based ACO to dynamic optimization problems. In: Dorigo, M., Caro, G., Sampels, M. (eds.) ANTS 2002. LNCS, vol. 2463, pp. 111–122. Springer, Heidelberg (2002). doi:10.1007/3-540-45724-0_10

15. Guntsch, M., Middendorf, M.: Solving multi-criteria optimization problems with population-based ACO. In: Fonseca, C.M., Fleming, P.J., Zitzler, E., Thiele, L., Deb, K. (eds.) EMO 2003. LNCS, vol. 2632, pp. 464–478. Springer, Heidelberg (2003). doi:10.1007/3-540-36970-8_33

16. Izzo, D.: PyGMO and PyKEP: Open source tools for massively parallel optimization in astrodynamics (the case of interplanetary trajectory optimization). In: 5th International Conference on Astrodynamics Tools and Techniques (ICATT) (2012)

17. Izzo, D.: Revisiting Lambert's problem. Celest. Mech. Dyn. Astron. **121**(1), 1–15 (2015)

18. Izzo, D., Getzner, I., Hennes, D., Simões, L.F.: Evolving solutions to TSP variants for active space debris removal. In: Proceedings of the 2015 Annual Conference on Genetic and Evolutionary Computation, pp. 1207–1214. ACM (2015)

19. Izzo, D., Hennes, D., Simões, L.F., Märtens, M.: Designing complex interplanetary trajectories for the global trajectory optimization competitions. In: Fasano, G., Pintér, J.D. (eds.) Space Engineering: Modeling and Optimization with Case Studies, pp. 151–176. Springer, Heidelberg (2016)

20. Izzo, D., Simões, L.F., Märtens, M., de Croon, G.C., Heritier, A., Yam, C.H.: Search for a grand tour of the jupiter galilean moons. In: Genetic and Evolutionary Computation Conference (GECCO 2013), pp. 1301–1308. ACM (2013)

21. Izzo, D., Simões, L.F., Yam, C.H., Biscani, F., Di Lorenzo, D., Addis, B., Cassioli, A.: GTOC5: results from the European Space Agency and University of Florence. Acta Futura **8**, 45–55 (2014)

22. Maron, O., Moore, A.W.: The racing algorithm: model selection for lazy learners. In: Aha, D.W. (ed.) Lazy Learning, pp. 193–225. Springer, Dordrecht (1997)

23. Oliveira, S., Hussin, M.S., Stützle, T., Roli, A., Dorigo, M.: A detailed analysis of the population-based ant colony optimization algorithm for the TSP and the QAP. Technical report, TR/IRIDIA/2011-006, IRIDIA February 2011

24. Petropoulos, A.E., Bonfiglio, E.P., Grebow, D.J., Lam, T., Parker, J.S., Arrieta, J., Landau, D.F., Anderson, R.L., Gustafson, E.D., Whiffen, G.J., Finlayson, P.A., Sims, J.A.: GTOC5: results from the Jet Propulsion Laboratory. Acta Futura **8**, 21–27 (2014)

25. Sinha, A., Malo, P., Deb, K.: Evolutionary bilevel optimization: an introduction and recent advances. In: Bechikh, S., Datta, R., Gupta, A. (eds.) Recent Advances in Evolutionary Multi-objective Optimization. ALO, vol. 20, pp. 71–103. Springer, Heidelberg (2017). doi:10.1007/978-3-319-42978-6_3

26. Stuart, J.R., Howell, K.C., Wilson, R.S.: Design of end-to-end trojan asteroid rendezvous tours incorporating scientific value. J. Spacecr. Rockets **53**(2), 278–288 (2016)

27. Weise, T., Chiong, R., Lässig, J., Tang, K., Tsutsui, S., Chen, W., Michalewicz, Z., Yao, X.: Benchmarking optimization algorithms: an open source framework for the traveling salesman problem. IEEE Comput. Intell. Mag. **9**(3), 40–52 (2014)

28. Wilt, C.M., Thayer, J.T., Ruml, W.: A comparison of greedy search algorithms. In: Proceedings of the Third Annual Symposium on Combinatorial Search (SOCS 2010), pp. 129–136 (2010)

29. Zitzler, E., Thiele, L., Laumanns, M., Fonseca, C.M., Da Fonseca, V.G.: Performance assessment of multiobjective optimizers: an analysis and review. IEEE Trans. Evol. Comput. **7**(2), 117–132 (2003)

Optimizing Charging Station Locations
for Electric Car-Sharing Systems

Benjamin Biesinger$^{(\boxtimes)}$, Bin Hu, Martin Stubenschrott, Ulrike Ritzinger,
and Matthias Prandtstetter

AIT Austrian Institute of Technology, Center for Mobility Systems – Dynamic
Transportation Systems, Giefinggasse 2, 1210 Vienna, Austria
{benjamin.biesinger,bin.hu,martin.stubenschrott,ulrike.ritzinger,
matthias.prandtstetter}@ait.ac.at

Abstract. This paper is about strategic decisions required for running
an urban station-based electric car-sharing system. In such a system,
users can rent and return publicly available electric cars from charging
stations. We approach the problem of deciding on the location and size
of these stations and on the total number of cars in such a system using
a bi-level model. The first level of the model identifies the number of
rental stations, the number of slots at each station, and the total number
of cars to be acquired. Then, such a generated solution is evaluated by
computing which trips can be accepted by the system using a path-based
heuristic on a time-expanded location network. This path-based heuristic
iteratively finds paths for the cars through this network. We compare
three different pathfinder methods, which are all based on the concept of
tree search using a greedy criterion. The algorithm is evaluated on a set
of benchmark instances which are based on real-world data from Vienna,
Austria using a demand model derived from taxi data of about 3500 taxis
operating in Vienna. Computational tests show that for smaller instances
the algorithm is able to find near optimal solutions and that it scales well
for larger instances.

Keywords: Location problem · Car-sharing · Electric cars · Variable
neighborhood search

1 Introduction

Urban transportation as a major consumer of energy and a contributor to air
pollution raises challenges to local governments and companies that must be
solved in the near future. One of those challenges is to reduce the usage of
conventional cars in urban areas. It can already be observed that the market
for alternatives to combustion engine powered vehicles, especially (full) battery

This work has been partially funded by the Austrian Federal Ministry for Transport,
Innovation and Technology (bmvit) in the JPI Urban Europe programme under grant
number 847350 (e4share).

B. Hu and M. López-Ibáñez (Eds.): EvoCOP 2017, LNCS 10197, pp. 157–172, 2017.
DOI: 10.1007/978-3-319-55453-2_11

electric cars, is steadily growing. Moreover, car-sharing systems as an alternative or an addition to public transport is getting increasingly popular and in many larger cities such systems are already installed and in use. As the main drawbacks of electric cars—high acquisition costs, limited battery range, and high infrastructure costs—is mitigated when used in an urban shared environment, electric car-sharing systems have attracted increased attention over the last years. Such car-sharing systems provide a fleet of vehicles in a defined area of operation in which users can rent and return cars. Specifically for electric cars, however, charging stations have to be installed to recharge the battery of the vehicles. We assume a one-way station-based system in which the users can rent and return the cars only at these stations in contrast to free-floating systems where the customers can return the cars in any free parking space within the operational area. Therefore, we consider stationary electric car-sharing systems where the decision is where to place these stations and how many charging slots to install. This is crucial for the functionality and success of the whole system. In Fig. 1 an illustration of such a system is given using, as an example, the inner districts of Vienna, Austria. A good location for a station naturally depends on the customer demand of the nearby area and therefore we use a demand model to evaluate station locations. The demand model is given by a forecast of a set of trip requests in which each request has an individual estimated profit. The evaluation method maximizes the total profit of the acceptable requests.

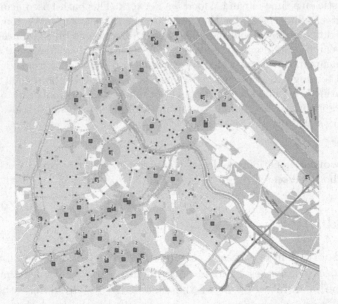

Fig. 1. Illustration of station locations (green rectangles) with their area of attraction and chosen number of installed charging slots in the inner city of Vienna, Austria. Red circles show possible station locations which are not chosen. (Color figure online)

This work considers the problem of deciding the charging locations, their size, and the total number of cars in the system as combinatorial optimization problem that is heuristically solved using a two-stage solution algorithm. In the first stage, the decision variables are fixed using a variable neighborhood search (VNS) [13]. Each generated solution candidate is evaluated in the second stage using an iterated procedure based on a greedy criterion using a demand forecast. Specifically, a greedy, a PILOT, and a beam search algorithm is used to iteratively find paths for the used cars through the system. The presented algorithm is evaluated on a set of benchmark instances based on real-world data of Vienna, Austria. First, the problem is formally defined in Sect. 2. Then, related work is described in Sect. 3 followed by the description of the algorithm in Sect. 4. Computational results are presented in Sect. 5 and finally, conclusions are drawn and possible future research directions are given in Sect. 6.

2 Problem Definition

The charging station location problem (CSLP) is defined on a road network $G = (V, A)$ given by a set of vertices V and a set of directed arcs A representing road segments. Each arc $a = (i, j) \in A$, $i, j \in V$ has a length l_{ij} and an associated weight representing the travel time δ_{ij} needed to travel from vertex i to j. Possible charging stations $S \subseteq V$ are given by a subset of the vertices and each potential station $i \in S$ has an associated opening cost $F_i \geq 0$, a capacity $C_i \in \mathbb{N}$, and a cost per slot $Q_i \geq 0$. The maximum number of cars is given by H, and each car has the same acquisition cost F_c, battery capacity $B^{\max}$, and charging rate per time unit ρ.

Furthermore, a set of requested trips K is given, which corresponds to a demand forecast within the time horizon $T = \{0, \ldots, T_{\max}\}$ and is the basis of the solution evaluation. Each request $k \in K$ has a starting $s_k \in T$ and ending time $e_k \in T$ with $e_k > s_k$, an origin $o_k \in V$, and a destination $d_k \in V$. Further, a duration δ_k, an estimated battery consumption b_k, and a profit p_k is given for each request $k \in K$. It is assumed that the customers are willing to walk to a nearby station if the walking time does not exceed a pre-defined maximum duration β^w. Based on this maximum walking time we have a set of potential starting $N(o_k)$ and ending stations $N(d_k)$ for each request $k \in K$. If there are no potential stations within the walking range, i.e., $N(o_k) = \emptyset$ or $N(d_k) = \emptyset$ then the request is unrealizable and therefore not considered. Figure 2 illustrates the CSLP on a small road network.

The goal of the CSLP is to find the set of stations to open $S' \subseteq S$, the number of slots to use for each open station, and the total number of cars $H' \leq H$ in the system with a limited budget W such that the total profit of all accepted trips is maximized. To compute the obtained profit each car $c = 0, \ldots, H' - 1$ must be assigned a set of feasible trips $K'_c \subseteq K$ where each trip $k \in K'_c, \forall c = 0, \ldots, H' - 1$ can only be assigned to at most one car c. Each car $c = 0, \ldots, H' - 1$ must be able to perform its assigned trips K'_c by ensuring *capacity feasibility*, *battery feasibility*, and *connectivity* of the car route:

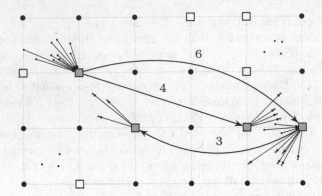

Fig. 2. Example of a problem instance and a specific solution in a small graph. The road network is illustrated by the big circles as crossings, the rectangles as possible station locations, and the streets between them. The starting points of the trip requests are shown as small circles and the respective ending points as red diamond-shaped rectangles. The filled rectangles correspond to the chosen locations to which the nearby starting and ending points of the acceptable requests are assigned. The arrows between the stations indicate the travel paths of the cars which are routed through the network. The number of cars traveling along these routes is shown by the number above these arcs. In the upper right and lower left corner some requests are shown which are unfulfillable because there is no station in the vicinity of their starting points.

- *Capacity feasibility* is given when at each time-step $t \in T$ there are no more cars in station $s \in S$ than the available number of slots.
- *Battery feasibility* is given if the battery capacity of the car is sufficient for performing the requested trip taking potential preceding battery charging into account. More formally, the solution is battery feasible if between two consecutive trips $k^1, k^2 \in K$ starting/ending at station i of a car $\min\{(s_{k^2} - e_{k^1})\rho + B^{k^1}, B^{\max}\} \geq b_{k^2}$ is valid, where B^{k^1} is the remaining battery capacity of the car after performing trip k^1.
- *Connectivity* is given when the ending station of a trip k is equal to the starting station of the next trip.

Then, the total profit of a solution S is the total sum over the profits of all accepted trips: $p(S) = \sum_{c=0}^{H'-1} \sum_{k \in K_c'} p_k$. We are aware that in practice it might be unrealistic to plan the accepted trips in such a way, but this academic problem formulation allows us to obtain an upper bound for the total profit based on a given demand forecast.

3   Related Work

The charging station location problem in the presented form is a relatively new research topic and was introduced by Brandtstätter et al. [5]. In their work the authors defined the problem, proved its NP-hardness and described some polynomially solvable cases. Furthermore, they modeled the CSLP as integer linear program in several ways and compared the efficiency of their models on a set of artificially created and real-world instances. The best formulations are able to solve real-world instances with up to 480 trips, 50 cars, and 153 possible station locations to proven optimality. In another work Brandstätter et al. [4] considered stochastic aspects of the demand and presented two-stage stochastic models and a heuristic for solving the stochastic CSLP.

The literature of optimization problems arising in (electric) car-sharing systems is, however, broader and a recent literature survey can be found in [3]. A similar problem concerning the location of charging stations for electric car-sharing systems was described by Boyacı et al. [2]. Additionally, they also considered operational decisions regarding relocation of cars from stations in low to stations in high demand areas to balance the system as a whole. The topic of relocating cars is also considered by Weikl and Bogenberger [16] who investigated different relocation strategies for free floating conventional car-sharing systems. Related problems in the domain of exact methods and heuristics for optimization problems arising in electric car-sharing systems are described, e.g., by Hess et al. [12], Ge et al. [11], Cavadas et al. [6], and Frade et al. [9]. Another related problem of placing charging stations for electric taxis in urban areas was approached by Asamer et al. [1]. The authors propose a decision support system which identifies promising regions in which charging stations should be placed instead of actual locations.

4   Algorithm Description

We approach this problem using a two-stage solution algorithm. In the first (or upper) level it is fixed which stations are opened, how many slots are built at each open station, and how many cars are purchased. Then, in the second (or lower) level we compute which requests can be accepted by the system using the solution of the upper level problem. As both the upper level and the lower level problem are computationally demanding, we solve both of them heuristically. Therefore, a variable neighborhood search (VNS) [13] is used for the upper level and three different procedures based on a greedy criterion are used for solving the second stage. In the next sections we will describe the algorithms in detail.

4.1   Variable Neighborhood Search for the Upper Level Problem

A solution to the upper level problem is represented by an integer vector z of size $|S|$ and each element z_i, $\forall i = 0, \ldots, |S| - 1$ can take values $0, \ldots, C_i$. In our approach we determine the total number of cars of a solution candidate implicitly by first fixing the stations and number of slots and then using as many cars as possible with the remaining budget bounded by H.

The first step of the algorithm is to create an initial solution which is then passed on to a subsequent variable neighborhood search. The initial solution generation is a greedy construction heuristic based on following greedy criterion:

Each potential station is assigned an *attractiveness* value which determines the order of the stations which are considered to be opened. This attractiveness value is the number of requests which can start or end at this station and thereby represents a heuristic guidance which stations are useful for the given set of requests. Then, the construction algorithm iterates over the set of stations in descending order with regard to their attractiveness values, opens station i if the remaining budget is sufficient, and chooses the number of slots out of the budget feasible values of $\{1, \ldots, C_i\}$ uniformly at random. The algorithm ensures that there is enough remaining budget to purchase at least one car and terminates if no station can be opened anymore.

This starting solution is then passed to a general variable neighborhood search, which was introduced by Mladenović and Hansen [13]. The underlying variable neighborhood descent (VND) uses four different neighborhood structures (NBs) which are searched in the following order (sorted by their estimated complexity in ascending order) in a best improvement fashion:

- *Close station:* In this neighborhood structure a randomly chosen station out of the currently open ones is closed. As a consequence, this move primarily increases the number of purchased cars and can therefore improve the objective value. Here, we do not only accept improving moves but also accept moves which do not change the objective value. The idea behind this strategy is to have as much budget available for the next neighborhood structures as possible.
- *Open station:* A move in this neighborhood structure opens a previously closed station with as many slots as possible.
- *Change slots:* An already open station i is chosen, the number of slots z_i is set to all feasible values out of $\{1, \ldots, C_i\}$ and the best result is taken.
- *Swap slots:* For each pair of stations $i, j \in S$ for which $z_i \neq z_j$ the values of z_i and z_j are swapped. Resulting infeasibilities are repaired by iteratively reducing the number of slots of station i or j (chosen uniformly at random) by one until the solution is feasible again.

The VNS uses four shaking neighborhood structures which are based on the NBs described above. The first two shaking NBs are based on the *swap slots* neighborhood structure and performs two and four swaps, respectively. The third and fourth NBs perform two and four moves in the *close station* neighborhood structure.

4.2   Path-Based Heuristic

After fixing the solution (z, H') of the upper level problem, in the lower level problem we now have to compute the trips that are accepted by the system so that the total profit is maximized. Therefore, we use a variant of the path-based heuristic (PBH) introduced by Brandstätter et al. [5]. The PBH finds iteratively paths for each car through space and time in an adaptable time-expanded location network (TELN). The TELN is a time-discretized directed acyclic multigraph basically consisting of one node for each station and time unit and arcs between each two consecutive time slots. In the following we will formally define the TELN $G^T = (N, A')$:

First, the set of open stations is defined as $S' = \{i \in S \mid z_i > 0\}$. Then, for each station $i \in S'$ and each time unit $t \in T$ a node i_t is introduced forming the set of nodes N. Additionally an artificial source r^s and target node r^t are created. The set of arcs A' is divided into three disjoint subsets of arcs A^I, A^W, A^T and each arc a has an associated profit p_a and a battery consumption b_a.

- The set of *initialization arcs* $A^I = \{(r^s, i_0) \mid i \in S'\} \cup \{(i_{t_{\max}}, r^t) \mid i \in S'\}$ are the arcs leaving the artificial source node to each station node of the first time slot and entering the artificial target node from each station node of the last time slot $t_{\max}$. All initialization arcs have a profit and battery consumption of zero.
- *Waiting arcs* A^W are added between two time slots of any station, i.e., in the simplest case $A^W = \{(i_t, i_{t'}) \mid i \in S', t \in T \setminus t_{\max}, t' = t + 1\}$. This set, however, can be reduced as we will see later. These arcs also have a profit of zero but a negative battery consumption of $-\phi_i$ which corresponds to battery charging.
- Finally, the set of *trip arcs* A^T corresponds to the requested trips and contains arcs for each trip that can be accepted (denoted by $K' \subseteq K$) with the given set of stations S'. For each such trip $k \in K'$ an arc is introduced for all possible starting and ending station combinations, i.e., $A^T = \bigcup_{k \in K'} A_k^T$, with $A_k^T = \{(i_{s_k}, j_{e_k}) \mid i \in N(o_k), j \in N(d_k)\}, \forall k \in K'$. Each trip arc a has battery consumption b_k and profit p_k, where $k \in K'$ is the corresponding request.

An illustrative example of a TELN with four stations is shown in Fig. 3. In this figure the solid lines are initialization or waiting arcs and the dashed lines are trip arcs corresponding to a request.

In this example we see a reduced version of the graph which omits many of the unnecessary waiting arcs. Let $N^i \subseteq N = \{n_0^i, \ldots, n_{|N_i|}^i\}$ be the set of all starting/ending nodes of all trips at station $i \in S'$ sorted in ascending order of their starting/ending time. Then, the set of actual introduced waiting arcs is the following: $A^W = \{(i_0, n_0^i)\} \cup \{(n_{|N_i|}^i, i_{t_{\max}})\} \cup \{(n_t^i, n_{t+1}^i) \mid t = 0, \ldots, |N_i| - 1\}$.

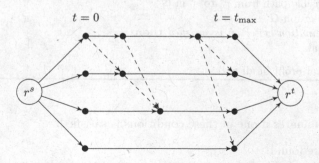

Fig. 3. Example of a time-expanded location network with four open stations.

After building the graph G^T, as Algorithm 1 shows, the PBH consists of iteratively finding paths from r^s to r^t in the TELN and updating the graph based on the found path. The resulting objective value of the upper level solution is then the total profit of all found paths where the profit of a path is defined as the total profit of all requests fulfilled by this path. The path finding step corresponds to solving a resource constrained shortest path problem with the battery as resource and the negative profits as the arc weights to minimize. Note that here we do not have to consider the station capacities as it is assured by the update procedure of the graph, which is described later, that they can never be violated. As this problem is NP-hard [10], we use several heuristics for solving it. All of these heuristics construct iteratively a path starting from r^s using a greedy criterion based on the concept of a *potential profit* p' of a node. For each node $i_t \in N$ a potential profit $p'(i_t)$ is calculated by computing shortest paths from i_t to the target node r^t relaxing the battery constraints. In the PBH this is done by computing shortest paths from r^t to all other nodes in the arc-reversed graph. Then, each time a path P is found through the TELN, the graph is updated for the next iteration. Therefore, for each served request $k \in K'$ on P, all corresponding trip arcs A_k^T are deleted from G^T which ensures that one request can only be fulfilled by at most one car. Furthermore, we have to ensure capacity feasibility of the paths and therefore we introduce global variables u_{it}, $\forall i \in S', t \in T$ which are initially set to zero. These variables store the current number of cars at each station in each time step over all iterations. For each used arc $a = (i_t, j_{t'}) \in P, i, j \in S'$, we increase $u_{jt'}$ by one and check if the fixed number of slots of j is reached. If $u_{jt'} = z_j$, then all incoming arcs of node $j_{t'}$ are deleted from G^T which ensures that in later iterations this node cannot be reached anymore. Finally, also the potential profits of the nodes have to updated and the shortest path values are recomputed.

Algorithm 1. Path-based Heuristic

Input : Upper level solution (z, H')
Output: Approximated profit
build time-expanded location graph G^T
for $i = 0, \ldots, H'$ **do**
    find feasible path from r^s to r^t in G^T
    update graph G^T
    if *termination criterion is satisfied* **then**
        break;
return total profit of all paths

The PBH terminates if one of these conditions is satisfied:

- H' paths are found.
- The potential profit of r^s is zero or r^t is unreachable.

- The path finding algorithm could not find a path from r^s to r^t. This can happen if it gets stuck at a node with either no outgoing arc or with only outgoing arcs with a battery consumption greater than the remaining capacity.
- The profit of the found path is zero. Although this condition is not needed for the PBH to be valid it is introduced for improving its efficiency by reducing the number of generated paths with zero profit.

In the following we will describe three heuristics we designed, implemented, and evaluated for finding feasible paths. They are described in order of their complexity and time consumption and a comparison of their performance is given in Sect. 5.3.

Greedy Pathfinder. The greedy pathfinder always appends at the current node i_t a battery-feasible arc $a = (i_t, j_t')$ for which $p_a + p(j_t')$ is maximum and ties are broken randomly.

PILOT Pathfinder. The PILOT pathfinder extends the greedy algorithm by incorporating a look-ahead mechanism as described in [8]. Instead of always extending the current partial solution with the arc $a = (i_t, j_t')$ for which $p_a + p(j_t')$ is maximum, for each possible extension a lower bound on the objective is computed (using the greedy algorithm described above). Then, the extension is performed which results in the highest value and ties are, again, broken randomly.

Beam Search Pathfinder. The beam search pathfinder is based on the beam search algorithm [14]. Beam search uses a set of partial solutions of size k_{bw}, called the beam, and evaluates the k_{ext} most promising extensions to these. For this problem we choose the extensions based on the greedy criterion as explained before. In the end, all partial solutions of the beam are complete and the best one among those is taken as the final solution.

5 Computational Results

The developed algorithm is tested on a set of benchmark instances based on real-world data, which are described in Sect. 5.1. First, computational results on a set of instances are shown, which previously also have been approached by exact algorithms [5]. Then, in the second set of experiments, the developed pathfinder methods are compared on a different set of instances with stricter battery constraints to better highlight the differences of these methods.

5.1 Instance Description

The basis of all instances is the road network of Vienna, Austria obtained with OpenStreetMap data[1]. We assume potential locations for stations at supermarkets, parking lots, and subway stations. The number of slots for each station is chosen uniformly at random between 1 and 10. Station opening and

---

[1] https://www.openstreetmap.org/.

slot costs are chosen uniformly at random from $F_i \in \{9000, \ldots, 64000\}$ and $Q_i \in \{22000, \ldots, 32000\}$ Euro, respectively. We further assume fast charging slots with a maximum charging rate per slot of 50 kW. The set of homogeneous cars is based on the data of the Smart ED car with an acquisition cost of 20000 Euro, a battery capacity of 17.6 kWh, and a maximum charging rate of 17.6 kW. The data for the requested trips is based on real Taxi data of one week in spring provided by a Taxi provider of Vienna. For each trip we are given the starting and ending point, and the duration and battery consumption is computed using the routing framework by Prandtstetter et al. [15]. Furthermore, we are given a starting and ending time, which is discretized in time intervals of 15 min, resulting in 672 time periods. The profit is computed by assuming a rate of 0.3 Euro/minute which approximately corresponds to the rate of local car-sharing systems. We assume that each trip can start and end at the three closest stations within a walking distance of 5 min. Therefore, the original road network G is extended by introducing an arc $a' = (j, i)$ for each arc $a = (i, j) \in A$ of the original network and setting $l_{ji} = l_{ij}$. We assume a walking speed of 1.34 m/s and compute a shortest path using Dijkstra's algorithm [7] from the starting and the ending point of each request to all stations using the walking time as arc weight. The instance contains 693 potential station locations and 37965 trips.

The size of the whole instance is very large and especially the outer areas are not economically relevant for practical applications. Therefore, we filtered the instance and use only a subset of the trips and stations for the algorithm evaluation. Figure 4 shows four subsets of instances $I_1, \ldots, I_4$ based on the political districts of Vienna.

The first instance I_1 uses only the potential stations and requests starting and ending in one of the five red districts. For I_2 the yellow district is added and for I_3 the yellow and the blue districts. Finally, the largest instance I_4 uses

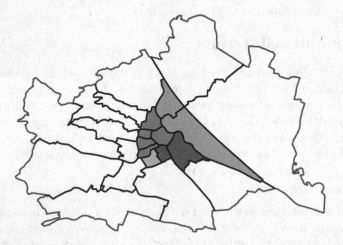

Fig. 4. The 23 districts of Vienna, Austria with the highlighted subset of districts which are used as instances. (Color figure online)

all ten colored districts. Table 1 shows the properties for each instance, where r_K is the average number of requests per hour, and $\overline{b_K}$ is the average energy consumption of the trips relative to the battery capacity of the Smart ED car in percent.

Table 1. Additional instance data.

Instance	$|S|$	$|K|$	r_K	$\overline{b_K}$
I_1	105	108	0,68	3,10
I_2	131	209	1,29	3,28
I_3	198	719	4,31	3,37
I_4	280	1347	8,04	3,27

Each such created instance is further split into specific instances by specifying a maximum budget and a maximum number of cars. These instances are preprocessed in two ways:

1. Stations in which no requests can start or end are not considered.
2. The times of the requested trips are shifted forward in time so that the first fulfillable request starts at $t = 0$ and t_{max} is given by the ending time of the last fulfillable request. Thereby, the size of T and hence the size of the TELN decreases which results in a faster solution evaluation.

5.2 Comparison to Exact Algorithms

First, we compare the results of the developed algorithm with exact methods based on integer linear programming (ILP) models developed by Brandstätter et al. [5]. They proposed several models and three of them (C1, C2, F1) turned out to perform best on the considered instances. Model C1 uses connectivity cuts and continuous battery tracking, C2 uses connectivity cuts and battery-infeasible path cuts, and model F1 is a multi-commodity flow formulation with continuous battery tracking. Table 2 shows the results of our proposed VNS with the greedy pathfinder compared to the results obtained from the exact models. We used instances with different budget constraints out of $W \in \{1000000, 2000000, 3000000, 4000000, 5000000\}$ Euro and the number of cars H is restricted by either 10, 25, or 50. Here we used only the district subsets of I_1 and I_2 to be able to make comparisons. For instances I_3 and I_4 the state-of-the-art exact approaches are not able to produce reasonable results in reasonable time. Table 2 shows the average objective value over 30 independent runs of the VNS $(\overline{\text{obj}})$, their associated standard deviation (sd), and the best objective value over these 30 runs (obj^b). Each run was performed on an Intel Xeon E5-2643 processor and terminated after 3600 s. The best solution, however, was usually found earlier (depending on the size of the instance) as the algorithm converged to its final solution. For the models the table shows the lower bound

Table 2. Comparison to exact integer linear programming approaches

Instance			VNS			C1			C2			F1		
W	H	I	obj	sd	obj^b	obj^{lb}	obj^{ub}	gap	obj^{lb}	obj^{ub}	gap	obj^{lb}	obj^{ub}	gap
1M	10	I_1	13559,0	303,7	13769	14045	14045,0	0,0	14045	14046,3	0,0	14045	14045,0	0,0
2M	10	I_1	19563,2	489,6	20444	21899	21901,0	0,0	21899	21899,0	0,0	21899	21901,1	0,0
3M	10	I_1	21411,1	220,2	21557	21956	21956,0	0,0	21956	21956	0,0	21956	21956,0	0,0
4M	10	I_1	21430,4	198,1	21557	21956	21956,0	0,0	21956	21956,0	0,0	21956	21956,0	0,0
5M	10	I_1	21493,1	123,1	21557	21956	21956,0	0,0	21956	21956,0	0,0	21956	21956,0	0,0
1M	25	I_1	13542,8	338,5	13769	14045	14045,0	0,0	14045	14046,2	0,0	14045	14046,2	0,0
2M	25	I_1	23467,8	320,2	24017	18302	25279,0	38,1	25279	25281,2	0,0	25279	25281,4	0,0
3M	25	I_1	28089,8	538,9	29168	30845	30845,0	0,0	30845	30845,0	0,0	30845	30848,0	0,0
4M	25	I_1	29807,1	359,4	30441	31215	31215,0	0,0	31215	31215,0	0,0	31215	31215,0	0,0
5M	25	I_1	30410,1	190,7	30686	31215	31215,0	0,0	31215	31215,0	0,0	31215	31215,0	0,0
1M	50	I_1	13625,1	201,0	13769	14045	14045,0	0,0	14045	14045,0	0,0	14045	14299,0	1,8
2M	50	I_1	23511,4	342,7	24086	25147	25291,3	0,6	25279	25279,0	0,0	25186	25645,5	1,8
3M	50	I_1	28783,9	524,6	29858	26057	31140,2	19,5	26837	31135,5	16,0	31131	31134,1	0,0
4M	50	I_1	31545,2	410,6	32403	30948	33971,0	9,8	33708	33952,8	0,7	33877	34197,0	0,9
5M	50	I_1	33154,4	425,9	33754	34093	34197,0	0,3	34197	34197,0	0,0	34197	34197,0	0,0
1M	10	I_2	23142,6	389,6	23515	24403	24403,0	0,0	24403	24403,0	0,0	24403	24403,0	0,0
2M	10	I_2	32032,7	748,9	33351	36916	37277,0	1,0	36846	37277,0	1,2	37277	37280,5	0,0
3M	10	I_2	37041,1	873,7	38431	40822	40822,0	0,0	40822	40822,0	0,0	40822	40822,0	0,0
4M	10	I_2	38967,6	347,9	39459	40822	40822,0	0,0	40822	40822,0	0,0	40822	40822,0	0,0
5M	10	I_2	38967,2	320,1	39496	40822	40822,0	0,0	40822	40822,0	0,0	40822	40822,0	0,0
1M	25	I_2	23141,2	436,7	23542	24403	24403,0	0,0	24403	24403,0	0,0	22675	27100,2	19,5
2M	25	I_2	37457,9	818,4	38415	27340	40952,2	49,8	24809	40957,0	65,1	39667	43608,0	9,9
3M	25	I_2	46500,5	1003,7	48301	36783	53172,5	44,6	52506	53166,0	1,3	52392	54470,6	4,0
4M	25	I_2	49988,0	1092,5	51987	58219	58253,0	0,1	58160	58253,0	0,2	58155	58253,0	0,2
5M	25	I_2	53639,6	981,8	55736	58253	58253,0	0,0	58253	58253,0	0,0	58253	58253,0	0,0
1M	50	I_2	22962,8	591,7	23542	24403	24403,0	0,0	22895	24430,1	6,7	23180	27745,0	19,7
2M	50	I_2	37349,1	825,6	38536	24852	40985,6	64,9	24031	40987,0	70,6	38940	43853,8	12,6
3M	50	I_2	47953,0	837,4	49381	36783	53564,4	45,6	36783	53502,5	45,5	49906	55732,9	11,7
4M	50	I_2	54131,0	1266,0	57060	47877	62012,2	29,5	47877	62014,9	29,5	61343	62958,9	2,6
5M	50	I_2	59358,4	1453,9	61943	57238	66494,4	16,2	57238	66494,4	16,2	65783	66971,0	1,8

(obj^{lb}), the upper bound (obj^{ub}), and the optimality gap (gap). The models are solved using CPLEX 12.6.3 with a time limit of 6 h.

The results in Table 2 show that the VNS is barely able to reach the results (obj^{lb}) of the exact methods when the gap is below 20%. For the larger instances i_1 with $H = 50$, however, the VNS starts finding better results than model C1 and C2. The average gap of the VNS to the optimal value of the optimally solved instances is about 6.1%. After an analysis of the results we observed that the main reason for the worse performance on some instances lies in the inaccuracy of the solution evaluation. Although the greedy pathfinder is often able to find the optimal path through the TELN, the overall evaluation procedure is just an approximation since the path of each car is evaluated separately in succession. Therefore, even if the VNS would generate an optimal solution, the pathfinder heuristic would not be able to identify it as such because it might be assigned a worse objective value from the procedure which calculates the accepted trips.

5.3   Pathfinder Results for Larger Instances

In the second set of experiments the impact of the used pathfinder (greedy (G), PILOT (P), and beam search (B)) on the final results is investigated in more detail. The instances, as described in Sect. 5.1 have a low trip density and average energy consumption per trip. Preliminary results showed that for these instances the greedy pathfinder always performed best because it is the fastest algorithm and the PILOT and beam search method were never able to find better paths. Therefore, we now assume an adapted demand forecast in which the system is more utilized and the requested trips are longer. So, we change the instances in the following way:

- The starting times of all requests of the instance are scaled in such a way that they all start within an 8 h period.
- The battery consumption of each request is increased by a factor of 6.

By applying these adaptions the instance becomes, on the one hand, more *crowded* so that more trips are requested in a short time period and, on the other hand, it becomes more battery restricted. Therefore, the used pathfinder is more crucial to the solution evaluation. The results of the different algorithms are shown in Table 3 and, again, 30 runs per instance are performed with a maximum run-time of 3600 s.

Table 3. Results of the different pathfinders for the altered instances aggregated by used subset of districts i and number of available cars H.

Instance		obj_g			#best			$\#G>_{sig}$		$\#P>_{sig}$		$\#B>_{sig}$	
i	H	G	P	B	G	P	B	P	B	G	B	G	P
I_1	10	17800,5	18040,9	17965,7	1	3	1	1	1	4	4	3	0
	25	23752,6	23933,8	23886,5	1	4	0	1	1	4	0	4	0
	50	25332,2	25229,1	25102,8	3	2	0	3	3	2	3	0	0
I_2	10	28529,0	28829,0	28684,0	0	5	0	0	1	1	2	0	0
	25	39973,8	39680,7	39220,2	5	0	0	3	5	0	2	0	0
	50	42788,8	42317,0	42019,2	5	0	0	4	5	0	1	0	0
I_3	10	46528,5	47068,0	46257,9	1	4	0	0	2	2	4	0	0
	25	71937,7	71052,8	69499,3	4	1	0	3	4	1	4	0	0
	50	81131,1	78513,6	76170,1	4	1	0	4	4	1	4	0	0
I_4	10	46648,1	46499,7	45775,3	3	2	0	1	3	1	3	0	0
	25	71833,6	69268,5	66719,6	5	0	0	4	5	0	3	0	0
	50	81547,6	77482,9	75111,6	5	0	0	4	5	0	3	0	0

In this table the instances are grouped by the subset of districts $I_1, \ldots, I_3$ and the maximum number of cars $H \in \{10, 25, 50\}$ resulting in 5 instances with different budget constraints per row. For each row the geometric mean $(\overline{obj_g})$, the number of instances for which the algorithm yields the best average objective value over all 30 runs (#best), and the results of the statistical tests are given. For testing the statistical differences between the algorithms we performed one-sided Wilcoxon rank sum tests using an error level of 5%. The entries of the table corresponds to the number of instances for which algorithm A yields statistically significant better results than algorithm B, i.e., the entries in column P below $\#G>_{sig}$ are the number of instances the greedy pathfinder finds better results than the PILOT pathfinder.

Table 3 shows that for the smallest instance I_1 and for the instances with $H = 10$ both the PILOT and the beam search pathfinder tend to yield better results than the greedy pathfinder. As the instance size or the number of available cars increases, however, the greedy pathfinder starts performing better than the other two. Especially the beam search pathfinder is outperformed by the greedy and PILOT pathfinder, as it could not find significantly better results than the latter for any of our test instances. This behavior can be explained by the higher run-time complexity for the PILOT and beam search pathfinder as the instance size grows. When the greedy pathfinder is used, the algorithm can perform more iterations overall and is therefore able to find better solutions.

6   Conclusions and Future Work

In this work a heuristic algorithm for finding and designing charging stations for electric vehicles is presented. It is based on a bi-level model formulation in which the first level decides on the station locations, number of slots per station, and the total number of cars by using a variable neighborhood search. Then, for each generated solution candidate of the first level, a path-based heuristic is used for the evaluation, i.e., deciding which trips can be accepted by the system. For the path-based heuristic three different path-finders are implemented and compared to each other. The results show that for smaller instances and instances with a smaller number of available cars that have a high trip density and many trips with a high energy consumption, the PILOT pathfinder performs best in most of these instances. On the other hand, if the instance size is larger, the trips are shorter, or the trip density decreases, the greedy pathfinder outperforms the other two. The beam search pathfinder is, however, not able to consistently find any significantly better results than the greedy or the PILOT method. The implemented algorithm was further compared with several exact methods based on integer linear programming from the literature. The results show that for those instances which are hard to solve by exact approaches and thus have high optimality gaps, the VNS is able to yield better results. However, it was not able to find solutions with the same quality for those instances that could be solved to optimality. The main reason for this is that the solution evaluation using the path-based heuristic is not exact and therefore the algorithm is not able to find the optimal trips to be accepted. An exact evaluation might improve these results but the algorithm would not scale well with the instance size anymore.

Although we considered many aspects of a real-world electric car-sharing system, some operational decisions are neglected currently. As one-way car-sharing systems tend to fall out of balance over time because of the not uniformly distributed demand in the operational area some kind of re-distribution of the vehicles must be performed. These re-distributions can be user-based (e.g., by giving the users incentives) or system-based. When using the latter, the re-distribution is usually performed by dedicated relocators who move cars between stations. This step is important for the operational decisions of such an electric car-sharing system but can also be considered for the strategic decisions of the station planning.

The solution algorithm could potentially be improved by using a more advanced initial solution generation method. The path-based heuristic used for the solution evaluation can be extended to generate an initial solution by letting the heuristic decide which station to open or to extend if it is budget-feasible. For larger instances, however, the run-time requirements would increase because the whole set of available stations has to be considered for the time-expanded location network. Another possibility to improve the algorithm is to incorporate feedback from the lower level into the upper level. Data generated by the solution of the lower level problem, e.g., which stations have a too high/low capacity, could guide the search in the upper level into promising directions.

Another interesting direction for future work are free-floating systems in which the users can rent and return the cars anywhere within the operational area. When using electric cars in such a system, charging stations still have to be planned and the re-distribution may become even more important. The impact of using such a system on the trip acceptance rate, the user experience, and on the strategic planning is promising and relevant for future research.

References

1. Asamer, J., Reinthaler, M., Ruthmair, M., Straub, M., Puchinger, J.: Optimizing charging station locations for urban taxi providers. Transp. Res. Part A: Policy Pract. **85**, 233–246 (2016)
2. Boyacı, B., Zografos, K.G., Geroliminis, N.: An optimization framework for the development of efficient one-way car-sharing systems. Eur. J. Oper. Res. **240**(3), 718–733 (2015)
3. Brandstätter, G., Gambella, C., Leitner, M., Malaguti, E., Masini, F., Puchinger, J., Ruthmair, M., Vigo, D.: Overview of optimization problems in electric car-sharing system design and management. In: Dawid, H., Doerner, K.F., Feichtinger, G., Kort, P.M., Seidl, A. (eds.) Dynamic Perspectives on Managerial Decision Making. DMEEF, vol. 22, pp. 441–471. Springer, Heidelberg (2016). doi:10.1007/978-3-319-39120-5_24
4. Brandstätter, G., Kahr, M., Leitner, M.: Determining optimal locations for charging stations of electric car-sharing under stochastic demand (2016, submitted)
5. Brandstätter, G., Leitner, M., Ljubić, I.: Location of charging stations in electric car sharing systems (2016, submitted)

6. Cavadas, J., de Almeida Correia, G.H., Gouveia, J.: A MIP model for locating slowcharging stations for electric vehicles in urban areas accounting for driver tours. Transp. Res. Part E: Logist. Transp. Rev. **75**, 188–201 (2015)
7. Dijkstra, E.W.: A note on two problems in connexion with graphs. Numer. Math. **1**(1), 269–271 (1959)
8. Duin, C., Voß, S., et al.: The pilot method: a strategy for heuristic repetition with application to the steiner problem in graphs. Networks **34**(3), 181–191 (1999)
9. Frade, I., Ribeiro, A., Gonçalves, G., Antunes, A.: Optimal location of charging stations for electric vehicles in a neighborhood in Lisbon, Portugal. Transp. Res. Rec.: J. Transp. Res. Board **2252**, 91–98 (2011)
10. Garey, M.R., Johnson, D.S.: Computers and Intractability: A Guide to the Theory of NP-Completeness. W. H. Freeman & Co., New York (1979)
11. Ge, S., Feng, L., Liu, H.: The planning of electric vehicle charging station based on grid partition method. In: 2011 International Conference on Electrical and Control Engineering (ICECE), pp. 2726–2730. IEEE (2011)
12. Hess, A., Malandrino, F., Reinhardt, M.B., Casetti, C., Hummel, K.A., Barceló-Ordinas, J.M.: Optimal deployment of charging stations for electric vehicular networks. In: Proceedings of the First Workshop on Urban Networking, pp. 1–6. ACM (2012)
13. Mladenović, N., Hansen, P.: Variable neighborhood search. Comput. Oper. Res. **24**(11), 1097–1100 (1997)
14. Ow, P.S., Morton, T.E.: Filtered beam search in scheduling. Int. J. Prod. Res. **26**(1), 35–62 (1988)
15. Prandtstetter, M., Straub, M., Puchinger, J.: On the way to a multi-modal energy-efficient route. In: IECON 2013–39th Annual Conference of the IEEE Industrial Electronics Society, pp. 4779–4784. IEEE (2013)
16. Weikl, S., Bogenberger, K.: Relocation strategies and algorithms for free-floating car sharing systems. IEEE Intell. Transp. Syst. Mag. **5**(4), 100–111 (2013)

Selection of Auxiliary Objectives Using Landscape Features and Offline Learned Classifier

Anton Bassin[✉] and Arina Buzdalova

ITMO University, 49 Kronverksky Pr., Saint-Petersburg, Russia 197101
anton.bassin@gmail.com, abuzdalova@gmail.com

Abstract. In order to increase the performance of an evolutionary algorithm, additional auxiliary optimization objectives may be added. It is hard to predict which auxiliary objectives will be the most efficient at different stages of optimization. Thus, the problem of dynamic selection between auxiliary objectives appears. This paper proposes a new method for efficient selection of auxiliary objectives, which uses fitness landscape information and problem meta-features. An offline learned meta-classifier is used to dynamically predict the most efficient auxiliary objective during the main optimization run performed by an evolutionary algorithm. An empirical evaluation on two benchmark combinatorial optimization problems (Traveling Salesman and Job Shop Scheduling problems) shows that the proposed approach outperforms similar known methods of auxiliary objective selection.

Keywords: Evolutionary algorithms · Multi-objective optimization · Auxiliary objectives · Fitness landscape features

1 Introduction

Evolutionary algorithms (EAs) are generic meta-heuristic optimization algorithms. An EA searches solution candidates based on the current state whilst previously reached historical states are not taken into account during the runtime. The information derived from the fitness landscape and from the optimization problem instance may be used to determine the state of an evolutionary algorithm. In these terms, the optimization problem transforms into searching the EA states which correspond to global optima on the fitness landscape. Auxiliary objectives may be used instead of the target objective or along with the target objective. Auxiliary objectives serve to multi-objectivise a single objective problem. In some cases this transformation may increase efficiency of an EA.

An auxiliary objective is efficient if its usage leads to decrease of the time needed to find the optimum of the target objective. Different auxiliary objectives have different efficiency on various stages of optimization. For example, at the stagnation point of the EA the most aggressive auxiliary objective may move the optimization process away from getting stuck in a local optimum. Inversely,

B. Hu and M. López-Ibáñez (Eds.): EvoCOP 2017, LNCS 10197, pp. 173–188, 2017.
DOI: 10.1007/978-3-319-55453-2_12

if the current algorithm state corresponds to the situation where solution candidates are located near the global optimum, we would want to use less aggressive auxiliary objectives.

At present time, researchers are looking for new ways of parameterizing fitness landscape features and applying received techniques in the evolutionary computation field [1,2]. Authors of paper [3] suggested a method to multi-objectivise single objective problems by using an elementary landscape decomposition of their objective function. However, to the best of our knowledge, landscape features have never been used to guide dynamical selection of auxiliary objectives. The existing objective selection algorithms use random selection based on the number of iterations [4] or reinforcement learning based on differences in target objective values [5].

One of the first approaches to transform a single-objective problem into a multi-objective one was proposed by Knowles et al. in [6]. The authors suggest decomposing the target objective into several components, which should be independent. The decomposed objectives are optimized simultaneously. Another method belongs to Jensen [4]. The idea is to use auxiliary objectives in combination with the target objective. Furthermore, the auxiliary objectives used in paper [4] are changed dynamically. The author concluded from the obtained experimental results that using one auxiliary objective at a time is the best approach, but he also underlined questions on when to change the auxiliary objective and to which objective it should be changed. There also exists an adaptive auxiliary objective selection method based on reinforcement learning called EA+RL [5]. The main idea is to use the reinforcement learning to train online (during the EA runtime) an agent, which tries to predict the most efficient auxiliary objective on each evolutionary iteration. The aforementioned method was improved by Petrova et al. in [7].

We tested the efficiency of the method that we propose in the present paper on multiple instances of two benchmark problems: The Traveling Salesman Problem (TSP) and The Job Shop Scheduling Problem (JSSP). However, the proposed method is not designed for solving any specific optimization problem, it is a general approach for selection of the most efficient auxiliary objective during EA runtime. Therefore, we compared the proposed method with other approaches of objective selection. Additionally, to confirm the reliability of the obtained results we checked the corresponding values for statistical distinguishability.

The rest of the paper is organized as follows. In Sect. 2 discusses the main aspects of the proposed method of auxiliary objective selection. Section 3 presents experiment results of solving TSP. Section 4 presents the results for the JSSP, and we conclude in Sect. 5.

2   The OLHP Method

We propose a new auxiliary objective selection strategy named *The Offline Learned Helper Picker (OLHP)*. The term *Offline* is used because the meta-classifier for auxiliary objective selection is trained offline with machine learning methods. The learning dataset is gathered from the training EA runs on training instances of an optimization problem. The learning dataset vector contains properties of the problem instance and feature values of the fitness landscape of the current EA population. The term *Helper* is used as a synonym to "auxiliary objective". The OLHP consists of two main phases: the meta-classifier learning and objective selection during the EA runtime. The implementation of OLHP, along with experimental evaluation results, is available at Bitbucket repository[1].

2.1   Learning the Meta-Classifier Phase

The meta-classifier is learned only once for each optimization problem T. Input parameters of this phase are: **1.** A set of training problem instances $L \subset T$, L consists of any $\ell \in T$, where T is the set of all possible instances of a particular optimization problem. **2.** A set of rules for auxiliary objectives generation $G = \{g(i)\}, g(i) = h_i : \mathbb{N} \to H$, where H is the set of all possible auxiliary objectives for the considered optimization problem.

At the considered phase we construct the dataset of learning samples and then build the meta-classifier. To do this we need to perform n runs of the EA for each training problem instance. The value of the constant n may be manually tuned to find better meta-classifier metrics.

The current EA state is described by two components: static meta-features of the problem instance and features of the target objective landscape at the current population, which are extracted dynamically during the runtime. From each particular EA state st_j we make k runs ($k = |H_k|$ is the total number of used auxiliary objectives, $G(H) = H_k = \{h_i\}$ is a set of generated auxiliary objectives) in parallel for $n_{\text{iter}} = \frac{I_{\text{max}}}{k}$ iterations, where I_{max} corresponds to the maximum number of iterations in a training run of the EA. Thus, the maximum number of considered EA states is $j = k$. Accordingly, there is a chance for each auxiliary objective to be picked at each EA state. All parallel EA threads optimize different auxiliary objective function h_i simultaneously with the target objective. After performing latter evaluations we can make an assumption on which auxiliary objective h_i would be the most efficient if using it from the EA state st_j for n_{iter} EA iterations.

We identify an efficiency of an auxiliary objective by comparing the values of target objective for the best solution candidate in the beginning of each EA thread and in the end of its work. If several threads showed an equal increase in the target fitness value of the best solution, then the best auxiliary objective for the EA state st_j is selected from the first thread. Thereby, we have one learning dataset vector which is comprised of the meta-features of the problem instance and the fitness landscape features corresponding to the EA state st_j. The target value (or class value) of this vector is number i, which corresponds to the most efficient auxiliary objective h_i.

After performing the training evaluations from the EA state st_j, we move forward to a new state st_{j+1}. As the state st_{j+1} we pick an EA state after using the most efficient auxiliary objective h_j. The steps described above are processed until the maximum number of EA iterations I_{max} is reached. The final thing to

---

[1] https://bitbucket.org/BASSIN/2017-olhp-tsp-jssp/src.

do is to train and save the meta-classifier for selection of auxiliary objectives. Algorithm 1 presents the pseudocode of the meta-classifier learning phase.

Algorithm 1. Learning of the meta-classifier

1: **procedure** LEARNSELECTOR($Insts_{tr}, G$)   ▷ $Insts_{tr}$ is a set of training instances, G - set of auxiliary objective generating rules
2:    $r \leftarrow$ the number of runs for each training instance
3:    **for all** tr in $Insts_{tr}$ **do**
4:       $metaFeatures \leftarrow extractMetaFeatures(tr)$
5:       $H_k \leftarrow G(tr)$ ▷ Generate auxiliary objectives for train problem instance tr
6:       $i \leftarrow 0$
7:       **while** $i < r$ **do**
8:          $I_{max} \leftarrow$ calculateMaxIterationNumber(tr)
9:          $n_{iter} \leftarrow \frac{I_{max}}{|H_k|}$ ▷ Setting iterations number between algorithm states
10:         $EA.initialize(tr)$ ▷ Initialize evolutionary algorithm with train instance tr
11:         $j \leftarrow 0$
12:         **while** $j < I_{max}$ **do**
13:            $st_j \leftarrow saveState(EA, metaFeatures)$
14:            **for all** h in H_k **do**   ▷ For each auxiliary objective
15:               $EA.run(n_{iter}, h)$ ▷ Run EA with auxiliary objective h for n_{iter} iterations
16:               $fitnessRaise_h \leftarrow calculateFitnessDiff(EA)$
17:               $st_h \leftarrow saveState(EA)$
18:            **end for**
19:            $h_{best} \leftarrow findBestHelper(\forall fitnessRaise_h)$ ▷ Identify the most efficient auxiliary objective
20:            $dataset.put(h_{best}, st_j.getMetaFeatures(), st_j.getLandscapeFeatures())$
21:            $EA.setState(st_{h_{best}})$ ▷ Set EA state to the state after using the best auxiliary objective
22:            $j \leftarrow j + n_{iter}$
23:         **end while**
24:         $i \leftarrow i + 1$
25:      **end while**
26:   **end for**
27:   $classifier.train(dataset)$ ▷ Learn the meta-classifier for objective selection
28:   $classifier.serialize()$ ▷ Serialize trained model for future usages
29: **end procedure**

2.2   Objective Selection During the EA Runtime

This subsection describes the algorithm that dynamically selects and applies auxiliary objectives during EA runtime. The input parameters for this $OLHP$ phase are: **1.** An instance of the optimization problem we want to solve $\ell \in T$, where T is a space of all possible instances of a particular optimization problem; **2.** A set of rules for generation of auxiliary objectives $G = \{g(i)\}, g(i) = h_i$:

$\mathbb{N} \to H$, where H is the set of all possible auxiliary objective functions we are working with. **3.** A meta-classifier learned beforehand on problem instances from T: $predict(v) = k, g(k) = h_{best}$, where v is the vector of fitness landscape features of the current population and static meta-features of the problem instance.

The first step of this phase is initialization of all required structures and deserialization of the meta-classifier for considered optimization problem. At the next step, we start an evolutionary algorithm. Each $n_{iter} = \frac{I_{max}}{k}$ ($k = |H_k| = |G(H)|$) iteration period we predict the most efficient auxiliary objective basing on the problem instance static features and fitness landscape features of the current EA population. The predicted objective h_{best} is used by the EA simultaneously with the target objective for the next n_{iter} iterations. Restriction on optimization by only one additional objective is made for making it possible to learn and solve many problem instances of various sizes for a reasonable computational time. Algorithm 2 presents the detailed pseudocode for this OLHP phase.

Algorithm 2 Solving the problem instance

1: **procedure** SOLVEPROBLEM(inst, G, cl)   ▷ inst is a problem instance to optimize, cl – learned classifier
2:     $cl.deserialize()$
3:     $metaFeatures \leftarrow extractMetaFeatures(\text{inst})$
4:     $H_k \leftarrow G(inst)$ ▷ Generate auxiliary objectives for the problem instance
5:     $I_{max} \leftarrow \text{calculateMaxIterationNumber}(\text{inst})$
6:     $n_{iter} \leftarrow \frac{I_{max}}{|H_k|}$ ▷ Setting iterations number between algorithm states
7:     $EA.initialize(\text{inst})$ ▷ Initialize EA with optimization problem instance
8:     $j \leftarrow 0$
9:     **while** $j < I_{max}$ **do**
10:       **if** $j \bmod n_{iter} = 0$ **then**   ▷ Time for switching an auxiliary objective
11:          $st_j \leftarrow getState(EA, metaFeatures, getLandscapeFeatures(EA))$
12:          $h_{predicted} \leftarrow cl.predict(st_j)$
13:          $EA.setHelper(h_{predicted})$
14:       **end if**
15:       $EA.runIteration()$ ▷ Make one iteration of EA
16:       $j \leftarrow j + 1$
17:     **end while**
18:     $EA.saveBestSolution()$ ▷ Save the optimization result
19: **end procedure**

2.3 Fitness Landscape Features

We use generic fitness landscape features of an EA population which are eligible for almost any optimization problem. A population of size p may be represented as a set of random variables $P = \{s_i\}, i = 1..p$. On each EA iteration we can obtain values of random variables s_i. Hence, we can calculate statistical metrics of set P.

Since we are solving meta-classification task for any instance of optimization problem T, we need to normalize the values of received landscape features. For such a normalization, we propose to use the ratio of the target objective value on current solution candidate to the best known solution candidate value: $\frac{G(s_i)}{G(s_{\text{best}})} = x_i$, where $G(s)$ is the target objective function. The set of random variables $\{x_i\} = P_{\text{normalized}}$ is normalized for all instances of problem T. Thus, we suggest to use the following statistical metrics calculated on the set $P_{\text{normalized}}$ as fitness landscape features:

1. Med – the median value.
2. $\bar{x} = \frac{1}{p} \sum_{i=1}^{p} x_i$ – the arithmetical mean.
3. $H_{\text{mean}} = \frac{p}{\sum_{i=1}^{p} \frac{1}{x_i}}, x_i > 0$ – the harmonic mean.
4. Dev – the standard deviation.
5. $Q_{\text{mean}} = \sqrt{\frac{1}{p} \sum_{i=1}^{p} (x_i - \bar{x})^2}$ – the sample variance.

3   Applying OLHP to Traveling Salesman Problem

The Traveling Salesman Problem (TSP) is a classic NP-hard problem in combinatorial optimization. Each TSP instance may be described by a set of cities $\{c_i\}, \forall i = 1 \ldots N$ and a distance matrix M of size $N \times N$. Elements of M represent the distance between a pair of cities. For example, $M(c_i, c_j)$ is the distance between the cities c_i and c_j. The TSP asks the following question: "What is the shortest possible route that visits each city once and returns to the origin city?". In other words, we need to find a Hamiltonian path with the lowest total distance. For the path vector $\rho = (\rho_1, \rho_2, \ldots, \rho_N)$ we can calculate the total distance cost using (1):

$$D(\rho) = \sum_{i=1}^{N} M(c_{\rho_{[i]}}, c_{\rho_{[i \oplus 1]}}),$$

$$\text{where } i \oplus 1 = \begin{cases} i+1, & \text{if } i < N \\ 1, & \text{if } i = N \end{cases}$$

(1)

In experimental evaluations, we use symmetric TSP problem instances. In the symmetric TSP problem, the value of a path from one city to another is equal to the value of the reverse path: $M(c_i, c_j) = M(c_j, c_i)$. More detailed explanation of the TSP may be found in [8].

3.1   TSP Meta-Features

We need to specify TSP meta-features which we would use as a part of machine learning vector. Meta-features should contain information, which represents the properties of a particular TSP instance. The following features meet this requirement:

1. V_{num} – the number of cities.
2. E_{min} – the minimum distance between a pair of cities.
3. E_{max} – the maximum distance.
4. E_{avg} – the average distance.
5. E_{med} – the median distance.
6. Dev_E – the standard deviation of distances.
7. QE_{-avg} – the number of distances shorter than E_{avg}.
8. Sum_{min_E} – the sum of V_{num} minimal distances.

The latter TSP meta-features were successfully used by Kanda et al. in [9] to classify Traveling Salesman Problems and in [10] to recommend meta-heuristics for solving TSP.

3.2   TSP Auxiliary Objectives Generation

In the OLHP method, an auxiliary objective is required to have some property, which depends only on the problem instance, but not on the individual or the iteration number. Unfortunately, the existing approaches of auxiliary objective generation [4,6] do not provide us objectives with such kind of a property, because in these approaches objectives are generated using randomly picked cities.

To generate auxiliary objectives which depend only on the problem instance, we propose a new method of auxiliary objective generation inspired by the k-nearest neighbor classification algorithm [11].

In [4], the following auxiliary objective function was proposed:

$$h(\rho, s) = \sum_{i=1}^{|s|} (M(c_{\rho[\rho^{-1}[s[i]\ominus 1]}, c_{s[i]}) + M(c_{s[i]}, c_{[\rho^{-1}[s[i]]\oplus 1]}), \tag{2}$$

where s is the subset of the set of cities $C = \{1, 2, \ldots, N\}$, $\rho^{-1}(x)$ is the position of x in ρ, $\ominus 1$ is the reverse operator to $\oplus 1$.

In our approach, we use Eq. (2) to generate auxiliary objectives. We generate subsets of cities (the s parameter in (2)) by partitioning the set of cities $C = \{c_i\}, i = 1 \ldots N$ using the following algorithm:

1. Sort the set of all cities C by the following criteria:

$$Knn(C) \rightarrow S_C = \{c_1, c_2, \ldots, c_N\}, \tag{3}$$

where $Knn(C)$ is the sorting operator, S_C – is the ordered set.
For each pair of elements from S_C:

$$c_i \preceq_{knn} c_j, \tag{4}$$

where $i, j = 1 \ldots N$ and the relation $\preceq_{knn}$ is true when the total distance from city c_i to k nearest neighbor cities is less or equal to the corresponding total distance for the city c_j.

2. Divide the ordered set S_C into subsets $s_j \subset S_C, j = 1 \ldots r$ of equal cardinality. The number of elements in the last subset s_r may be less than the number of elements in the other subsets.

Note that $s_i \cap s_j = \emptyset$, $\forall i, j = 1 \ldots r, i \neq j$. This fact means that the generated auxiliary objectives $h(\rho, s_j)$ have disjoint properties on a problem instance. This should make possible for each auxiliary objective to be the most efficient objective at different stages of optimization process.

3.3   Experimental Evaluation on TSP

We compare the OLHP method with the following approaches of optimizing the target objective with auxiliary objectives. First, we consider the method proposed by Jensen [4], where the auxiliary objective is dynamically reselected after a fixed number of EA iterations. In the second considered approach [7], named Multi-Objective Evolutionary Algorithm + Reinforcement Learning (MOEA+RL), RL agent learns to select auxiliary objectives during the EA runtime and non-stationarity of the environment is taken into account. The last considered approach is a modified combination of two known algorithms. Two auxiliary objectives are composed in the manner described by Jähne et al. in [12] and the first selected auxiliary objective is switched to another one at the half of the EA runtime as suggested in [4].

The TSP instances for the experimental evaluation were taken from TSPLIB[2] library. The crossover and mutation operators were identical to the corresponding operators from the papers mentioned above. Also, the *2-opt* local search heuristic [13] was used.

The crossover probability was equal to 40%. The population size was $P_{\text{size}} = 100$. The limit of target objective evaluations for each EA run was calculated in the manner proposed in [12]: $\text{ev}_{\max}(N, m) = \sqrt{N^3} * m$, where N is the total number of cities, m is a manually chosen parameter (in our experiments we used $m = 10$). The number of EA runs for each training problem instance was $n = 4$.

The auxiliary objectives for the methods proposed by Jensen and Jähne were generated using the rules, which are provided in the related papers. The results for both the OLHP and MOEA+RL algorithms were obtained using the same auxiliary objectives, namely the *K-nn* auxiliary objectives proposed in Sect. 3.2. We used the following parameters to generate auxiliary objectives: $k = 5$, $r = 5$.

The comparison with MOEA+RL was intended to evaluate the efficiency of the proposed objective selection scheme without the influence of the objective generation approach, as the both OLHP and MOEA+RL algorithms are in equal conditions in terms of the used auxiliary objectives. At the same time, in the Jensen and Jähne/Jensen methods, the original auxiliary objectives from the corresponding works were applied, so the comparison with these methods was performed to evaluate the efficiency of the entire proposed approach for the TSP optimization.

---

[2] http://comopt.ifi.uni-heidelberg.de/software/TSPLIB95.

The Non-dominated Sorting Genetic Algorithm II (NSGA-II) [14] was used as the base evolutionary algorithm in all experiments. The program code was written in Java and Groovy languages. The following frameworks were used: Watchmaker[3] – for evolutionary computations, Weka[4] – for the OLHP machine learning operations.

Learning of Meta-Classifier. The Random Forest classifier was used for building the OLHP auxiliary objective selector. The Random Forest parameters had default[5] values from the Weka framework.

To obtain train and test sets, TSP instances from the TSPLIB were divided into subsets. We were guided by the idea that train and test instances should have items with similar meta-properties. The train TSP instances are listed in Appendix A.

To estimate how accurately our predictive model would perform in practice, we cross-validated our classifier. For the same limitation on similarity of train and test problem instances we were forced to use a special set of problems for cross-validation. We made 10-fold cross-validation on instances listed in Appendix A. The performance metric values of our classification model on TSP after the cross-validation were: Estimated Error Rate = 0.16, Precision = 0.90, Recall = 0.96, F-measure = 0.93. The latter values confirm that there exists a correlation between the EA state features and selection of the most efficient auxiliary objective.

Results of Solving TSP. We used 44 TSP instances to perform experiments. Further, the final solution results for each method of auxiliary objective selection were obtained $\eta = 40$ times and averaged.

Table 1 shows mean and standard deviation of the best obtained value of the target objective for the OLHP, MOEA+RL, Jensen and Jensen/Jähne methods. Cells with the best values are marked with bold text. The last row of Table 1 shows the total number of instances, on which the particular method has outperformed other approaches. Note that the sum of the values in the last row is not equal to the number of total considered test instances. It is explained by the situation when several methods showed the best mean target value. In this case, we increment the corresponding counters for each such method. To summarize, it can be concluded from Table 1 that the newly proposed OLHP method outperformed the considered approaches of auxiliary objective selection on the set of test TSP instances.

Statistical Testing. According to [15], we used the Wilcoxon signed-rank test to detect significant differences in behavior of two algorithms. The pairwise statistical test was applied on the average results obtained on test instances for each pair of the considered approaches. In order to perform multiple comparisons and

---

[3] http://watchmaker.uncommons.org.

[4] http://www.cs.waikato.ac.nz/ml/weka.

[5] http://weka.sourceforge.net/doc.dev/weka/classifiers/trees/RandomForest.html.

Table 1. Mean and standard deviation of resulting fitness (TSP)

Problem	Best	OLHP	MOEA+RL	Jensen	Jen/Jäh
a280	2579	2597.9 ± 13.1	2599.4 ± 13.4	2597.2 ± 16.3	**2597.2 ± 11.8**
ali535	202339	204624.4 ± 1587.0	**204382.0 ± 1482.5**	205909.5 ± 1505.1	204853.0 ± 1463.7
att48	10628	**10628.5 ± 2.2**	10629.1 ± 4.9	10631.4 ± 6.5	10645.1 ± 17.7
bier127	118282	118337.3 ± 116.0	118436.2 ± 259.4	**118320.6 ± 83.0**	118358.1 ± 187.9
brg180	1950	1952.8 ± 4.5	1952.5 ± 4.3	**1950.0 ± 0.0**	**1950.0 ± 0.0**
ch150	6528	6548.0 ± 12.4	6550.8 ± 12.7	**6544.4 ± 12.6**	6547.6 ± 12.9
d1291	50801	51585.9 ± 210.2	51594.4 ± 240.7	51592.7 ± 191.1	**51560.5 ± 270.5**
d657	48912	49292.3 ± 151.9	49353.6 ± 141.5	**49285.6 ± 127.6**	49358.0 ± 131.0
dsj1000	18659688	**18860887.9 ±45886.0**	18870762.9 ±50593.3	18861694.6 ±60582.9	18887428.3 ±61706.0
eil76	538	**544.4 ± 0.0**	544.5 ± 0.5	544.7 ± 0.9	544.4 ± 0.1
fl417	11861	11927.0 ± 5.2	**11926.8 ± 6.0**	11942.1 ± 13.9	11939.7 ± 22.4
gr137	69853	**69862.6 ± 28.5**	69878.8 ± 57.9	69878.8 ± 44.4	69864.0 ± 29.3
gr229	134602	135130.0 ± 400.5	135087.3 ± 346.0	**135048.9 ± 346.7**	135051.8 ± 267.1
gr666	294358	297940.3 ± 1536.3	298189.9 ± 1454.3	**297663.3 ± 1336.9**	297826.6 ± 1426.5
gr96	55209	**55305.2 ± 58.1**	55312.6 ± 77.0	55352.1 ± 60.6	55326.9 ± 70.2
kroA100	21282	21287.0 ± 9.5	21287.0 ± 10.0	21287.0 ± 10.0	**21285.4 ± 0.0**
kroA200	29368	**29393.5 ± 59.2**	29398.2 ± 67.6	29426.1 ± 111.9	29395.7 ± 62.3
kroB100	22141	**22140.9 ± 11.1**	22148.2 ± 28.7	22141.3 ± 13.3	22145.9 ± 21.6
kroB150	26130	**26148.2 ± 49.5**	26190.0 ± 64.9	26162.5 ± 60.7	26149.8 ± 51.1
kroC100	20749	**20750.8 ± 0.0**	**20750.8 ± 0.0**	**20750.8 ± 0.0**	20750.8 ± 0.5
kroD100	21294	**21311.2 ± 30.1**	21333.5 ± 45.0	21374.0 ± 31.8	21314.5 ± 33.2
lin105	14379	**14383.0 ± 0.0**	**14383.0 ± 0.0**	**14383.0 ± 0.0**	**14383.0 ± 0.0**
lin318	42029	42321.6 ± 175.3	42323.7 ± 180.6	42314.6 ± 154.1	**42297.7 ± 167.8**
pcb1173	56892	57909.2 ± 166.2	57982.5 ± 250.3	**57885.9 ± 199.6**	57909.2 ± 201.1
pcb442	50778	**51263.1 ± 132.1**	51305.7 ± 199.0	51312.8 ± 202.1	51307.3 ± 167.3
pr1002	259045	**262779.5 ± 868.7**	263100.1 ± 1115.0	263427.5 ± 786.9	263178.0 ± 942.0
pr107	44303	**44328.9 ± 39.9**	44337.7 ± 50.8	44372.1 ± 74.5	44336.6 ± 50.0
pr124	59030	**59030.7 ± 0.0**	**59030.7 ± 0.0**	59032.9 ± 9.6	**59030.7 ± 0.0**
pr144	58537	**58535.2 ± 0.0**	58536.1 ± 5.2	58562.5 ± 14.8	58565.3 ± 19.9
pr152	73682	73697.4 ± 41.3	73711.2 ± 55.0	73783.6 ± 61.2	**73691.6 ± 49.9**
pr226	80369	80374.6 ± 12.4	**80374.4 ± 12.4**	80382.9 ± 38.8	80385.8 ± 29.7
pr299	48191	**48320.1 ± 111.8**	48370.7 ± 165.5	48434.7 ± 210.8	48398.6 ± 134.6
pr439	107217	107600.7 ± 424.3	**107597.9 ± 354.4**	107666.0 ± 498.6	107875.6 ± 510.4
rat195	2323	**2340.6 ± 5.5**	2344.9 ± 6.3	2343.1 ± 5.2	2342.9 ± 6.1
rat783	8806	**8932.3 ± 23.8**	8949.3 ± 25.4	8946.2 ± 25.5	8941.6 ± 22.8
rd100	7910	7912.3 ± 8.3	**7911.9 ± 5.4**	7914.7 ± 12.5	7914.7 ± 11.4
rd400	15281	15394.0 ± 54.6	15428.7 ± 58.8	15386.8 ± 54.3	**15377.9 ± 55.8**
si1032	92650	**92650.0 ± 0.2**	92656.6 ± 19.8	92720.3 ± 43.9	92673.4 ± 22.2
si535	48450	48496.4 ± 17.4	48496.2 ± 22.6	48543.2 ± 33.1	**48487.8 ± 16.2**
tsp225	3916	3876.3 ± 21.1	3877.8 ± 21.1	3873.1 ± 21.5	**3869.2 ± 16.8**
u1060	224094	226731.1 ± 671.0	**226627.1 ± 721.0**	226825.6 ± 690.4	226962.5 ± 687.4
u159	42080	**42075.7 ± 0.0**	**42075.7 ± 0.0**	**42075.7 ± 0.0**	**42075.7 ± 0.0**
u724	41910	42298.9 ± 137.3	42337.1 ± 119.3	**42247.3 ± 96.4**	42280.5 ± 121.7
vm1084	239297	241691.7 ± 962.0	241953.6 ± 898.0	**241432.7 ± 704.0**	241822.2 ± 770.0
Total best		**21**	10	12	12

control the family-wise error rate we adjusted the obtained p-values by using the Holm–Bonferroni correction method.

The *adjusted p-values* for pairs of methods of auxiliary objective selection were the following: OLHP – MOEA+RL = 1.0e-03, OLHP – Jensen = 4.2e-02, OLHP – Jensen/Jähne = 4.2e-02. Therefore, the OLHP method shows a significant improvement over MOEA+RL, Jensen and Jensen/Jähne approaches, with the level of significance $\alpha = 0.05$.

4   Applying OLHP to Job Shop Scheduling Problem

The OLHP method was applied to the Job Shop Scheduling Problem (JSSP) for further verification of its efficiency. Similarly to TSP, JSSP is a well-known NP-hard combinatorial optimization problem.

A JSSP instance of size $n \times m$ consists of n jobs $\{J_1, J_2, \ldots, J_n\} = J$ and m machines $\{M_1, M_2, \ldots, M_m\} = M$. Each job J_i contains a sequence of m operations $(o_{i1}, o_{i2}, \ldots, o_{im})$. Jobs and machines have mutual constraints, because an operation o_{ij} may be processed only on the corresponding machine M_j. Each operation o_{ij} takes the corresponding processing time $\tau_{ij} \in \mathbb{N}$. All jobs from the set J need to be scheduled properly on the given machines, while trying to minimize the amount of spent time resources. We consider the following JSSP variation: each machine may process only one operation at the same time, operations related to one job can not be processed concurrently and processing of an operation can not be interrupted.

There are several types of target objective which can be used in evolutionary computations for the JSSP. We minimize the total flow-time of the schedule S:

$$F(S) = \sum_{i=1}^{n} (S(o_{\max_i}) + \tau_{o_{\max_i}}), \tag{5}$$

where $o_{\max_i}$ is the operation of the job J_i with the maximum start time in the schedule S, $S(o_{\max_i})$ defines the start time value of operation $o_{\max_i}$ and $\tau_{o_{\max_i}}$ is the processing time of operation $o_{\max_i}$.

4.1   JSSP Meta-Features

The JSSP meta-features were developed similarly to the TSP meta-features from Sect. 3.1. The following features present the JSSP instance properties:

1. M_{num} – the number of machines.
2. J_{num} – the number of jobs.
3. $MJ_{ratio} = \frac{M_{num}}{J_{num}}$ – the ratio between the number of machines and the number of jobs.
4. τ_{min} – the minimum operation processing time.
5. τ_{max} – the maximum operation processing time.
6. τ_{mean} – the mean operation processing time.
7. Dev_τ – the standard deviation of the operation processing time.
8. $\tau_{avg_M} = \frac{\sum_{j=1}^{m} \frac{\sum_{i=1}^{n} o_{ij}}{n}}{m}$ – the average processing time, which is also averaged per machine.

4.2   JSSP Auxiliary Objectives Generation

The restriction mentioned in the Sect. 3.2 (an auxiliary objective should have some property defined by the problem instance) is reached by the *The Shortest*

Job First (SJF) auxiliary objective generating method proposed by Lochtefeld et al. in [16].

Each particular job J_i has a minimum processing time, which can be calculated by the following equation: $F_{\min}(J_i) = \sum_{j=1}^{m} o_{ij}$. The next step is to define the subset of jobs H_k, which would be used in the k-th auxiliary objective. Then, the Eq. (5) evaluates the value of each auxiliary objective. The following algorithm is used to determine a subset of jobs H_k of the particular auxiliary objective:

1. The minimum processing times of all jobs $F_{\min_i}, i = 1 \ldots n$ are calculated.
2. The set of all jobs J is sorted with respect to the minimum processing time:

$$\text{Sort}(J) \rightarrow S_J = J_1, J_2, \ldots, J_n, \text{where} F_{\min}(J_i) \le F_{\min}(J_j). \tag{6}$$

3. The sorted set of all jobs S_J is divided into r subsets with equal number of elements. Such a subset defines the auxiliary objective.

More details about the SJF auxiliary objectives can be found at [17].

The subsets H_k for the objectives can also be formed in a random way. Such a technique is used by Jensen in [4]. We test performance of the random composed subsets of jobs and the subsets generated by the SJF algorithm.

4.3   Experimental Evaluation on JSSP

In [4], Jensen also suggests using random auxiliary objectives for the JSSP problem. In [18] Petrova et al. apply the MOEA + RL method to this problem. We also consider the problem specific approach proposed by Lochtefeld et al. [16], based on job prioritization for auxiliary objective selection order. The OLHP method was compared with the aforementioned algorithms.

The JSSP instances were taken from the *Beasley's OR Library*[6]. The Generalized Order Crossover (GOX) [19] and the Position Based Mutation (PBM) were used in all the considered algorithms. EA solution candidates were represented as an ordered permutation list of different operations with repetitions. For example, an individual for a 2×3 problem may be encoded as $(1, 2, 2, 1, 1, 2)$, where the first "1" is the first operation of the job J_1, the second "1" is the second operation of J_1 and so forth. Furthermore, the Giffler-Thompson schedule builder [20] is used to transform genome to the proper solution candidate.

The crossover probability was set to 80%. The population size was $P_{\text{size}} = 100$. The stopping condition was whether an EA reached the limit of iterations. This limit was calculated as follows: $\text{iter}_{\max}(N, M) = N * M * 2$, where N was the total number of machines, M was the total number of jobs in the problem instance. Likewise in the TSP experiments, we used NSGAII algorithm. The number of EA runs for each training instance was $n = 4$. For the OLHP, Lochtefeld's, MOEA+RL methods we used 4 different SJF auxiliary objectives (remember that only one auxiliary objective was optimized simultaneously with

---

[6] http://people.brunel.ac.uk/~mastjjb/jeb/.

Table 2. Mean and standard deviation of resulting fitness (JSSP)

Problem	Best	OLHP	MOEA+RL	Lochtefeld	Jensen
abz6	7808	**8180.2 ± 143.8**	8261.3 ± 137.6	8213.2 ± 134.3	8244.6 ± 125.9
abz7	12561	**13116.1 ± 167.3**	13272.3 ± 180.7	13193.5 ± 188.9	13150.7 ± 179.7
abz9	12813	**13441.6 ± 187.7**	13574.6 ± 172.9	13475.9 ± 191.0	13463.0 ± 216.9
ft06	265	272.7 ± 7.0	272.0 ± 7.1	272.2 ± 7.0	**270.4 ± 6.8**
ft20	14279	**16047.1 ± 516.4**	16892.9 ± 542.8	16370.6 ± 499.5	16426.7 ± 590.5
la01	4832	**4989.6 ± 81.7**	5043.2 ± 96.6	5015.2 ± 87.6	5003.3 ± 71.3
la03	4175	**4236.4 ± 68.2**	4328.6 ± 70.5	4271.8 ± 64.6	4280.6 ± 73.9
la05	4094	**4189.0 ± 59.5**	4270.0 ± 61.6	4215.3 ± 63.5	4263.4 ± 66.1
la06	8694	**9298.3 ± 162.3**	9693.7 ± 201.4	9442.8 ± 215.1	9435.1 ± 189.1
la08	8176	**8708.9 ± 202.6**	9175.4 ± 194.6	8929.1 ± 221.8	8947.6 ± 229.2
la09	9452	**9777.7 ± 140.6**	10166.5 ± 170.2	9932.6 ± 170.4	9952.9 ± 211.5
la10	9230	**9557.7 ± 161.7**	10015.0 ± 214.6	9698.0 ± 176.8	9762.9 ± 189.8
la11	14801	**15840.5 ± 321.2**	16786.1 ± 367.3	16197.8 ± 339.1	16253.5 ± 434.1
la12	12484	**13449.3 ± 263.5**	14319.7 ± 301.4	13783.4 ± 339.3	13928.9 ± 377.5
la14	15595	**16199.8 ± 268.9**	17119.5 ± 300.9	16568.6 ± 338.3	16624.7 ± 341.3
la16	7393	**7836.2 ± 136.7**	7984.5 ± 151.4	7893.9 ± 158.1	7872.4 ± 145.4
la17	6555	**6847.5 ± 99.5**	6907.5 ± 96.3	6850.9 ± 88.1	6866.2 ± 99.5
la20	7427	**7751.9 ± 117.6**	7832.4 ± 144.0	7794.4 ± 124.8	7773.9 ± 119.5
la21	12953	**13940.4 ± 203.1**	14272.6 ± 200.6	14083.7 ± 253.4	14068.1 ± 219.9
la22	12106	13120.0 ± 221.2	13251.0 ± 213.5	**13094.6 ± 208.3**	13132.4 ± 223.9
la25	12465	**13154.3 ± 222.3**	13445.0 ± 243.1	13344.7 ± 260.6	13246.0 ± 214.8
la26	20234	**22351.7 ± 309.0**	22823.0 ± 312.0	22467.0 ± 272.3	22449.4 ± 311.8
la29	20404	**21498.7 ± 394.6**	22047.7 ± 386.5	21707.8 ± 383.1	21733.0 ± 392.2
la30	22333	**23725.8 ± 472.4**	24323.1 ± 382.8	24026.2 ± 471.2	23948.2 ± 419.7
la31	39007	**44400.9 ± 619.9**	45201.8 ± 541.6	44548.8 ± 580.2	44524.4 ± 614.9
la35	44059	**46014.8 ± 638.6**	46843.0 ± 633.2	46382.7 ± 679.1	46161.3 ± 784.9
la36	17073	**18461.3 ± 289.8**	18565.4 ± 244.1	18485.6 ± 272.2	18484.2 ± 270.3
la38	16621	**17346.9 ± 290.6**	17595.4 ± 280.3	17474.5 ± 280.0	17439.8 ± 282.9
la40	16618	**17667.4 ± 264.3**	17894.0 ± 270.7	17771.0 ± 267.2	17726.0 ± 274.6
orb02	7353	**7684.6 ± 123.8**	7753.5 ± 118.9	7739.5 ± 125.5	7708.2 ± 125.8
orb03	8280	**8772.8 ± 170.3**	8895.3 ± 194.0	8774.7 ± 189.7	8784.9 ± 235.5
orb06	8418	**8950.3 ± 205.2**	9113.8 ± 221.3	8979.7 ± 202.7	8950.6 ± 203.8
orb07	3296	**3505.6 ± 61.5**	3551.1 ± 74.9	3523.7 ± 63.2	3510.5 ± 74.1
orb09	7582	**8025.0 ± 180.5**	8231.8 ± 233.7	8062.8 ± 162.9	8149.8 ± 198.5
orb10	8043	**8335.7 ± 125.8**	8419.0 ± 146.9	8358.3 ± 147.4	8367.5 ± 134.5
swv01	20688	**24859.5 ± 711.2**	25710.4 ± 649.8	25441.2 ± 630.2	25461.3 ± 709.7
swv03	23266	**24617.6 ± 623.1**	25547.2 ± 633.0	25023.4 ± 668.7	25150.6 ± 652.5
swv04	24271	**25665.5 ± 574.4**	26457.5 ± 543.8	25978.6 ± 642.5	25957.5 ± 660.7
swv07	27385	**32738.5 ± 710.4**	33407.3 ± 652.0	32744.2 ± 711.8	32843.7 ± 755.0
swv08	32976	**36043.0 ± 775.3**	36655.9 ± 765.7	36136.3 ± 852.7	36265.8 ± 724.1
swv09	31841	**33783.7 ± 696.4**	34350.2 ± 744.4	34037.0 ± 855.6	33920.3 ± 788.5
swv11	108842	**140735.9 ± 2939.1**	145240.3 ± 3350.8	142351.2 ± 3360.4	141638.6 ± 3386.9
swv12	109128	**140695.3 ± 3173.4**	145495.7 ± 3093.0	142674.0 ± 2961.3	141281.9 ± 3247.0
swv14	126333	**137137.8 ± 3101.6**	141119.4 ± 3524.9	137850.9 ± 2779.5	137362.5 ± 3225.5
swv15	131037	**139467.0 ± 3991.3**	143849.0 ± 3676.6	140435.0 ± 3224.6	140211.6 ± 3476.1
swv16	113398	**117369.9 ± 1303.0**	119494.0 ± 1008.8	117937.3 ± 1045.1	117466.7 ± 1194.2
swv17	110145	**113689.1 ± 1020.0**	115666.0 ± 1042.6	114240.4 ± 981.0	113760.8 ± 1444.7
swv20	109742	**112866.7 ± 1060.6**	115285.1 ± 977.5	113277.0 ± 1038.2	113184.2 ± 1043.8
yn1	17317	**18199.5 ± 234.9**	18397.8 ± 212.6	18236.6 ± 210.0	18229.1 ± 227.3
yn4	19107	**19916.3 ± 253.7**	20115.0 ± 220.2	19997.4 ± 244.7	19959.7 ± 222.8
Total best		48	0	1	1

the target objective). Jensen's method was compared with the approach based on random generating of auxiliary objectives. We used the same OLHP implementation and the same frameworks as for the TSP.

Meta-Classifier Learning. The meta-classifier for selection of auxiliary objectives was trained on 32 JSSP instances (see Appendix A). The considered training instances were not so diverse as the training instances for the TSP. So we could cross-validate the learned classifier model on the training set of JSSP instances with high probability not to have a situation, when we would try to validate on some data, which was not considered in the learning process. The performance measure values of our classification model on JSSP after the 10-fold cross-validation were: Estimated Error Rate = 0.21, Precision = 0.72, Recall = 0.74, F-measure = 0.73. The correlation between the EA state features and selection of the most efficient auxiliary objective also exists for this benchmark problem.

Results of Solving the JSSP. There were 50 various JSSP test instances. Each instance was solved $\eta = 100$ times with NSGAII and each auxiliary objective selection method. We present comparison results in the same way as for the TSP problem. The averaged results for each method of objective selection with the standard deviations are listed in Table 2. The calculated comparison results show that the OLHP method has significantly outperformed all the other methods of auxiliary objective selection. It is also worth mentioning that we were not able to find sources with the best known solutions for the *Beasley's OR Library* problems. So we run a single objective EA 1000 times on each instance and gathered the best found results.

Statistical Testing. As in the TSP problem experiments, we used the Wilcoxon signed-rank test and the Holm-Bonferroni correction method for statistical verification of the results. We obtained the following adjusted p-values for pairs of objective selection methods: OLHP – MOEA+RL = 2.3e-09, OLHP – Lochtefeld = 2.3e-09, OLHP – Jensen = 2.3e-09. The aforementioned adjusted p-values show that average performance of the OLHP method and the other considered algorithms was significantly different. Moreover, the confidence level of this fact is more than 99%.

5   Conclusion and Future Work

The new method for selection of the most efficient auxiliary objective named *The Offline Learned Helper Picker* is proposed in this paper. The OLHP approach consists of two stages. At first, training instances of an optimization problem are used to build a meta-classifier for selection of auxiliary objectives. Properties of a problem instance and features derived from the fitness landscape of the current EA population compose the state of the evolutionary algorithm, i.e. the data vector for machine learning. Further, the trained meta-classifier is used to predict the most efficient auxiliary objective at different EA runtime points. Specifically, the selected auxiliary objective is predicted and optimized simultaneously with the target objective during a number of EA iterations.

The OLHP method was compared with similar approaches of objective selection on two NP-hard combinatorial problems: The Traveling Salesman Problem and The Job Shop Scheduling Problem. The newly proposed method outperformed the considered algorithms. Statistical significance of the obtained results was confirmed by the Wilcoxon signed-rank test followed by the Holm-Bonferroni correction.

In the future work we plan to use additional well-known statistical, probabilistic and informational measures for fitness landscapes. This should increase performance of the meta-classifier used for selection of auxiliary objectives. It is also desirable to use more computational power for obtaining experimental evaluation on real world problems. Better computational performance will also provide an opportunity to automatically find and use the most efficient parameters for EA with OLHP and other auxiliary objective selection methods. For example, the most efficient algorithm settings may be found with tools such as the irace package [21].

A   Appendix: TSP and JSSP Instances Lists

TSP Train: att532, bays29, brazil58, ch130, d198, d493, eil101, gil262, gr120, gr202, gr24, gr431, hk48, kroA150, kroB200, kroE100, p654, pa561, pr136, pr264, rat575, rat99, si175, st70, ts225, u574, ulysses22. **TSP Cross-validate:** a280, att48, bayg29, bays29, berlin52, bier127, brazil58, brg180, burma14, ch130, ch150, d198, d493, dantzig42, eil101, eil51, eil76, fl417, fri26, gil262, gr17, gr21, gr24, gr48, gr96, gr120, gr137, gr202, gr229, gr431, hk48, kroA100, kroA150, kroA200, kroB100, kroB150, kroB200, kroC100, kroD100, kroE100, lin105, lin318, pcb442, pr76, pr107, pr124, pr136, pr144, pr152, pr226, pr264, pr299, pr439, rat195, rat99, rd100, rd400, si175, st70, swiss42, ts225, tsp225, u159, ulysses16, ulysses22.

JSSP Train and Cross-validate: abz5, abz8, ft10, la02, la04, la07, la13, la15, la18, la19, la23, la24, la27, la28, la32, la33, la34, la37, la39, orb01, orb04, orb05, orb08, swv02, swv05, swv06, swv10, swv13, swv18, swv19, yn2, yn3.

References

1. Pitzer, E., Affenzeller, M.: A comprehensive survey on fitness landscape analysis. In: Fodor, J., Klempous, R., Araujo, C.P.S. (eds.) Recent Adv. Intell. Eng. Syst., pp. 161–191. Springer, Heidelberg (2012)
2. Picek, S., Jakobovic, D.: From fitness landscape to crossover operator choice. In: Proceedings of the 2014 Annual Conference on Genetic and Evolutionary Computation, pp. 815–822. ACM (2014)
3. Ceberio, J., Calvo, B., Mendiburu, A., Lozano, J.A.: Multi-objectivising the quadratic assignment problem by means of an elementary landscape decomposition. In: Puerta, J.M., Gámez, J.A., Dorronsoro, B., Barrenechea, E., Troncoso, A., Baruque, B., Galar, M. (eds.) CAEPIA 2015. LNCS (LNAI), vol. 9422, pp. 289–300. Springer, Heidelberg (2015). doi:10.1007/978-3-319-24598-0_26
4. Jensen, T.: Helper-objectives: using multi-objective evolutionary algorithms for single-objective optimisation. J. Math. Model. Algorithms 3(4), 323–347 (2005)

5. Buzdalova, A., Buzdalov, M.: Increasing efficiency of evolutionary algorithms by choosing between auxiliary fitness functions with reinforcement learning. In: 2012 11th International Conference on Machine Learning and Applications (ICMLA), vol. 1, pp. 150–155. IEEE (2012)

6. Knowles, J.D., Watson, R.A., Corne, D.W.: Reducing local optima in single-objective problems by multi-objectivization. In: Zitzler, E., Thiele, L., Deb, K., Coello Coello, C.A., Corne, D. (eds.) EMO 2001. LNCS, vol. 1993, pp. 269–283. Springer, Heidelberg (2001). doi:10.1007/3-540-44719-9_19

7. Petrova, I., Buzdalova, A., Buzdalov, M.: Improved selection of auxiliary objectives using reinforcement learning in non-stationary environment. In: 2014 13th International Conference on Machine Learning and Applications (ICMLA), pp. 580–583. IEEE (2014)

8. Applegate, D.L., Bixby, R.E., Chvatal, V., Cook, W.J.: The Traveling Salesman Problem: A Computational Study. Princeton University Press, Princeton (2011)

9. Kanda, J., Carvalho, A., Hruschka, E., Soares, C.: Using meta-learning to classify traveling salesman problems. In: 2010 Eleventh Brazilian Symposium on Neural Networks, pp. 73–78. IEEE (2010)

10. Kanda, J.Y., de Carvalho, A.C., Hruschka, E.R., Soares, C.: Using meta-learning to recommend meta-heuristics for the traveling salesman problem. In: 2011 10th International Conference on Machine Learning and Applications and Workshops (ICMLA), pp. 346–351. IEEE (2011)

11. Cover, T., Hart, P.: Nearest neighbor pattern classification. IEEE Trans. Inf. Theory **13**(1), 21–27 (1967)

12. Jähne, M., Li, X., Branke, J.: Evolutionary algorithms and multi-objectivization for the travelling salesman problem. In: Proceedings of the 11th Annual conference on Genetic and evolutionary computation, pp. 595–602. ACM (2009)

13. Johnson, D.S., McGeoch, L.A.: The traveling salesman problem: a case in study local optimization. Local Search Comb. Optim. **1**, 215–310 (1997)

14. Deb, K., Pratap, A., Agarwal, S., Meyarivan, T.: A fast and elitist multiobjective genetic algorithm: NSGA-II. IEEE Trans. Evol. Comput. **6**(2), 182–197 (2002)

15. Derrac, J., García, S., Molina, D., Herrera, F.: A practical tutorial on the use of nonparametric statistical tests as a methodology for comparing evolutionary and swarm intelligence algorithms. Swarm Evol. Comput. **1**(1), 3–18 (2011)

16. Lochtefeld, D.F., Ciarallo, F.W.: Deterministic helper-objective sequences applied to job-shop scheduling. In: Proceedings of the 12th Annual Conference on Genetic and Evolutionary Computation, pp. 431–438. ACM (2010)

17. Lochtefeld, D.F., Ciarallo, W.: Helper-objective optimization strategies for the job-shop scheduling problem. Appl. Soft Comput. **11**(6), 4161–4174 (2011)

18. Petrova, I., Buzdalova, A., Buzdalov, M.: Improved helper-objective optimization strategy for job-shop scheduling problem. In: 2013 12th International Conference on Machine Learning and Applications (ICMLA), pp. 374–377, vol 2. IEEE (2013)

19. Bierwirth, C.: A generalized permutation approach to job shop scheduling with genetic algorithms. Oper. -Res. -Spektrum **17**(2–3), 87–92 (1995)

20. Giffler, B., Thompson, G.L.: Algorithms for solving production-scheduling problems. Oper. Res. **8**(4), 487–503 (1960)

21. López-Ibáñez, M., Dubois-Lacoste, J., Stützle, T., Birattari, M.: The irace package, iterated race for automatic algorithm configuration. Technical report TR/IRIDIA/2011-004, IRIDIA, Université Libre de Bruxelles, Belgium (2011)

Sparse, Continuous Policy Representations for Uniform Online Bin Packing via Regression of Interpolants

John H. Drake[1]([✉]), Jerry Swan[2], Geoff Neumann[3], and Ender Özcan[4]

[1] Operational Research Group, Queen Mary University of London,
Mile End Road, London E1 4NS, UK
j.drake@qmul.ac.uk

[2] Department of Computer Science, University of York,
York YO10 5GH, UK
jerry.swan@york.ac.uk

[3] Computing Science and Mathematics, University of Stirling,
Stirling FK9 4LA, UK
gkn@cs.stir.ac.uk

[4] School of Computer Science, University of Nottingham,
Jubilee Campus, Wollaton Road, Nottingham NG8 1BB, UK
exo@cs.nott.ac.uk

Abstract. Online bin packing is a classic optimisation problem, widely tackled by heuristic methods. In addition to human-designed heuristic packing policies (e.g. first- or best- fit), there has been interest over the last decade in the automatic generation of policies. One of the main limitations of some previously-used policy representations is the trade-off between locality and granularity in the associated search space. In this article, we adopt an interpolation-based representation which has the jointly-desirable properties of being sparse and continuous (i.e. exhibits good genotype-to-phenotype locality). In contrast to previous approaches, the policy space is searchable via real-valued optimization methods. Packing policies using five different interpolation methods are comprehensively compared against a range of existing methods from the literature, and it is determined that the proposed method scales to larger instances than those in the literature.

Keywords: Hyper-heuristics · Online bin packing · CMA-ES · Heuristic generation · Sparse policy representations · Metaheuristics · Optimisation

1 Introduction

Bin-packing is a well-known NP-hard problem in combinatorial optimization, in which the goal is to pack a set of items into the smallest possible number of fixed-capacity bins [1]. It has been extensively studied in both its online and

© Springer International Publishing AG 2017
B. Hu and M. López-Ibáñez (Eds.): EvoCOP 2017, LNCS 10197, pp. 189–200, 2017.
DOI: 10.1007/978-3-319-55453-2_13

offline forms. Whereas the sizes of all of the items to be packed are known in advance in the offline case, online bin packing [2,3] requires each piece from a 'lengthy' sequence of items to be considered in individually, with no knowledge of the sizes of the following pieces to pack. A packing policy is a heuristic defining how to pack items of different sizes depending on the currently available space in the set of open bins. Here we consider the one dimensional variant of the online bin packing problem, where items have a fixed width and vary in size in only a single dimension [4]. This problem has a wide range of practical applications in industry, e.g. in stock cutting, where a length of fixed width stock material needs to be cut into shorter segments with the minimum waste [5].

Our approach uses the method of *generative hyper heuristics* [6]. These methods seek to generate new heuristics, operating over a search space of heuristics rather than directly over a space of solutions (e.g. [7,8]). A number of generative hyper-heuristic approaches exist in the online bin-packing literature, with previous work focussing on generating packing policies using different representations. Some previous methods have used Genetic Programming (GP) to represent a packing policy [9,10], evolving a scoring metric to rank each choice of bin for the current item under consideration. Other work used a matrix representation to define a packing policy [11]. When using a matrix-based representation, each row of the matrix corresponds to a particular item size and each column to a particular remaining bin capacity. Entries for each (*size, capacity*) combination define the score for packing an item of that size into a bin with that remaining capacity.

These two representations for packing policies suffer from opposing limitations of the search space they present. Typically, GP suffers from a poor genotype-to-phenotype locality, meaning that small changes to a GP program lead to large changes in the solution and the search landscape is correspondingly rugged. Conversely, the use of a matrix representation suffers from being too *dense*: a large number of changes to the representation are required in order to make a significant difference to its phenotypic expression, tending to necessitate a correspondingly large number of evaluations of the objective function.

In this article, we describe an alternative representation of bin packing policies using *interpolants* that we claim does not suffer from defects present in both GP and matrix-based representations. Interpolants are mathematical functions defined by a set of *control points*, with an associated deterministic formula for values between these points. Interpolants are sparsely represented by their control points, and are constructed specifically so that they exhibit good locality. Searching the space of control points, a vector of real-valued parameters, we test five different interpolation methods to define packing polices for the online one-dimensional bin packing problem. We compare our approach to a number of previous approaches from the literature over 12 sets of instances for this problem.

2 Previous Approaches for Online One-Dimensional Bin Packing

Given a set of n items, with each item j having an associated weight w_j, and a set of n bins with capacity c, Martello and Toth [1] formulated the bin packing problem as follows:

$$\text{Minimise} \quad \sum_{i=1}^{n} y_i \tag{1}$$

$$\text{Subject to} \quad \sum_{j=1}^{n} w_j x_{ij} \le c y_i, \qquad i \in N = \{1,...,n\} \tag{2}$$

$$\sum_{i=1}^{n} x_{ij} = 1, \qquad j \in N \tag{3}$$

$$\text{with} \quad y_i \in \{0,1\}, \qquad i \in N \tag{4}$$

$$x_{ij} \in \{0,1\}, \qquad i \in N, j \in N \tag{5}$$

where y_i denotes whether or not bin i has pieces packed in it, x_{ij} denotes whether or not item j has been packed into bin i. The objective function (Expression 1) minimises the total number of bins used, Expression 2 ensures that the fixed capacity of each bin is respected and Expression 3 ensures that each item is only packed once. The online bin packing variant considers the packing of a 'large' number of items which arrive one at a time and a decision regarding which open bin to place each item needs to be made immediately.

Traditionally, online bin packing problems were solved using deterministic heuristics such as Best Fit (BF) and First Fit (FF) [11]. In FF bins are placed into a fixed order and each item is placed into the first bin with sufficient space [12]. The intention is that bins early on in the sequence will be quickly filled and removed from consideration [11]. However, this method relies on an ordering of the bins and this is not possible in the online case. In BF each item is placed into the fullest bin which has room for it. Where ties occur this algorithm operates like FF [12]. Lee and Lee [4] introduced a Harmonic heuristic, which normalises item sizes, and then separates this interval from (0,1] into non-uniform partitions, each representing a certain type and restricting the number of items than can be placed.

A disadvantage of all of these methods is that they assume that the relationship between the preferable choice of bin on one hand, and space/item size on the other, is smooth. A recent study by Özcan and Parkes found that good (optimal) policies could actually be 'spiky' and complex [11,13]. Recent research in bin packing has tended to focus on metaheuristic strategies capable of automatically devising policies which are more complex than FF or BF and better suited to solving the problem [11,13,14].

One metaheuristic often employed is Genetic Programming (GP). An example of a GP solution to the bin packing problem can be found in the work of

Burke et al. [15]. In this work the trees evolved by GP are used to assign a score to each open bin, indicating the desirability of packing the current item into that bin. This technique was able to automatically generate human-developed policies such as FF, as well as a wide range of alternative policies. Further work by Burke et al. [16] showed that the evolved policies were able to scale effectively to instances much larger than those on which they were trained. Burke et al. [9] evolved heuristics for specific sets of bin packing instances and were able to outperform the classic BF heuristic in some cases. Although this method was able to gain some results comparable to BF, crucially, it was not able to consistently outperform it on a regular basis.

Ross et al. developed a hyper heuristic approach using the XCS learning classifier system [17]. Motivated by the fact that traditional metaheuristics such as GAs generate a single heuristic policy that will likely not adapt if the nature of the problem changes, their approach instead evolved a set of rules through which low-level heuristics can be adapted to a changing problem. As more bins were packed, the state of the problem was analysed and matched to appropriate policies using the rule set. This approach performed well on a range of data sets.

Özcan and Parkes [11] used an approach in which policies were represented as two dimensional matrices, with rows corresponding to remaining bin capacity and columns corresponding to item size. The desirability of placing an item of size s into a bin with remaining capacity r, is provided in each matrix at column s and row r. Each item is then packed into the bin with the highest desirability. Matrices were evolved using a Genetic Algorithm. Unlike the previously discussed approaches based on GP, policies evolved using this representation were able to outperform the BF heuristic. This approach was expanded on in a later paper in which each matrix was viewed as a heuristic with a high number of parameters [14]. A heuristic configuration method called the Iterated Racing Algorithm [18] was then used to tune these parameters. Even though the number of parameters was greater than the number usually found in the problems to which iterated racing is applied, it still managed to improve upon human made heuristics such as BF. The original, Genetic Algorithm based approach was still found to be the more successful of the two approaches. In developing these approaches Özcan and Parkes found that the ideal solution was often one which could not easily be expressed through via an arithmetic function. This demonstrates an advantage over GP which is designed to find solutions, expressed through arithmetic functions [11,13]. Moreover, it was observed that GP mutations often correspond to large moves within the space of policy matrices [19].

3   Learning Mechanisms for Packing Policies

A packing policy can be implemented as a function of the incoming item size s, by assigning an ordinal value to each bin of remaining capacity r (and also to the empty bin). This can either be a bivariate function $p_2(s, r)$ or else, as is the case here, a univariate function $p_1(r - s)$.

As discussed in Sect. 1, policy representations (e.g. matrix, GP as above) may be characterized as *dense* or *sparse*, according to the minimum granularity of possible changes to the representation. Since individual matrix elements are independent, the dense matrix representation is clearly maximally fine-grained. Additionally, we can characterize representations as *continuous*[1] if small changes to the input produce correspondingly small changes to the output. The matrix representation is therefore both dense and continuous, whereas the GP representation is comparatively more sparse and less continuous. While continuity of representation is clearly desirable, density is not, since many invocations of the objective function are required in order to learn the corresponding policy. This suggests that it may be advantageous to consider alternative representations that are both sparse and continuous. One such possibility are the variety of function interpolation schemes used in numerical analysis. A function interpolator is an (invariably sparse) representation parametrically defined by a set of *control points*. Notably, this includes splines: piecewise polynomial functions which are continuous by construction.

Our approach is therefore to perform a hyper-parameter search over the vector of control points of a univariate function interpolator, which is then used to implement the packing policy. The hyper-heuristic search space is given by $\mathbb{R}^k$, where k is the number of control points. A candidate solution (represented at the hyper-level by a point in $\mathbb{R}^k$) is used to generate a packing policy by using these k values as the y value of the control points (with corresponding x points equally-spaced across the input domain $[0, c]$ of packing policies). Each value along the along the x-axis corresponds to a potential packing (i.e. $r - s$). A packing policy then ranks the desirability of placing the current item into each bin, using the y value defined by the interpolation scheme for the corresponding x value of $r - s$ for that bin. The interpolation schemes considered are:

- **Linear**: Piecewise linear function.
- **Cubic Spline**: Piecewise degree 3 polynomial function, which is continuous and twice differentiable.
- **Divided Difference**: Interpolation via Newton's method of divided differences, expressing the interpolating polynomial as a linear combination of Newton basis polynomials [20].
- **LOESS**: Piecewise polynomial function obtained via locally weighted least squares [21].
- **Neville**: Polynomial function with degree one less than the number of control points which passes exactly through them [22]. The construction uses Newton polynomials via the method of divided differences.

Figure 1 plots the values from different interpolation schemes. For the control points given, the plot for Neville visibly coincides with divided difference, however, the resulting function values do exhibit small differences. LOESS is parameterized here by a vector of random weights of length equal to the number of control points—if all weights were the same, then LOESS would coincide

---

[1] The term 'locality' is often used in this context in evolutionary computation.

with cubic. The piecewise description of interpolators also helps to overcome one of the limitations of expressing packing policies through purely arithmetic functions (i.e. lack of conditional statements) [11].

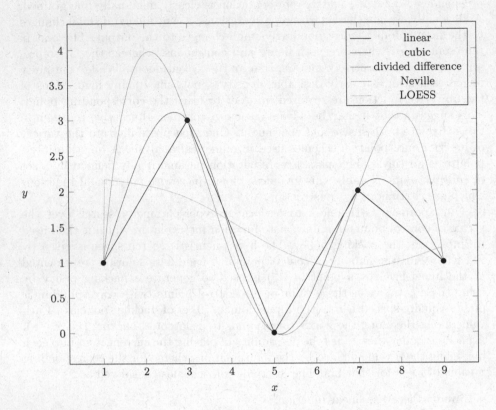

Fig. 1. Interpolation from control points $\{(1,1),(3,3),(5,0),(7,2),(9,1)\}$

As seen in Fig. 1, alternative interpolation schemes define rather different functions. For our purposes, this is not an issue: since we only require good genotype-to-phenotype locality, i.e. from the control points to the corresponding univariate function.

The components used in the hyper-parameter search are as follows:

– **Representation**: For k x-values equally-spaced across the domain of the packing policy function, the solution representation is then a vector in $\mathbb{R}^k$, denoting the corresponding y-values of the k control points.
– **Fitness**: the sum of the *average generic fullness* value, taken over each UBP instance in the training set, where average fullness f for each instance is calculated as:

$$f = \frac{1}{B} \sum_t f_t \tag{6}$$

where B is the number of bins used and f_t is the fullness of bin t.
- **Perturbation**: Solution vector elements are modified according to the mechanism of CMA-ES.

CMA-ES [23] is a well-known and effective metaheuristic for search in $\mathbb{R}^k$. It is one of the most widely-used gradient-free approaches (partly because it requires minimal parameter tuning by the user), and is of particular value when applied to problems with rugged search landscapes. CMA-ES is based upon the foundational Evolutionary Strategies method due to Rechenberg [24], which maintains a companion vector σ of k real values, denoting the mutation step-size to be applied to the corresponding element of the solution vector. CMA-ES further develops the ES approach by adaptively updating mutation step-size via a covariance matrix. Details of the CMA-ES implementation we use in this paper are given in the following section.

4 Experimental Framework and Results

Our experimental design follows the same methodology used by Asta et al. [13] to enable us to fairly compare against both their technique and the state of the art. Each algorithm compared was tested on a set of progressively larger configurations of the Uniform Bin Packing problem, referred to as *UBP's*. Each UBP is defined by three parameters, maximum bin capacity, minimum item size and maximum item size, denoted as UBP(*maxCapacity*, *minItemSize,maxItemSize*) herein. An additional parameter, the total number of items, was kept constant at 10^5 in every test. The first 10 UBP problems that we have used were taken directly from the work of Asta et al. [13]. An 11th and 12th have also been introduced in this paper in order to demonstrate the scalability of our technique. These two UBPs, UBP(225,30,150) and UBP(300,40,200), were produced by multiplying each parameter in UBP(150,20,100) by 1.5 and 2 respectively.

For each UBP a two step testing process was used. First of all, a training set consisting of 10 instances of the UBP is randomly generated and candidate packing policies are evolved in one evolutionary run on these training instances. The best packing policy generated is then tested on 100 instances of the UBP, the testing set. Separate training and testing sets ensures that the policies obtained generalise to new problem instances and are not simply obtained by overfitting. Randomization, both of the evolutionary process and of the generation of problem instances, is achieved through the use of the Mersenne Twister random number generator, known to produce a good distribution of random values [25]. In accordance with the recommendations of Luke [26], a set of random seeds is first generated. Each seed is used to generate a separate Mersenne Twister for each UBP. Each interpolant variant is run using the same seeds and tested on the same training and test sets.

As discussed in the previous section, the search over the vector of control points was performed via CMA-ES, a widely-used evolutionary algorithm. The CMA-ES implementation used here was Apache Commons Math, using default parameter settings[2]. After some initial experimentation, the number of control points k was set to 15, with each training run allowed a maximum of 7,500 fitness evaluations. The possible input range for each function is the amount of space left in a bin after adding the current item under consideration, that is $\in [0, (maxCapacity - minItemSize)]$. For each possible bin, the interpolated functions generate a real-valued output score, representing the preference value for that bin.

4.1   Comparison Between Interpolants and Previous Results in the Literature

The results in Table 1 show the %-average (mean) fullness of the best solution for the interpolant techniques compared to other approaches over 100 instances for each UBP type. Existing methods used for comparison are the classic Best Fit (BF), First Fit (FF), Worst Fit (WF) and Harmonic [4] heuristics as given by Asta et al. [13], in addition to more recent methods: policy matrix based approaches of Yarimcam et al. [14], Asta et al. [13] and Asta and Özcan [27].

From this table, we can see that the performance of different interpolation methods varies depending on the size of the instance considered. In general, we can observe that the interpolant methods offer very good results on the largest instances tested whilst still offering good performance on smaller instances. Linear interpolation is particularly strong in the larger instances, outperforming all other methods on 3 of the 5 largest instance sets. For the largest 3 instance sets we can compare to previous automated policy generation methods: UBP(75,10,50), UBP(80,10,50) and UBP(150,20,100), the best method is always one of the interpolants (twice linear and once divided difference). LOESS and Neville's method of interpolation perform well on the smallest instances, particularly UBP(6,2,3) and UBP(20,5,10) compared to the traditional heuristic methods, however in general they are outperformed by the policy matrix approaches.

Interestingly, the interpolant methods seem to struggle on some mid-size instances when compared to matrix based approaches, particularly in the case of UBP(40,10,20) and UBP(60,15,25). For both of these instance sets, the maximally dense $GA_{ORIGINAL}$ method outperforms all others. Asta et al. [13] observed that in the smaller instances tested (particularly from UBP(6,2,3) to UBP(15,5,10)), there are disconnected neutralities (plateaus) in the rugged search space of policies for GA to traverse. It could be the case that these mid-sized instances also exhibit this behaviour and are relatively easy for $GA_{ORIGINAL}$ to traverse. In the case of the interpolant methods, performance might be improved by increasing the number of control points used, creating a denser, more fine-grained space of policies.

It is clear that the policy matrix approaches perform well on the smallest instances, however as the size of the problem increases, the interpolant methods

---

[2] https://commons.apache.org/proper/commons-math/javadocs/api-3.6.1/index.html.

Table 1. Results obtained using different interpolant-based representations compared to existing methods from the literature in terms of %-average fullness over 100 instances of each UBP type

Methods	UBP (6,2,3)	UBP (15,5,10)	UBP (20,5,10)	UBP (30,4,20)	UBP (30,4,25)	UBP (40,10,20)	UBP (60,15,25)	UBP (75,10,50)	UBP (80,10,50)	UBP(150, 20,100)	UBP(225, 30,150)	UBP(300, 40,200)
Cubic spline	92.26	94.49	91.64	99.69	90.19	96.77	91.90	98.94	93.97	99.04	98.97	98.63
Linear	92.26	95.37	91.28	96.80	98.27	90.21	90.94	99.33	96.35	99.26	99.09	93.37
Divided difference	92.26	99.49	91.58	99.68	98.39	90.26	92.53	98.80	99.49	98.89	98.86	96.67
LOESS	96.56	99.00	97.80	93.16	98.81	90.04	92.28	99.04	98.96	99.09	98.66	98.97
Neville	99.99	96.56	96.93	99.51	98.14	90.21	92.53	97.07	99.47	98.93	98.75	98.87
BF	92.30	99.62	91.55	96.84	98.38	90.23	92.55	96.08	96.39	95.82	95.73	95.68
FF	92.30	99.55	91.54	96.68	97.93	90.22	92.55	95.91	96.29	95.64	95.55	95.50
WF	91.70	86.58	90.54	88.61	84.10	88.66	90.80	87.94	89.25	87.73	87.64	87.60
Harmonic [4]	-	74.24	90.04	73.82	74.21	89.10	85.18	71.59	72.96	71.97	-	-
$(\pi_r, 10^4)$ [14]	99.99	99.60	97.65	99.43	98.46	96.22	99.01	97.17	97.51	-	-	-
$(\pi_r, 10^5)$ [14]	99.99	99.66	98.21	99.09	99.51	96.81	99.49	96.93	97.36	-	-	-
$GA_{Original}$ [13]	99.99	99.63	98.18	99.41	98.39	96.99	99.68	98.22	98.54	97.88	-	-
GA_{Binary} [13]	99.99	99.61	98.42	99.58	99.55	96.75	96.96	98.45	98.46	97.63	-	-
GA+TA [27]	99.99	99.62	98.28	99.53	99.53	96.27	99.47	98.53	98.66	98.22	-	-

begin to outperform these approaches. As mentioned previously, matrix-based representation is maximally dense, as a desirability score for each (s, r) pair is maintained explicitly. This property restricts this representation in terms of scalability, as an increasingly large number of independent variables must be maintained as the problem size grows. This leads to an incredibly large search space in the case of large instances, which is subsequently much more difficult to search effectively. The search space of policies expressed using interpolant methods is constant irrespective of the problem instance size, as long as the number of control points and the range of values each point can take is fixed.

5   Conclusions

In this paper we have presented a new method of representing packing policies for online bin packing using function interpolation. A policy is defined as a function of the remaining space in a given bin after adding the current item to be packed, providing a score for the desirability of packing the item that bin. Such policies are represented using a set of 'control points', fixed along the input axis, with the exact nature of the function determined by the interpolation method used. Search takes place in hyper-parameter space, across the locations of each control point on the output axis, consisting of a vector of real-valued variables. Unlike previously proposed representations, policies defined using this approach are both sparse and exhibit good locality. Our experiments have shown that policies generated by CMA-ES using this representation can yield better results than both traditional heuristics and state-of-the-art 'policy matrix' approaches, particularly in the case of larger problem instances.

As a result of this work, a number of potential avenues for further research have emerged. One of the limitations of this work is that a fixed number of control points are used. It may be the case that the best choice in terms of number of control points is dependent on the size of the instance being solved, or even differ within a particular instance set depending on the interpolation method used. We intend to explore the relationship between the number of control points used and the number of possible item sizes in an instance and different interpolation methods. Additionally, although here we have chosen to use CMA-ES to search the hyper-parameter space, other continuous optimisation methods such as Genetic Algorithms or Differential Evolution could have been used. Future work will focus on applying other continuous optimisation methods to this problem, assessing their ability to search the hyper-parameter space effectively.

References

1. Martello, S., Toth, P.: Knapsack Problems: Algorithms and Computer Implementations. Wiley, Hoboken.(1990)
2. Csirik, J., Woeginger, G.J.: On-line packing and covering problems. In: Fiat, A., Woeginger, G.J. (eds.) Online Algorithms. LNCS, vol. 1442, pp. 147–177. Springer, Heidelberg (1998). doi:10.1007/BFb0029568

3. Coffman Jr., E.G., Csirik, J., Galambos, G., Martello, S., Vigo, D.: Bin packing approximation algorithms: survey and classification. In: Pardalos, P.M., Du, D.Z., Graham, R.L. (eds.) Handbook of Combinatorial Optimization, pp. 455–531. Springer, New York (2013)
4. Lee, C.C., Lee, D.T.: A simple on-line bin-packing algorithm. J. ACM **32**(3), 562–572 (1985)
5. Sinuany-Stern, Z., Weiner, I.: The one dimensional cutting stock problem using two objectives. J. Oper. Res. Soc. **45**(2), 231–236 (1994)
6. Burke, E.K., Hyde, M., Kendall, G., Ochoa, G., Özcan, E., Woodward, J.R.: A classification of hyper-heuristic approaches. In: Gendreau, M., Potvin, J.Y. (eds.) Handbook of Metaheuristics, vol. 146, pp. 449–468. Springer, Heidelberg (2010)
7. Woodward, J.R., Swan, J.: The automatic generation of mutation operators for genetic algorithms. In: Proceedings of the Genetic and Evolutionary Computation Conference (GECCO 2012), pp. 67–74. ACM (2012)
8. Drake, J.H., Hyde, M., Ibrahim, K., Ozcan, E.: A genetic programming hyper-heuristic for the multidimensional knapsack problem. Kybernetes **43**(9/10), 1500–1511 (2014)
9. Burke, E.K., Hyde, M.R., Kendall, G., Woodward, J.: Automatic heuristic generation with genetic programming: evolving a jack-of-all-trades or a master of one. In: Proceedings of the Genetic and Evolutionary Computation Conference (GECCO 2007), pp. 1559–1565. ACM (2007)
10. Burke, E.K., Hyde, M.R., Kendall, G., Woodward, J.: Automating the packing heuristic design process with genetic programming. Evol. Comput. **20**(1), 63–89 (2012)
11. Özcan, E., Parkes, A.J.: Policy matrix evolution for generation of heuristics. In: Proceedings of the Genetic and Evolutionary Computation Conference (GECCO 2011), pp. 2011–2018. ACM (2011)
12. Johnson, D.S., Demers, A., Ullman, J.D., Garey, M.R., Graham, R.L.: Worst-case performance bounds for simple one-dimensional packing algorithms. SIAM J. Comput. **3**(4), 299–325 (1974)
13. Asta, S., Özcan, E., Parkes, A.J.: CHAMP: creating heuristics via many parameters for online bin packing. Expert Syst. Appl. **63**, 208–221 (2016)
14. Yarimcam, A., Asta, S., Özcan, E., Parkes, A.J.: Heuristic generation via parameter tuning for online bin packing. In: IEEE Symposium on Evolving and Autonomous Learning Systems (EALS 2014), pp. 102–108. IEEE (2014)
15. Burke, E.K., Hyde, M.R., Kendall, G.: Evolving bin packing heuristics with genetic programming. In: Runarsson, T.P., Beyer, H.-G., Burke, E., Merelo-Guervós, J.J., Whitley, L.D., Yao, X. (eds.) PPSN 2006. LNCS, vol. 4193, pp. 860–869. Springer, Heidelberg (2006). doi:10.1007/11844297_87
16. Burke, E.K., Hyde, M.R., Kendall, G., Woodward, J.R.: The scalability of evolved on line bin packing heuristics. In: 2007 IEEE Congress on Evolutionary Computation, pp. 2530–2537. IEEE (2007)
17. Ross, P., Schulenburg, S., Marín-Blázquez, J.G., Hart, E.: Hyper-heuristics: learning to combine simple heuristics in bin-packing problems. In: Proceedings of the Genetic and Evolutionary Computation Conference (GECCO 2002), pp. 942–948 (2002)
18. López-Ibáñez, M., Dubois-Lacoste, J., Cáceres, L.P., Birattari, M., Stützle, T.: The irace package: iterated racing for automatic algorithm configuration. Oper. Res. Perspect. **3**, 43–58 (2016)

19. Parkes, A.J., Özcan, E., Hyde, M.R.: Matrix analysis of genetic programming mutation. In: Moraglio, A., Silva, S., Krawiec, K., Machado, P., Cotta, C. (eds.) EuroGP 2012. LNCS, vol. 7244, pp. 158–169. Springer, Heidelberg (2012). doi:10. 1007/978-3-642-29139-5_14
20. Abramowitz, M., Stegun, I.: Handbook of Mathematical Functions. Dover Publications, New York (1965)
21. Cleveland, W.S.: Robust locally weighted regression and smoothing scatterplots. J. Am. Stat. Assoc. **74**(368), 829–836 (1979)
22. Stoer, J., Bulirsch, R.: Introduction to Numerical Analysis. Texts in Applied Mathematics. Springer, Heidelberg (2002)
23. Hansen, N., Ostermeier, A.: Completely derandomized self-adaptation in evolution strategies. Evol. Comput. **9**(2), 159–195 (2001)
24. Rechenberg, I.: Evolutionsstrategie: optimierung technischer systeme nach prinzipien der biologischen evolution. Number 15 in Problemata. Frommann-Holzboog, Stuttgart-Bad Cannstatt (1973)
25. Matsumoto, M., Nishimura, T.: Mersenne twister: a 623-dimensionally equidistributed uniform pseudo-random number generator. ACM Trans. Model. Comput. Simul. (TOMACS) **8**(1), 3–30 (1998)
26. Luke, S.: Essentials of Metaheuristics, 2nd edn. Lulu, Raleigh (2013)
27. Asta, S., Özcan, E.: A tensor analysis improved genetic algorithm for online bin packing. In: Proceedings of the 2015 Annual Conference on Genetic and Evolutionary Computation, pp. 799–806. ACM, New York (2015)

The Weighted Independent Domination Problem: ILP Model and Algorithmic Approaches

Pedro Pinacho Davidson[1,2], Christian Blum[3(✉)], and José A. Lozano[1,4]

[1] Department of Computer Science and Artifical Intelligence,
University of the Basque Country UPV/EHU, San Sebastian, Spain
ppinacho@santotomas.cl, ja.lozano@ehu.eus
[2] Escuela de Informática, Universidad Santo Tomás, Concepción, Chile
[3] Artificial Intelligence Research Institute (IIIA-CSIC),
Campus UAB, Bellaterra, Spain
christian.blum@iiia.csic.es
[4] Basque Center for Applied Mathematics (BCAM), Bilbao, Spain

Abstract. This work deals with the so-called weighted independent domination problem, which is an NP-hard combinatorial optimization problem in graphs. In contrast to previous theoretical work from the literature, this paper considers the problem from an algorithmic perspective. The first contribution consists in the development of an integer linear programming model and a heuristic that makes use of this model. Second, two greedy heuristics are proposed. Finally, the last contribution is a population-based iterated greedy algorithm that takes profit from the better one of the two developed greedy heuristics. The results of the compared algorithmic approaches show that small problem instances based on random graphs are best solved by an efficient integer linear programming solver such as CPLEX. Larger problem instances are best tackled by the population-based iterated greedy algorithm. The experimental evaluation considers random graphs of different sizes, densities, and ways of generating the node and edge weights.

1 Introduction

The so-called weighted independent domination (WID) problem is a combinatorial optimization problem that was introduced in [1]. The problem is an extension of the well-known independent domination (ID) problem. Given an undirected graph $G = (V, E)$, V is the set of nodes and E refers to the set of edges. An edge $e \in E$ that connects nodes $u \neq v \in V$ is equally denoted by (u, v) and by (v, u). The *neighborhood* $N(v)$ of a node $v \in V$ is defined as $N(v) := \{u \in V \mid (v, u) \in E\}$, the *closed neighborhood* $N[v]$ of a node $v \in V$ is defined as $N[v] := N(v) \cup \{v\}$, and the set of edges incident to a node $v \in V$ is

This work was supported by project TIN2012-37930-C02-02 (Spanish Ministry for Economy and Competitiveness, FEDER funds from the European Union).

© Springer International Publishing AG 2017
B. Hu and M. López-Ibáñez (Eds.): EvoCOP 2017, LNCS 10197, pp. 201–214, 2017.
DOI: 10.1007/978-3-319-55453-2_14

defined as $\delta(v) := \{e = (v, u) \in E\}$. Given an undirected graph $G = (V, E)$, a subset $D \subseteq V$ of the nodes is called a *dominating set* if every node $v \in V \setminus D$ is adjacent to at least one node from D, that is, if for every node $v \in V \setminus D$ exists at least one node $u \in D$ such that $v \in N(u)$. Furthermore, a set $I \subseteq V$ is called an *independent set* if no two nodes from I are adjacent to each other. Correspondingly, a subset $D \subseteq V$ is called an *independent dominating set* if D is both an independent set and a dominating set. Finally, given an independent dominating set $D \in V$, for all $v \in V \setminus D$ we define the *D-restricted neighborhood* $N(v \mid D)$ as $N(v \mid D) := N(v) \cap D$, that is, the neighborhood of v is restricted to all its neighbors that are in D.

In the WID problem we are given an undirected graph $G = (V, E)$ with node and edge weights. More specifically, for each $v \in V$, respectively $e \in E$, we are given an integer weight $w(v) \geq 0$, respectively $w(e) \geq 0$. The WID problem consists in finding an independent dominating set D in G that minimizes the following cost function:

$$f(D) := \sum_{u \in D} w(u) + \sum_{v \in V \setminus D} \min\{w(v, u) \mid u \in N(v \mid D)\} \tag{1}$$

In words, the objective function value of D is obtained by the sum of the weights of the nodes in D plus the sum of the weights of the minimum-weight edges that connect the nodes that are not in D to nodes that are in D. As an example consider the graphics in Fig. 1. The node weights are indicated inside the nodes and the edge weights are provided besides the edges. A possible input graph is shown in Fig. 1a. An optimal *minimum weight dominating set* (the set of gray nodes) is shown in Fig. 1b. However, note that this set is not an independent set because the two nodes that form the set are adjacent to each other. An optimal *minimum weight independent dominating set*[1] is given in Fig. 1c. Note that for both, the minimum weight dominating set problem and the minimum weight independent dominating set problem, the edge weights are not considered. Finally, the optimal solution to the WID problem is shown in Fig. 1d. The minimum weight edges that are chosen to connect nodes not in D to nodes in D are indicated with bold lines. The objective function value of this solution is 13, which is composed of the nodes weights $(2 + 1 + 2)$ and the edge weights $(4 + 1 + 3)$.

1.1   Our Contribution

So far, the WID problem has only been considered from a theoretical perspective. It is easy to see that the problem is NP-hard. This is because with $w(v) = 1$ for all $v \in V$ and $w(e) = 0$ for all $e \in E$ the problem reduces to the independent domination problem which was shown to be NP-hard in [2]. A linear time algorithm for the WID problem in series-parallel graphs was proposed in [1].

---

[1] In this problem, given an undirected graph with node weights, the goal is to find an independent dominating set for which the sum of the weights of the nodes is minimal.

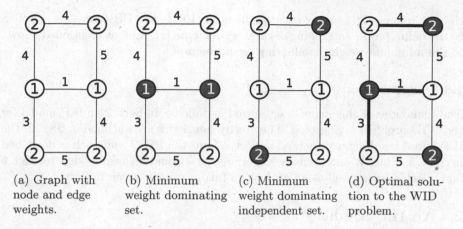

(a) Graph with node and edge weights.

(b) Minimum weight dominating set.

(c) Minimum weight dominating independent set.

(d) Optimal solution to the WID problem.

Fig. 1. Example that relates the WID problem with the *minimum weight dominating set* problem and with the *minimum weight independent dominating set* problem.

In this work we consider the WID problem in general graphs from an algorithmic perspective. Our contributions are as follows. First, we present an integer linear programming (ILP) model for the WID problem, together with an ILP-based heuristic. Second, we propose two different greedy heuristics for solving the problem. The first one is known from the minimum weight independent dominating set problem, while the second one is specifically developed for the WID problem. Finally, we propose a so-called population-based iterated greedy (PBIG) algorithm. This algorithm employs an iterated greedy metaheuristic in a population-based fashion, and can therefore be seen as a hybrid between methods based on local search and population-based methods.

1.2   Related Work

On one side, there is related work for problems similar to the one considered in this work. The *minimum independent dominating set* problem, for example, has recently been tackled by a greedy randomized adaptive search procedure (GRASP) in [3]. Another related problem is the *minimum weight dominating set* problem. This problem has been quite popular in recent years as a test case for metaheuristics. The most recent research efforts for this problem have led to the development of an ant colony optimization approach and a genetic algorithm in [4], a hybrid evolutionary algorithm in [5], a hybrid approach combining iterated greedy algorithms and an ILP solver in a sequential way in [6], and a memetic algorithm in [7].

On the other side, there is related work concerning the employed optimization technique, that is, PBIG. In general, iterated greedy (IG) algorithms have shown to be able to work very well in the context of problems for which a good and fast greedy heuristic is known. Prime examples include those to various scheduling problems such as [8,9]. The first PBIG approach was proposed in the context of the minimum weight vertex cover problem in [10]. Later, PBIG was also applied to the delimitation and zoning of rural settlements [11] and, as mentioned above, to the minimum weight dominating set problem [6].

1.3 Organization

The remainder of this paper is organized as follows. In Sect. 2 an ILP model for the WID problem is proposed. The greedy heuristics are outlined in Sect. 3, the ILP-based heuristic is presented in Sect. 4, and the PBIG approach is described in Sect. 5. Finally, an extensive experimental evaluation is provided in Sect. 6 and conclusions as well as an outlook to future work is given in Sect. 7.

2   An ILP Model

The proposed ILP uses three sets of binary variables. For each node $v \in V$ it uses a binary variable x_v. Moreover, for each edge $e \in E$ the model uses a binary variable y_e and a binary variable z_e. Hereby, x_v indicates if v is chosen for the solution. Moreover, z_e indicates if $e \in E$ is selected for connecting a non-chosen node to a chosen one. Variable y_e is an indicator variable, which indicates if e is choosable, or not.

$$(\text{ILP}) \quad \min \quad \sum_{v \in V} x_v w(v) + \sum_{e \in E} z_e w(e) \tag{2}$$

$$\text{s.t.} \quad x_v + x_u \leq 1 \qquad \qquad \text{for } e = (u,v) \in E \quad (3)$$

$$x_v + x_u = y_e \qquad \qquad \text{for } e = (u,v) \in E \quad (4)$$

$$z_e \leq y_e \qquad \qquad \text{for } e \in E \quad (5)$$

$$x_v + \sum_{u \in N(v)} x_u \geq 1 \qquad \text{for } v \in V \quad (6)$$

$$x_v + \sum_{e \in \delta(v)} z_e \geq 1 \qquad \text{for } v \in V \quad (7)$$

$$x_v \in \{0,1\} \qquad \qquad \text{for } v \in V$$

$$y_e \in \{0,1\} \qquad \qquad \text{for } e \in E$$

$$z_e \in \{0,1\} \qquad \qquad \text{for } e \in E$$

Hereby, constraints (3) are the independent set constraints, that is, they make sure that no two adjacent nodes can form part of the solution. Constraints (4) ensure the proper setting of the indicator variables. Note that edges that contribute to the objective function value must always connect a node that is not chosen for the solution with a node that is in the solution. Therefore, if—concerning an edge $e = (u,v)$—either v or u is in the solution, variable y_e is forced to take value one, which indicates that this edge is choosable. Constraints (5) relate the indicator variables with the variables that actually show which edges are chosen. In particular, if an indicator variable y_e has value zero, z_e is forced to take value zero, which means e cannot be chosen. Constraints (6) are the dominating set constraints. They ensure that for each node $v \in V$, either the node itself or at least one of its neighbors must form part of the solution. Finally, constraints (7) ensure that each node $v \in V$ that does not form part of the solution—that is, when $x_v = 0$—is connected by an edge to a node that forms part of the solution. Due to the fact that the optimization goal concerns minimization, the edge with the lowest weight is chosen for this purpose.

3   Greedy Heuristics

The first one of two different greedy heuristics developed in this work is a simple extension of a well-known heuristic for the minimum weight independent dominating set problem. Given an input graph G, this heuristic starts with an empty solution $S = \emptyset$ and adds, at each step, exactly one node from the remaining graph G' to S. Initally, the *remaining graph* G' is a copy of G. After adding a node $v \in V'$ to S, all nodes from $N[v \mid G']$—that is, from the closed neighborhood of v in G'—are removed from V'. Moreover all their incident edges are removed from E'. In this way, only those nodes that maintain the property of S being an independent set may be added to S. At each step, the node $v \in V'$ that maximizes $\frac{|N(v|G')|}{w(v)}$ is chosen to be added to S, where $N(v \mid G')$ refers to the neighborhood of v in G'. In other words, nodes with a high degree in the remaining graph G' and with a low node weight are preferred. Note that this greedy heuristic does not take the edge weights into account. They are only considered when calculating the objective function value of the final solution S. The pseude-code of this heuristic, henceforth referred to as GREEDY1, is shown in Algorithm 1.

In contrast to GREEDY1, the second greedy heuristic is designed to take into account the edge weights already during the process of constructing a solution. The algorithmic framework of this greedy heuristic—henceforth denoted by GREEDY2—is the same as the one of GREEDY1. However, the way in which a node is chosen at each step is different. For the description of this greedy heuristic the following notations are required. First, the maximum weight of any edge

---

Algorithm 1. Greedy Heuristic (GREEDY1)

1: **input:** a undirected graph $G = (V, E)$ with node and edge weights
2: $S := \emptyset$
3: $G' := G$
4: **while** $V' \neq \emptyset$ **do**
5:     $v^* := \text{argmax}\{\frac{|N(v|G')|}{w(v)} \mid v \in V'\}$
6:     $S := S \cup \{v^*\}$
7:     Remove from G' all nodes from $N[v \mid G']$ and their incident edges
8: **end while**
9: **output:** An independent dominating set S of G

in E is denoted by $w_{\max}$. Then, let $S \in V$ be a partial solution, that is, S is an independent set which is not yet a dominating set, but which can be extended to be a dominating set. The *auxiliary objective function value* $f^{\mathrm{aux}}(S)$ is defined as $\sum_{v \in V} c(v \mid S)$, where $c(v \mid S)$ is called the *contribution* of node v with respect to partial solution S. Given S, these contributions are defined as follows:

1. If $v \in S$: $c(v \mid S) := w(v)$
2. If $v \notin S$ and $N(v) \cap S = \emptyset$: $c(v \mid S) := w_{\max}$
3. If $v \notin S$ and $N(v) \cap S \neq \emptyset$: $c(v \mid S) := \min\{w(e) \mid e = (v, u), u \in S\}$

Note that in the case of S being a complete solution, it holds that $f(S) = f^{\mathrm{aux}}(S)$. Now, in order to obtain GREEDY2, line 5 of Algorithm 1 must be exchanged with the following one:

$$v^* := \operatorname{argmin}\{f^{\mathrm{aux}}(S \cup \{v\}) \mid v \in V'\} \tag{8}$$

4 Heuristic Based on the ILP Model

One possibility to take profit from the ILP model outlined in Sect. 2 is to devise a heuristic based on graph reduction. The main idea is to remove a certain percentage of the edges with the highest weights from the input graph G, which results in a reduced graph G'. Then, a general-purpose ILP solver such as CPLEX is used to solve the problem in G', forcing that the provided solution is also a feasible solution for G. However, this is not trivial, as indicated by the example in Fig. 2. In this example, the edge set E of the input graph G consists of all dashed and continuous lines. The edge set E' of the reduced graph G' only consists of the continuous lines. A feasible solution in the original graph consists of exactly one of the four nodes. As a consequence, the remaining three nodes must be connected to the chosen node. Observe that none of these solutions can be generated in the reduced graph. Therefore, for solving the problem in G' we devised the following ILP model, which makes use of additional binary variables p_v for all $v \in V$. Moreover, let w_v denote the weight of the edge with the highest weight of all those edges incident to v, that is, $w_v := \max\{w(e) \mid e \in \delta(v)\}$.

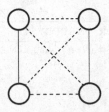

Fig. 2. Example for graph reduction. The edge set E of the input graph G consists of all dashed and continuous lines. The edge set E' of the reduced graph G' only consists of the continuous lines.

$$(\text{ILP2}) \quad \min \sum_{v \in V}(x_v w(v) + p_v w_v) + \sum_{e \in E'} z_e w(e) \tag{9}$$

$$\text{s.t.} \quad x_v + x_u \leq 1 \qquad\qquad\qquad \text{for } e = (u, v) \in E \tag{10}$$

$$x_v + x_u = y_e \qquad\qquad\qquad \text{for } e = (u, v) \in E' \tag{11}$$

$$z_e \leq y_e \qquad\qquad\qquad\qquad \text{for } e \in E' \tag{12}$$

$$x_v + \sum_{u \in N(v)} x_u \geq 1 \qquad\qquad \text{for } v \in V \tag{13}$$

$$x_v + p_v + \sum_{e \in \delta'(v)} z_e \geq 1 \qquad\quad \text{for } v \in V \tag{14}$$

$$x_v, p_v \in \{0, 1\} \qquad\qquad\qquad \text{for } v \in V$$

$$y_e \in \{0, 1\} \qquad\qquad\qquad\quad \text{for } e \in E'$$

$$z_e \in \{0, 1\} \qquad\qquad\qquad\quad \text{for } e \in E'$$

Note that constraints (10) and the neighborhood function N in constraints (13) are defined using input graph G. This is done such that the set of nodes chosen in any solution form a valid solution for the original input graph G. In contrast, constraints (11), (12) and the incidence function $\delta'()$ of constraints (14) refer to the edge set E' of the reduced graph G'. This is because for a solution of ILP2 only edges of the reduced graph may be chosen. In comparison to the original ILP, ILP2 has an objective function which is augmented by the term $\sum_{v \in V} p_v \cdot w_v$ and the left-hand-side of constraints (14) is augmented by summing p_v. This has the effect that, in those cases in which any feasible solution for G causes that node v cannot be connected to any chosen node using an edge from E', variable p_v is forced to take value one. This, in turn, results in summing the weight of the highest-weight edge from E which is incident to v to the objective function value.

In summary, the ILP-based heuristic—henceforth called ILP-HEURISTIC—works as follows. First, heuristic GREEDY2 is applied to G. Second, graph G is reduced by removing $X\%$ of the highest-weight edges, without removing any edges used by the solution of GREEDY2 and without removing more than $(100 - X)\%$ of the edges incident to any node in G. This is done by ordering all edges in E according to decreasing edge weight, and considering one edge after the other for removal, from left to right. This process results in a graph G'. Then, CPLEX is applied to G' using model ILP2. Moreover, the solution of GREEDY2 is provided as a warm-start to CPLEX. This process results in a set S' of chosen nodes. On the basis of S' we generate the corresponding solution in G by simply connecting any node in $V \setminus S'$ using the edge from E with the lowest weight to any of the nodes in S'. Note that by preventing any edges used in the GREEDY2 solution from being removed from G during the graph reduction step, the solution provided by ILP-HEURISTIC must always be at least as good as the one provided by GREEDY2.

'5 PBIG: Population-Based Iterated Greedy

A high level description of the implemented PBIG approach is given in Algorithm 2. Apart from the input graph G, PBIG requires values for five parameters: (1) the population size $p_{\text{size}} \in \mathbb{Z}^+$, (2) the lower bound (D^l) and the upper bound (D^u) for the degree of destruction applied to each solution of the population at each iteration, (3) the determinism rate $d_{\text{rate}} \in [0, 1]$, and (4) the candidate list size $l_{\text{size}} > 0$. The latter two parameters control the greediness of the probabilistic solution (re-)construction procedure. Moreover, note that for the values of the above-mentioned bounds it must hold that $0 \le D^l \le D^u \le 1$. For the following description, each solution S is a subset of the nodes of V, has an objective function value $f(S)$, and an individual, possibly dynamic, destruction rate D_S.

The algorithm works as follows. First, the p_{size} solutions of the initial population are generated by function GenerateInitialPopulation($p_{\text{size}}, d_{\text{rate}}, l_{\text{size}}$) (see line 2 of Algorithm 2). Afterwards, each iteration consists of the following steps. First, an empty population $\mathcal{P}_{\text{new}}$, called offspring population, is created. Then, each solution $S \in \mathcal{P}$ is partially destroyed using procedure DestroyPartially(S) (see line 6 of Algorithm 2). This results in a partial solution $\hat{S}$. On the basis of $\hat{S}$, a complete solution S' is then constructed using procedure Reconstruct($\hat{S}, d_{\text{rate}}, l_{\text{size}}$) (see line 7 of Algorithm 2). Then, the destruction rate D_S of solution S is adapted depending on the quality of solution S' in function AdaptDestructionRate(S, S'). Each newly obtained complete solution is stored in $\mathcal{P}_{\text{new}}$. Note that the two phases of destruction and re-construction are applied to all solutions from $\mathcal{P}$ independently of each other. When the iteration is completed, procedure Accept($\mathcal{P}, \mathcal{P}_{\text{new}}$) selects the best p_{size} solutions from $\mathcal{P} \cup \mathcal{P}_{\text{new}}$ for the population of the next iteration. In the case of two solutions from $\mathcal{P} \cup \mathcal{P}_{\text{new}}$ being equal, the criterion used for tie-breaking is based on the individual destruction rates. More specifically, the solution S with the highest individual destruction rate D_S is preferred over the other one. Finally, the algorithm terminates when a predefined CPU time limit is reached, and the best found solution is returned. The four procedures that form the core of PBIG are described in more detail in the following.

Algorithm 2. PBIG for the WID problem

1: **input:** input graph G, parameters $p_{\text{size}} > 0$, $D^l, D^u, d_{\text{rate}}, l_{\text{size}} \in [0, 1]$
2: $\mathcal{P} :=$ GenerateInitialPopulation($p_{\text{size}}, d_{\text{rate}}, l_{\text{size}}$)
3: **while** termination condition not satisfied **do**
4:      $\mathcal{P}_{\text{new}} := \emptyset$
5:      **for** each candidate solution $S \in \mathcal{P}$ **do**
6:          $\hat{S} :=$ DestroyPartially(S)
7:          $S' :=$ Reconstruct($\hat{S}, d_{\text{rate}}, l_{\text{size}}$)
8:          AdaptDestructionRate(S, S')
9:          $\mathcal{P}_{\text{new}} := \mathcal{P}_{\text{new}} \cup \{S'\}$
10:     **end for**
11:     $\mathcal{P} :=$ Accept($\mathcal{P}, \mathcal{P}_{\text{new}}$)
12: **end while**
13: **output:** argmin $\{f(S) \mid S \in \mathcal{P}\}$

GenerateInitialPopulation($p_{\text{size}}, d_{\text{rate}}, l_{\text{size}}$): This function generates p_{size} solutions for the initial population. For this purpose it uses the mechanism of GREEDY2[2] (see Sect. 3) in a probabilistic way. At each construction step, first, a random number $\delta \in [0, 1]$ is generated. In case $\delta \leq d_{\text{rate}}$, the best node according to the greedy function is chosen. Otherwise, a candidate list of size $\min\{|V'|, l_{\text{size}}\}$, where $V' \subseteq V$ are the nodes that can be selected at the current construction step, is generated, and one of the nodes from the candidate list is chosen uniformly at random. Note also that the initial destruction rate (D_S) of each solution S is set to the lower bound D^l for the destruction rates.

DestroyPartially(S): In this function, $\max\{3, \lfloor D_S \cdot |S| \rfloor\}$ randomly selected nodes are removed from S, where D_S is the current individual destruction rate of solution S.

Reconstruct($\hat{S}, d_{\text{rate}}, l_{\text{size}}$): Given as input a partial solution $\hat{S}$, this function reconstructs a complete solution S' in the same way in which solutions are probabilistically constructed in the context of generating the initial population (see above). Moreover, the initial destruction rate $D_{S'}$ of S' is set to D^l.

AdaptDestructionRate(S, S'): The individual destruction rate D_S of solution S (from which partial solution $\hat{S}$ was obtained) is updated on the basis of the lower bound D^l and the upper bound D^u as follows. If $f(S') < f(S)$, the value of D_S is set back to the lower bound D^l. Otherwise, the value of D_S is incremented by a certain amount. After initial experiments, we determined this amount to be 0.05. If the value of D_S, after this update, exceeds the upper bound D^u, it is set back to the lower bound D^l.

Note that the idea behind this way of dynamically changing the value of D_S is as follows. As long as the algorithm is able to improve a solution using a low destruction rate, this rate is kept low. In this way, the re-construction is faster. Only when the algorithm seems not to be able to improve over a solution, the individual destruction rate of this solution is increased in a step-wise manner.

6   Experimental Evaluation

The following five algorithmic approaches are evaluated on a variety of benchmark instances: (1) GREEDY1, (2) GREEDY2, considering edge-weights during the solution construction, (3) the application of the ILP solver CPLEX to the ILP model presented in Sect. 2 (CPLEX), (4) the ILP-based heuristic (ILP-HEURISTIC), and (5) PBIG. All techniques were implemented in ANSI C++ using GCC 4.6.3 for compiling the software. Moreover, we used CPLEX version 12.6 in single-threaded execution. The experimental results that are presented in the following were obtained on a cluster of computers with Intel® Xeon® CPU 5670 CPUs of 12 nuclei of 2933 MHz and (in total) 32 Gigabytes of RAM. For each run of CPLEX we allowed a maximum of 2 Gigabytes of RAM, which

---

[2] Note that GREEDY2 is chosen over GREEDY1 because, as it will be shown later, GREEDY2 generally works better than GREEDY1.

was never reached within the allotted computation time. In the following, first, the set of benchmark instances is described. Then, a detailed analysis of the experimental results is presented.

6.1 Benchmark Instances

For the evaluation of the proposed algorithms we used random graphs of various sizes and densities. In particular, we generated graphs of 100, 500 and 1000 nodes, that is, $|V| \in \{100, 500, 1000\}$. Edges between nodes were generated totally at random, with a given probability ep for each edge. This probability controls the density of the graph. In particular, we considered $ep \in \{0.05, 0.15, 0.25\}$. Three different schemes for generating the node and edge weights were considered. In the first scheme, both node and edge weights were drawn uniformly at random from $\{0, \ldots, 100\}$. Henceforth we call the resulting graphs *neutral graphs*. In the second scheme, node weights were drawn uniformly at random from $\{0, \ldots, 1000\}$ and edge weights were drawn uniformly at random from $\{0, \ldots, 10\}$. In these graphs, henceforth called *node-oriented* graphs, the choice of the nodes is presumably very important because of the high weights associated to the nodes. Finally, in the third scheme node weights were drawn uniformly at random from $\{0, \ldots, 10\}$ and edge-weights were drawn uniformly at random from $\{0, \ldots, 1000\}$. In these *edge-oriented* graphs, the choice of the nodes is important due to edges that are made available for connecting non-chosen nodes to chosen nodes. For each combination of graph size, edge probability, and weight generation scheme we produced 10 problem instances. This makes a total of 270 graphs.

6.2 Tuning of PBIG

The five concerned parameters are the following ones: p_{size}, D^l, D^u, d_{rate} and l_{size}. The automatic configuration tool irace [12] was applied separately for each combination of the number of nodes and the weight generation scheme. Note that no separate tuning was performed concerning the graph density (depending on ep). This is because, after initial runs, it was shown that the other parameters have a higher influence on the behavior of the algorithm. Summarizing, irace was applied 9 times with a budget of 1000 applications of PBIG per tuning run.

For each application of PBIG a time limit of $|V|/5$ CPU seconds was given. For each run of irace, two tuning instance were generated for each combination of number of nodes, graph density, and weight generation scheme. This gives a total of six tuning instances per run of irace. The following parameter value ranges were considered for each tuning run:

- $p_{size} \in \{1, 10, 50, 100\}$.
- For the lower and upper bound values of the destruction percentage, the following value combinations were considered: $(D^l, D^u) \in \{(10,10), (20,20), (30,30), (40,40), (50,50), (60,60), (70,70), (80,80), (90,90), (10,50), (30,70), (50,90)\}$. Note that in those cases in which both bounds have the same value, the percentage of deleted nodes is always the same.

– $d_{\text{rate}} \in \{0.0, 0.3, 0.5, 0.7, 0.9\}$.
– $l_{\text{size}} \in \{1, 3, 5, 10\}$.

The results of the tuning processes are shown in Table 1. The trends are very clear. The population size should be rather high, the determinism rate rather low, and the candidate list size rather high. Moreover, a dynamically changing value of the destruction rate does not seem to be necessary. In most cases a fixed value of around 0.5 is selected.

Table 1. Results of tuning PBIG with irace.

Weight scheme	$|V|$	p_{size}	(D^l, D^u)	d_{rate}	l_{size}
Neutral	100	100	(0.7, 0.7)	0.3	10
	500	100	(0.5, 0.9)	0.0	10
	1000	100	(0.4, 0.4)	0.3	10
	100	50	(0.5, 0.5)	0.3	10
Node-oriented	500	50	(0.6, 0.6)	0.5	10
	1000	100	(0.5, 0.5)	0.3	10
	100	100	(0.5, 0.5)	0.0	5
Edge-oriented	500	100	(0.5, 0.5)	0.0	10
	1000	100	(0.4, 0.4)	0.0	10

6.3 Numerical Results

The results are presented in numerical form in Table 2, which has the following format. The first three table columns indicate the number of nodes in the graph ($|V|$), the weight generation scheme, and the graph density in terms of the edge probability (ep). The results of GREEDY1, GREEDY2 and ILP-HEURISTIC are presented by means of two columns each. The first column presents in each row the average result obtained for the corresponding 10 problem instances. The second column provides the average computation times (in seconds). PBIG was applied with a computation time limit of $|V|/5$ seconds to each problem instance. We provide the average results in the first column and the average computation times at which these results were found in the second column. CPLEX was applied with two different computation time limits. In the columns with heading CPLEX we present the results that were obtained with the same computation time limit as PBIG, while the columns with heading CPLEX-L contain the results were the computation time limit was set to 3600 s per application. In both cases, the first one of the two columns presents the average of the objective function values of the best solutions found within the computation time limit for the 10 problem instances of each row. The second column indicates the average optimality gaps (in percent). Note that when the average optimality gap is zero,

Table 2. Computational results.

$|V|$	Weight scheme	ep	GREEDY1 mean	time	GREEDY2 mean	time	CPLEX mean	gap	CPLEX-L mean	gap	ILP-HEURISTIC mean	time	PBIG mean	time
100	neutral	0.05	3589.1	0.0	3519.1	0.0	3060.4	3.3	3049.8	0.0	3055.6	256.4	3051.0	1.3
		0.15	3014.4	0.0	2981.3	0.0	2470.1	34.4	2354.8	11.6	2343.4	2913.7	2093.0	2.7
		0.25	2883.5	0.0	2796.1	0.0	2195.2	41.1	2070.3	2.9	2069.3	1841.8	2070.9	0.1
	node-oriented	0.05	10465.6	0.0	11756.6	0.0	7715.4	0.0	7715.4	0.0	7715.4	0.5	7888.9	3.5
		0.15	4891.6	0.0	5845.4	0.0	3046.6	0.0	3046.6	0.0	3046.6	4.7	3155.8	1.4
		0.25	3297.5	0.0	3488.9	0.0	1808.4	0.0	1808.4	0.0	1808.4	7.8	1832.4	0.2
	edge-oriented	0.05	25698.7	0.0	22269.3	0.0	14423.9	3.1	14378.7	0.0	14435.2	102.1	14393.8	0.4
		0.15	27528.4	0.0	23404.5	0.0	16161.0	53.5	14634.7	12.9	14665.1	2813.2	14609.0	0.1
		0.25	25451.4	0.0	21770.0	0.0	16633.1	69.8	14841.2	27.8	14405.6	1913.2	14572.9	0.1
500	neutral	0.05	14143.1	0.0	13535.1	0.0	12955.5	56.2	11857.9	51.0	11780.8	3600.0	10154.6	73.9
		0.15	12268.5	0.0	11558.0	0.0	11734.5	73.8	10050.1	69.2	9734.1	3600.0	8283.0	11.4
		0.25	11630.3	0.1	10429.5	0.1	11462.4	79.0	10420.0	76.0	9287.1	3600.0	7689.6	14.6
	node-oriented	0.05	15501.5	0.0	18298.1	0.0	12567.6	53.4	12027.7	50.1	12547.5	3600.0	10264.8	71.6
		0.15	6496.3	0.1	7300.1	0.0	12863.7	82.8	5172.5	56.5	5152.6	3600.0	3718.3	20.3
		0.25	4212.4	0.1	4463.7	0.0	15011.7	86.5	3644.9	53.0	3447.8	3600.0	2667.1	7.9
	edge-oriented	0.05	125357.6	0.0	108178.0	0.0	100020.6	83.7	96350.0	82.4	94027.5	3600.0	69245.0	79.3
		0.15	114951.0	0.1	102365.1	0.0	107834.7	93.9	99377.3	92.4	90399.2	3600.0	64723.1	11.9
		0.25	111012.3	0.1	99018.2	0.0	599267.8	97.8	100750.3	94.6	84572.1	3600.0	64694.0	6.2
1000	neutral	0.05	25569.6	0.0	23489.7	0.0	25156.1	69.9	25156.1	69.4	21547.2	3600.0	17927.0	87.4
		0.15	20827.1	0.2	20689.1	0.2	129091.1	90.1	21117.8	81.0	19772.8	3600.0	15051.7	35.0
		0.25	20858.8	0.3	19280.5	0.3	229070.6	97.5	45126.9	86.5	18553.8	3600.0	14251.3	18.1
	node-oriented	0.05	18048.6	0.1	20142.3	0.1	35766.3	83.4	28123.1	76.4	16566.6	3600.0	11434.6	167.1
		0.15	7408.3	0.2	7987.4	0.2	40581.6	95.3	40581.6	93.2	7987.4	3600.0	4617.8	88.8
		0.25	4941.9	0.4	5566.6	0.3	35838.9	100.0	33023.7	93.4	5566.6	3600.0	3535.8	53.1
	edge-oriented	0.05	238600.0	0.1	202992.0	0.1	215587.9	91.9	209391.8	91.0	198508.6	3600.0	132097.2	149.5
		0.15	209709.3	0.2	182726.6	0.4	1242486.5	98.6	204521.2	96.7	179139.9	3600.0	127561.7	49.7
		0.25	198537.0	0.4	181150.0	0.3	1581837.5	99.0	903076.4	98.3	178003.7	3600.0	126145.5	21.3

all 10 corresponding instances were solved to optimality. Finally note that ILP-HEURISTIC was applied with a graph reduction of $X = 20\%$ and with the same computation time limit as CPLEX-L. The best result of each table row is shown with gray background.

The experimental results allow us to make the following observations:

- Concerning the comparison between GREEDY1 and GREEDY2 it can be observed that, generally, GREEDY2 outperforms GREEDY1 in the context of neutral graphs and edge-oriented graphs. Only in the context of node-oriented random graphs GREEDY1 outperforms GREEDY2. This shows that, generally, it is a good idea to take the edge weights already into account during the construction of a solution. Only when the edge weights are not important in comparison with the node weights—that is, in the context of node-oriented graphs—GREEDY1 has advantages.
- CPLEX is able to obtain very good results for the smallest instances with 100 nodes. However, for larger problem instances, it is not competitive anymore. Increasing the computation time limit to 3600 s (CPLEX-L) helps for some of the medium size problem instances, where the results in comparison to CPLEX improve considerably. However, when large problem instances are concerned, CPLEX-L is still not competitive.
- ILP-HEURISTIC improves in all but two cases over GREEDY2. However, this is at the cost of a huge increase in computation time. Moreover, it improves in most cases (especially for what concerns medium and large size instances) over CPLEX and CPLEX-L.
- PBIG is, overall, clearly the best-performing algorithm. It outperforms both greedy heuristics in all cases. Moreover, it outperforms both CPLEX variants and ILP-HEURISTIC for all problem instances with more than 100 nodes. Moreover, in those cases where PBIG is worse than CPLEX, it is only slightly worse. This is with the exception of two cases (node-oriented graphs on 100 nodes with $ep \in \{0.05, 0.15\}$) where the difference is more pronounced.
- Concerning the computation time requirements, the two greedy variants are clearly the fastest methods. However, even PBIG produces its best solutions in a very short computation time.

Summarizing, we can state that the algorithm of choice for small problem instances, no matter the graph density, is CPLEX, whereas for larger problem instances PBIG is clearly the best-performing approach.

7 Conclusions and Future Work

This paper has dealt with an NP-hard problem in graphs, the so-called weighted independent domination problem. We proposed the first integer linear programming model for this problem, together with a heuristic that makes use of this model. Additionally, we presented two different greedy heuristics, and a population-based iterated greedy algorithm which takes profit from the better one of the two greedy heuristics. The results have shown that small problem instances are best solved by applying a general-purpose integer linear programming solver. Medium and large scale instances, on the other side, are best solved by the population-based iterated greedy approach.

In the near future we plan to investigate if there are better ways to take profit from the developed ILP model in a heuristic way, for example, in the context of a large neighborhood search algorithm or another hybrid algorithm called construct, merge, solve and adapt.

References

1. Chang, S.C., Liu, J.J., Wang, Y.L.: The weighted independent domination problem in series-parallel graphs. In: Proceedings of ICS 2014 - The International Computer Symposium of Intelligent Systems and Applications, vol. 274, pp. 77–84. IOS Press (2015)
2. Chang, G.J.: The weighted independent domination problem is NP-complete for chordal graphs. Discrete Appl. Math. **143**(1), 351–352 (2004)
3. Wang, Y., Li, R., Zhou, Y., Yin, M.: A path cost-based GRASP for minimum independent dominating set problem. Neural Comput. Appl. (2016, in press)
4. Potluri, A., Singh, A.: Hybrid metaheuristic algorithms for minimum weight dominating set. Appl. Soft Comput. **13**(1), 76–88 (2013)
5. Chaurasia, S.N., Singh, A.: A hybrid evolutionary algorithm with guided mutation for minimum weight dominating set. Appl. Intell. **43**(3), 512–529 (2015)
6. Bouamama, S., Blum, C.: A hybrid algorithmic model for the minimum weight dominating set problem. Simul. Model. Practice Theor. **64**, 57–68 (2016)
7. Lin, G., Zhu, W., Ali, M.M.: An effective hybrid memetic algorithm for the minimum weight dominating set problem. IEEE Trans. Evol. Comput. (2016, in press)
8. Ruiz, R., Stützle, T.: A simple and effective iterated greedy algorithm for the permutation flowshop scheduling problem. Eur. J. Oper. Res. **177**(3), 2033–2049 (2007)
9. Fanjul-Peyro, L., Ruiz, R.: Iterated greedy local search methods for unrelated parallel machine scheduling. Eur. J. Oper. Res. **207**(1), 55–69 (2010)
10. Bouamama, S., Blum, C., Boukerram, A.: A population-based iterated greedy algorithm for the minimum weight vertex cover problem. Appl. Soft Comput. **12**(6), 1632–1639 (2012)
11. Porta, J., Parapar, J., Doallo, R., Barbosa, V., Santé, I., Crecente, R., Díaz, C.: A population-based iterated greedy algorithm for the delimitation and zoning of rural settlements. Comput. Environ. Urban Syst. **39**, 12–26 (2013)
12. López-Ibáñez, M., Dubois-Lacoste, J., Pérez Cáceres, L., Birattari, M., Stützle, T.: The irace package: iterated racing for automatic algorithm configuration. Oper. Res. Perspect. **3**, 43–58 (2016)

Towards Landscape-Aware Automatic Algorithm Configuration: Preliminary Experiments on Neutral and Rugged Landscapes

Arnaud Liefooghe[1,2]($\boxtimes$), Bilel Derbel[1,2], Sébastien Verel[3], Hernán Aguirre[4], and Kiyoshi Tanaka[4]

[1] Univ. Lille, CNRS, Centrale Lille, UMR 9189 – CRIStAL,
59000 Lille, France
arnaud.liefooghe@univ-lille1.fr
[2] Inria Lille – Nord Europe, 59650 Villeneuve d'Ascq, France
[3] Univ. Littoral Côte d'Opale, LISIC, 62100 Calais, France
[4] Faculty of Engineering, Shinshu University, Nagano, Japan

Abstract. The proper setting of algorithm parameters is a well-known issue that gave rise to recent research investigations from the (offline) automatic algorithm configuration perspective. Besides, the characteristics of the target optimization problem is also a key aspect to elicit the behavior of a dedicated algorithm, and as often considered from a landscape analysis perspective. In this paper, we show that fitness landscape analysis can open a whole set of new research opportunities for increasing the effectiveness of existing automatic algorithm configuration methods. Specifically, we show that using landscape features in iterated racing both (i) at the training phase, to compute multiple elite configurations explicitly mapped with different feature values, and (ii) at the production phase, to decide which configuration to use on a feature basis, provides significantly better results compared against the standard landscape-oblivious approach. Our first experimental investigations on NK-landscapes, considered as a benchmark family having controllable features in terms of ruggedness and neutrality, and tackled using a memetic algorithm with tunable population size and variation operators, show that a landscape-aware approach is a viable alternative to handle the heterogeneity of (black-box) combinatorial optimization problems.

1 Introduction

Following the advent of increasingly complex problems coming from different application fields, and implying optimization scenarios with different properties, the optimization community is continuously pushing towards the design of novel techniques that are both effective when tackling a particular problem instance, and as generic as possible in order to be flexibly adapted to a variety of problem classes. In particular, evolutionary algorithms are extremely effective to deal with a broad range of black-box optimization problems, which is one of the major reasons of their widespread uptake. Nonetheless, and despite the

© Springer International Publishing AG 2017
B. Hu and M. López-Ibáñez (Eds.): EvoCOP 2017, LNCS 10197, pp. 215–232, 2017.
DOI: 10.1007/978-3-319-55453-2_15

tremendous knowledge gained on the design of general-purpose techniques, this success can be seriously impacted by the choice of the algorithm components and parameters. For example, when designing a genetic algorithm, one has to specify what crossover and mutation rates to set in order to reach a good performance, as well as the choice of the variation operators. Moreover, it is a fact that the robustness of an algorithm, in terms of the best reachable performance, can be directly related to the characteristics of the problem instances being tackled. In this respect, a number of paradigms, techniques and dedicated software tools from automatic algorithm configuration have been proposed in order to alleviate the design of algorithms from the challenging and crucially important issue of setting their parameters [1–5]. Similarly, a huge body of literature from fitness landscape analysis was devoted to eliciting the features that make a problem instance fundamentally different from another, and to better grasp the behavior of evolutionary algorithms. In this paper, we aim at providing a first step in bridging automatic algorithm configuration with fitness landscape analysis, towards the achievement of a more powerful offline tuning framework.

Automatic Algorithm Configuration. Informally speaking, given a number of algorithm parameters (that might be numerical, discrete, or categorial), (offline) automatic algorithm configuration seeks a good configuration, that is a particular choice of the parameter values that best suits the solving of some *a priori* unknown instances [2]. Clearly, the motivation is not only to get rid from the burden of a manual calibration or the bias of personal and ad-hoc configuration processes, but more importantly to set up a principled approach for algorithm design, allowing to systematically explore their strengths and weaknesses when tackling a whole family of problems. In this context, several approaches have been proposed, ranging from racing [1,2] to statistics [3], experimental design [4], and heuristic search [5].

In this paper, we focus on the iterated racing method, which is gaining a lot of popularity, especially thanks to the flexibility of the user-friendly `irace` software [6]. Racing approaches, as most existing automated algorithm configuration methods, can be viewed from a machine learning perspective as operating in a training phase followed by a test or a production phase. Based on some given instances forming the training set, the training phase is intended to learn a good configuration that would hopefully perform well when experimented later, on some new unseen instances coming from the production phase. Roughly speaking, different configurations are first evaluated in parallel by racing, and those that are performing poorly are then discarded until one single configuration remains. Since the parameter space can be huge and an exhaustive search on the training set of instances prohibitive, a biased sampling procedure is typically implemented in order to cleverly select which configurations are to be evaluated. More specifically to iterated racing [6], the sampling distribution associated with each input parameter is updated at each iteration based on some statistical tests on the performance of running the considered configurations on some instances chosen from the input training set. It has been pointed out that the way the parameter sampling procedure and the statistical evaluation of the performance of different configurations plays a key role in guiding the iterated racing process towards the most promising configurations [6]. However, and as for any machine learning technique, the properties of the training set is a key issue in order to guarantee a high accuracy of the output configuration.

To our best knowledge, this issue has been studied only to a small extent in the context of automatic algorithm configuration. In fact, although one can safely claim that a set of available instances are already known *a priori* for a particular problem class, they might have fundamentally different structural properties, thus making them not homogeneous enough to be tackled using a single configuration. The heterogeneity of training instances was discussed briefly in [6] in the context of a tuning scenario implying SAT instances and `irace`. It was argued that such a scenario can constitute a real challenge for algorithm configuration. We also argue that a single output parameter configuration might not be suitable for the target algorithm to best suit a whole set of instances having different properties. In this paper, we rather advocate for the computation of a set of configurations, not a single one, that can then be mapped accurately with respect to the characteristics of an instance. Notice that, in iterated racing, a whole set of elite configurations can be provided as output – the set of configurations that were found to statistically have similar performance, which actually happens in many tuning scenarios, especially when the number of parameters is large. Nevertheless, it is still unclear which configuration has to be chosen in practice. Additionally, it often happens that the structural properties of a production instance, that is an instance on which the algorithm was not tuned beforehand, require a seemingly different parameter settings to reach optimal performance. This is for example typically the case in black-box optimization, where no assumption is made on the structure of the fitness function. This is precisely where fitness landscape analysis comes into play.

Fitness Landscape Analysis. When tackling black-box optimization problems, for which expert domain knowledge is typically not available, a fundamental issue is to understand what makes a problem instance difficult to solve. Similarly, it is essential to elicit the performance of a randomized search heuristic in light of the structural properties of the tackled problem. In this respect, fitness landscapes analysis [7,8] provides a set of general-purpose tools and a principled approach to systematically investigate the characteristics of an optimization problem in an attempt to guide algorithm designers towards a more in-depth understanding of the search behavior, and thus towards more effective algorithms. A typical issue addressed in fitness landscape analysis consists in studying how the performance of a given algorithm configuration can be impacted in light of insightful features from the considered problem instances. In particular, different general-purpose features were studied for this purpose [8], and such landscape features have prove their interest in successfully distinguishing between instances [9]. The general idea developed in this paper is that such features can actually serve to differentiate which parameter configuration can be more suitable for a particular problem instance, both during the training phase and during the production phase of automatic algorithm configuration. In other words, since it might be useless to search for just one single parameter configuration for an heterogeneous instance set, an alternative solution would be to consider a whole set of configurations that are explicitly associated with some elicited computable instance features. We, in fact, claim that such an idea is useful to enhance the robustness of the output configuration.

Contributions. The contributions of this paper can be stated following the next aspects:

- We adopt a landscape-oriented methodology to strengthen the accuracy of automated algorithm configuration. By partitioning the training set into different groups based on the value of landscape features, we conduct an independent training phase in parallel for each group, thus ending up with multiple algorithm configurations corresponding to the different groups. At the production phase, the appropriate configuration is selected based on the feature value of the considered instance. As a byproduct, we derive a novel landscape-aware methodology to complement existing automatic algorithm configuration in deciding on a suitable parameter setting.
- We validate the proposed landscape-aware methodology through an empirical study on the well-established benchmark family of NK-landscapes. This problem class allows us to model a black-box optimization scenario with a variety of problem instances coming from the same (pseudo-boolean) domain, but with seemingly different intrinsic characteristics. By construction, a number of features, that are often found to impact the performance of evolutionary algorithms, are in fact made controllable. This results in a particularly interesting adversary benchmark for studying the challenges that automated algorithm configuration has to face when tackling heterogenous instances. In particular, we focus on the behavior of iterated racing when tackling problems with a variable degree of ruggedness and neutrality.
- By fairly taking the extra computational cost induced by our methodology into account, we investigate the gain of deciding which parameter configuration to choose for an unseen production instance based on general-purpose low-cost computable features. Our empirical findings reveal that landscape-aware iterated racing is able to find better configurations when experimented in a conventional memetic algorithm with tunable population size, variation operators, crossover and mutation rates.

Positioning. Our work shares similarities with previous attempts from automatic configuration. In Hydra [10], a portfolio builder is used together with an automatic configuration method in order to construct a portfolio of algorithm configurations. The portfolio builder typically uses problem features to discard or add new configurations found by automatic configuration, and the method was proved effective when experimented with SAT specific tools. However, it requires both a suitable portfolio builder and a domain-specific knowledge, which can constitute a bottleneck in practice for black-box optimization. In SMAC [11], landscape features are used within the tuning process as a subset of input variables in order to construct a model predicting algorithm performance, but a single recommended algorithm configuration is returned for the whole instance set. In ISAC [12], features are used for instance-specific algorithm configuration, but the authors consider problem-specific features, whereas our proposal attempts to address black-box optimization problems.

Outline. For the sake of presentation and completeness, we first start by describing in Sect. 2 the rationale behind NK-landscapes, as well as by defining some general-purpose features that we shall use in order to empirically revisit the characteristics of NK-landscapes. In Sect. 3, which is the core of the paper, we describe the proposed landscape-aware methodology for automatic algorithm configuration and experimentally investigate its accuracy on NK-landscapes. In Sect. 4, we conclude the paper while providing some future research questions.

2 Initial Considerations on Pseudo-Boolean Landscapes

2.1 NK-, NK$_q$- and NK$_p$-Landscapes

The family of NK-landscapes constitutes a problem-independent model used for constructing multimodal benchmark instances with variable *ruggedness* [13]. The fitness function f is a pseudo-boolean function $f : \{0,1\}^N \to [0,1]$ to be maximized. Candidate solutions are binary strings of size N, i.e. the solution space is $X := \{0,1\}^N$. The fitness value $f(x)$ of a solution $x = (x_1, \ldots, x_i, \ldots, x_N)$ is an average value of the individual contributions associated with each variable x_i. Indeed, for each x_i, $i \in [\![1, N]\!]$, a component function $f_i : \{0,1\}^{K+1} \to 0,1$ assigns a positive contribution for every combination of x_i and its K *epistatic interactions* $\{x_{i_1}, \ldots, x_{i_K}\}$. Thus, the individual contribution of a variable x_i depends on the value of x_i, and on the values of $K < N$ other binary variables $\{x_{j_1}, \ldots, x_{j_K}\}$. The problem can be formalized as follows:

$$\arg\max_{x \in \{0,1\}^N} f(x) = \frac{1}{N} \sum_{i=1}^{N} f_i(x_i, x_{i_1}, \ldots, x_{i_K})$$

The epistatic interactions, i.e. the K variables that influence the contribution of x_i, are here set uniformly at random among the $(N-1)$ other variables, following the random model from [13]. By increasing the number of epistatic interactions K from 0 to $(N-1)$, NK-landscapes can be gradually set from smooth to rugged. It is worth noticing that this is intended to provide a family of black-box benchmark functions that allow to study challenging aspects that can make a practical combinatorial optimization problem instance difficult to solve, such as ruggedness or multimodality [7,13,14].

Moreover, NK-landscape were shown to be extendable to optimization scenarios in the presence of different degrees of *neutrality*, which is also a critical issue when dealing with combinatorial optimization problems [15,16]. Accordingly, Newman [17] and Barnett [18] introduced a controllable level of *neutrality* as follows. In the so-called quantized NK$_q$-landscapes [17], the f_i-values are generated following a discrete uniform distribution $[\![0, q-1]\!]$, and are scaled down by a factor of $\frac{1}{q-1}$. In the so-called probabilistic NK$_p$-landscapes [18], the f_i-values are set to 0 with a probability p, and otherwise generated as in the original NK-landscapes with a probability $(1 - p)$, where p is a benchmark parameter. To summarize, we shall consider $NK_{q|p}$-landscapes as described above, where it is expected that the larger K the higher the level of ruggedness, and that the smaller q (respectively the larger p) the higher the level of neutrality.

2.2 $NK_{q|p}$-Landscapes Features

As mentioned earlier, fitness landscape analysis aims at studying the topology of a combinatorial optimization problem by gathering important information such as ruggedness or multimodality [7,14]. It is important to remark that such an information is typically *not* available *a priori*, when effectively solving a given unseen problem instance. Actually, in a typical black-box optimization scenario, even the parameters that originate a particular problem instance might not be available. With respect to the $NK_{q|p}$-benchmark family, we might typically consider a configuration scenario where the instance parameter values such as K, p or q, are not known by the optimizer. In this context, a fitness landscape analysis might allow us to extract valuable information on the structural properties of an instance. For this purpose, we first report some general-purpose properties of the considered $NK_{q|p}$ benchmarks by taking inspiration from [18]. Our goal is also to provide empirical evidence that this benchmark family is rather heterogenous, and is indeed a good adversary candidate for evaluating the behavior of automatic algorithm configuration. We consider an instance dataset of 800 $NK_{q|p}$-landscapes with a problem size $N \in [\![500, 2\,000]\!]$, an epistatic degree $K \in [\![0, 10]\!]$, and a neutral degree $q \in [\![2, 10]\!]$ for NK_q-landscapes, respectively $p \in [0.60, 0.93]$ for NK_p-landscapes. The range of the parameters q and p have been chosen in order to obtain a similar range of neutral degrees on NK_q- and NK_p-landscapes. A total of 800 instances are considered, with one instance generated at random for each parameter combination. Half of the instances correspond to NK_q-landscapes, while the other half are NK_p-landscapes. The parameters have been generated from a design of experiments based on a latin hypercube sampling.

Formally, a fitness landscape is defined by a triplet $(X, \mathcal{N}, f)$, such that X is a set of admissible solutions (the search space), $\mathcal{N} : X \rightarrow 2^X$ is a neighborhood relation between solutions, and $f : X \rightarrow \mathbb{R}$ is a black-box fitness function, here assumed to be maximized. A simple sampling technique for examining features from the landscape is to perform a *random walk* over the landscape. More specifically, an infinite random walk is an ordered sequence $\langle x_0, x_1, \ldots \rangle$ of solutions such that $x_0 \in X$, and x_t is a neighboring solution selected uniformly at random from $\mathcal{N}(x_{t-1})$. In the same spirit than for the heterogeneous scenario mentioned in [6], a first feature that we might consider is the average fitness value of a random walk, which can be approximated by means of a finite random walk $\langle x_0, x_1, \ldots, x_\ell \rangle$ of length ℓ as follows: $\bar{f} = \frac{1}{\ell} \sum_{t=1}^{\ell} f(x_t)$. The average fitness value encountered along a random walk can actually be used to differentiate a given set of instances. This is exactly what we report in Fig. 1 for the $NK_{q|p}$-landscapes, where ℓ is set to $1\,000$. We can observe that NK_q-landscapes clearly differ from NK_p-landscapes, as the range of average fitness values is substantially different. While the instances generated with different $q-$values appear to

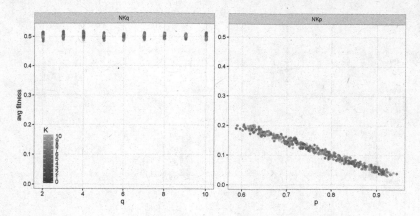

Fig. 1. Scatter plot of mufit (average fitness value) as a function of p and q for all instances.

be rather uniform in terms of average fitness value (independently of K), the average fitness value is in contrast decreasing linearly as a function of p. This provides a first hint on the differences that we might encounter in the landscape of different instances.

In order to go further in the analysis, the autocorrelation [14] between the fitness values of consecutive solutions in a random walk can be used to characterize an important feature of an instance, namely its ruggedness. We consider the following approximation to estimate the so-called autocorrelation coefficient $\hat{r}(k)$:

$$\hat{r}(k) = \frac{\sum_{t=1}^{\ell-k}(f(x_t) - \bar{f}) \cdot (f(x_{t+k}) - \bar{f})}{\sum_{t=1}^{\ell}(f(x_t) - \bar{f})^2}$$

We use the first autocorrelation coefficient $r(1)$ to characterize ruggedness: the larger $r(1)$, the smoother the landscape [14]. We report in Fig. 2 this coefficient as a function of K. As expected, we can observe that the first autocorrelation coefficient tends to decrease with the degree of non-linearity. This means that the larger K, the more likely to fall into a local optimum. Notice that this tendency is the same for both NK_q- and NK_p-landscapes.

At last, we shall examine a feature capturing the degree of neutrality, which explicitly relates to parameters p and q in $NK_{q|p}$-landscapes. Given a solution x, we denote a neighboring solution $x' \in \mathcal{N}(x)$ as a *neutral neighbor* if it has the same fitness value: $f(x') = f(x)$ [15]. The *neutral degree* of a solution is then defined as the number of its neutral neighbors. Consequently, different statistics can be used to quantify the neutral degree of a given instance, following different sampling strategies that induce different computational costs. Since we shall fairly include the cost of computing such features later when addressing the effectiveness of an algorithm configuration method, we consider a new estimator that solely looks at consecutive solutions along a random walk. More specifically, let $NN = \{(x_i, x_{i+1}) \mid f(x_i) = f(x_{i+1}), i \in \{0, \ldots, \ell - 1\}\}$ be the set of pairs of solutions with the same fitness value in the random walk. We consider

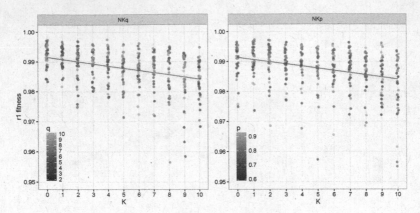

Fig. 2. Scatter plot of rho1fit (first autocorrelation coefficient) as a function of K for all instances.

the following low-cost feature to render neutrality: rateeq $= \frac{|NN|}{\ell}$, which is the proportion of pairs of neutral neighbors along the random walk. In Fig. 3, we report the neutral degree of the considered instances as a function of the different parameters K, q and p. The neutral degree decreases (resp. increases) with q (resp. p), which is with no surprise given the definition of these two parameters in $NK_{q|p}$-landscapes. However, a notable observation is that the neutral degree is relatively higher for NK_p-landscapes (up to 0.8) compared against NK_q-landscapes (up to 0.6), which is yet another interesting information about the heterogeneity of these instances. Interestingly, we clearly see that the neutral degree is not only dependent on parameters q or p, but also on the degree of non-linearity K, as previously pointed out in [18]. Actually, the higher the value of K, the lower the neutral degree. We also remark that for instances with a high level of non-linearity K, the difference in the range of neutrality between NK_p- and NK_q-landscapes decreases significantly, and the neutral degree appears to be roughly the same.

To conclude this section, let us emphasis that, although $NK_{q|p}$-landscapes belong to the same problem family, they are seemingly different as they expose different degrees of ruggedness and neutrality. This is likely to be the case in practice for other problem classes, where one can expect different instances to have different properties, and hence to expose different degrees of difficulty. In this respect, a reasonable hypothesis is that the optimal setting of the considered optimization algorithm depends on instance properties. This is precisely what we address in the remainder of this paper.

3   Feature-Based Algorithm Configuration

In this section, we describe a feature-based algorithm configuration methodology, and provide an empirical evidence of its benefits when tuning a standard memetic algorithm.

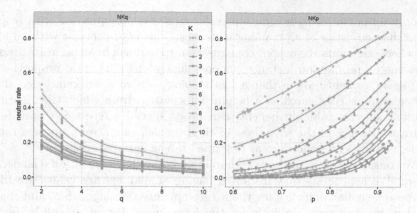

Fig. 3. Scatter plot of rateeq (neutral degree) as a function of p and q for all instances.

3.1 Feature-Aware Iterated Racing

For completeness, we first start recalling the main steps of conventional iterated racing as performed in `irace` [6]. Our interest in this approach stems from its successful application in tuning different optimization techniques for a rather wide range of optimization problems [6]. The input of `irace` is a set of parameters $\theta = \{x_1, \ldots, x_n\}$ from the algorithm to be configured, and a set of training instances $\mathcal{I} = \{I_1, \ldots, I_k\}$. The output is typically a set of elite configurations $\theta^* = \{\theta_1, \ldots, \theta_r\}$ that allow the target algorithm to perform at its best with respect to some performance metric. Notice that `irace` is actually a stochastic search process performing in the parameter space, and hence no guarantee is actually provided on the optimal performance of the output configuration. That said, `irace` consists in three main steps that are repeated sequentially as follows, until a termination condition is met. First, some configurations are sampled according to a particular probability distribution. The best configurations are then selected using a racing procedure [6]. More specifically, the sampled configurations are evaluated for a number of steps by executing the algorithm with the parameter setting mapping to those configurations. At each step of the race, one instance from $\mathcal{I}$ is considered. The configurations that were found to perform statistically worse than others are then discarded, and the race continues with the surviving configurations. Finally, the distribution from where the configurations are sampled from is updated in order to bias the search towards the most promising configurations found in previous iterations. As will be detailed later, we use a standard termination criterion which is a user-defined computational budget, in terms of a number of algorithm execution. The performance metric is simply the quality of the best solution found during an algorithm execution.

At this stage, it is important to remark that `irace` is intended to be a general-purpose tuning approach. In particular, no assumption is made from the set of input training instances $\mathcal{I}$. Following the same motivations from the no-free lunch theorem, the idea developed in this paper is precisely that there cannot exist a unique optimal configuration for a whole set of instances. Consequently, `irace` can only output a configuration representing a good compromise with respect to the characteristics of all training instances. This is to contrast with an ideal case where one wants the output configuration to perform in an accurate manner to an unseen production instance, independently of its intrinsic properties. In this paper, we hence argue that a methodology where some knowledge about the landscape is considered as a helpful information from which the algorithm configuration can valuably benefit, can be of special interest. To provide an empirical evidence of the soundness of the previous claim, we propose a rather simple, yet efficient, procedure as described in the next paragraph.

We consider that an instance is characterized by the value of some landscape feature. We hypothesis that instances having similar feature values are likely to expose a similar difficulty for the target optimization algorithm, and that it can then be configured similarly for those instances. Let us denote by $\mathsf{feat}(I)$ the value of feature feat for instance I. Since we might have numerical, discrete, or even categorial features, we assume for now that we are able to classify an instance I into a unique class according to its feature value $\mathsf{feat}(I)$. Let us assume as well that we have s such classes, where s is a pre-defined parameter. We then proceed as follows: (i) we partition the training set into s groups according to the feature values, i.e. $\mathcal{I} = \mathcal{I}^1 \cup \mathcal{I}^2 \cup \ldots \cup \mathcal{I}^s$, where $\mathcal{I}^i$ contains instances from the same class; and (ii) we run $\mathtt{irace}$ independently, using every partition $\mathcal{I}^i$ separately as an input training set. Since $\mathtt{irace}$ is then executed s times on the s training sets, we obtain as output s elites configurations: $\theta_1^* \cup \theta_2^* \cup \ldots \cup \theta_s^*$, where θ_j^* maps to instances of class $j \in \{1, \ldots, s\}$. Since these output configurations are hence explicitly related to the feature class, it becomes straightforward to decide which elite configuration to choose when experiencing a new unseen production instance. More specifically, given a new unseen test instance, we first compute its feature class j, and we simply consider the elite configuration θ_j^*, computed by $\mathtt{irace}$ beforehand, in order to effectively set the parameters of the optimization algorithm for this unseen instance. Designing insightful problem features is to be understood as a challenging issue in practice, and it is worth noticing that the general-purpose landscape features for black-box combinatorial optimization that we consider in this paper do not require any expert domain knowledge. The proposed methodology is to be viewed as a first step towards the design of more sophisticated approaches, as will be discussed in more details in the conclusions. Our main goal is in fact to study at which extent a landscape-aware automatic algorithm configuration methodology could be beneficial.

Up to now, we did not address the cost of computing the feature values, nor the computational effort devoted to the tuning task. This is an important issue when evaluating the proposed methodology. For fairness, we split the available budget B equally over the s runs of $\mathtt{irace}$, i.e. each run $j \in \{1, \ldots, s\}$ of $\mathtt{irace}$ with $\mathcal{I}^j$ uses as termination condition a maximum number of algorithm runs which is set to B/s. Additionally, we consider to subtract the cost of computing the feature from the computational effort devoted to execute the algorithm on a given instance, both at the training phase of $\mathtt{irace}$, but more importantly at the test or production phase, when computing the class of a new unseen instance. This is to be specified in more details in our experimental setup.

Table 1. Parameter space for tuning the Memetic Algorithm (MA) for $NK_{q|p}$-landscapes.

Parameter	Domain	Type
Population size	$\{1, 2, 4, 8, 16, 32, 64, 128, 256, 512, 1\,024, 2\,048\}$	Ordinal
Crossover operator	$\{$unif, 1-point, 2-point$\}$	Categorical
Crossover rate	$\{0.00, 0.05, 0.1, \ldots, 0.95, 1.00\}$	Ordinal
Mutation rate	c/N, s.t. $c \in \{0.0, 0.5, 1.0, \ldots, 9.5, 10.0\}$	Ordinal

3.2 Memetic Algorithm and Parameter Space

As a case study, and in order to highlight the relevance of the previously-described methodology, we consider the configuration of the main components of a memetic algorithm (MA) similar to [19] as one alternative to solve the class of $NK_{q|p}$-landscapes. The MA evolves a population of candidate solutions represented as binary strings. Starting from a randomly-generated population P of size μ, the MA proceed in consecutive iterations. At each iteration, two solutions from the current population are selected using a binary tournament selection, and a new offspring is created by means of crossover followed by mutation. The crossover is applied with a fixed probability r_c. The mutation consists in flipping each bit with a probability r_m. We then use a local search to enhance the so-obtained offspring. Specifically, a first-improvement hill-climbing algorithm is implemented. Solutions at hamming distance 1 are examined in a random order, and the first improving neighbor is selected until a local optimum is found. After a set of μ offspring solutions are created in this manner, a generational replacement is performed. The newly-generated solutions becomes the current population and the best individual from the old population replaces the worst solution if it is better than the best newly generated offspring. The algorithm terminates after a fixed number of fitness function evaluations.

The parameter space for the automatic design of the MA is given in Table 1. We consider to tune the population size, which is known to be a critical issue in evolutionary computation. We hence choose a set of values ranging from very small (1) to very large (2 048). For crossover, we consider three well-established binary string operators, namely one-point crossover, two-point crossover, and uniform crossover. The possible values for the crossover rate (r_c) ranges from 0 (no crossover) to 1 (crossover always performed). The possible values for the mutation rate (r_m) are set as a function of N (the bit-string size), and controls the number of bits that are flipped in average. Although some of these parameters could have been specified as real or integer parameters, we decided to discretize them in order to reduce the size of the parameter space in `irace`.

3.3 Experimental Setup

We use the `irace` R-package [6], that provides the reference implementation of iterated racing. As training instances, we consider the same set of 800 instances as described previously in Sect. 2. We consider two types of features: (i) the benchmark parameters from $NK_{q|p}$-landscapes: N, K, p or q, and the type of neutrality, where the first three are numerical and the last one is categorial (i.e. quantized or probabilistic), and (ii) the general-purpose features as discussed in Sect. 2, namely the average fitness mufit, the first autocorrelation coefficient rho1fit, and the neutral rate rateeq, all computed based on a random walk of budget $\ell = 1\,000$. In order to partition the training set, we consider a one-dimensional simple strategy that takes each feature separately, and then splits the instances into a fixed number of clusters with equal range of that feature values (see Table 2). This simple partitioning strategy is to be viewed as a first step towards more sophisticated clustering strategies involving more than one feature at a time, that is left for future research. Except for the feature involving the type of instance (and where the number of clusters is two), we choose to partition the training instances into four clusters. Notice also that since neutrality can be controlled independently by parameter p or q, we combine these parameters to constitute one feature denoted p|q, for which we also have four groups: two from NK_q- and two from NK_p-landscapes. For the test phase, we independently generate a test set of 200 instances, following the same experimental design discussed in Sect. 2. These additional instances are used to test the accuracy of the output configurations and are not available for irace during the training phase. As one can appreciate in Table 2, the instances from the training set and the test set are actually well balanced over the different clusters.

Following [6], we use irace with a tuning budget of 20 000 algorithm runs, where each run of the MA performs 100 000 calls to the fitness function. As previously mentioned, when the proposed feature-based methodology is experimented, we split the budget equally over the different clusters. Since we need to perform a random walk beforehand to compute the features mufit, rho1fit, rateeq, we subtract 1 000 fitness function calls from the overall MA budget, both during the training and the test phases, in order to tune the MA in production-like conditions. Notice that, although K, p and q are typically not available for the algorithm, we still include them in our experiments for the sake of illustrating the gain one can expect from the proposed methodology.

3.4   Experimental Results

In Table 2, we report the best configuration (the first one in the elite set) found when running irace with the whole set of training instances, which is considered as a baseline approach (first row in the Table). We thereby report the best configurations found when combining irace with the proposed feature-based methodology. The most notable observation at this stage of the analysis is that a uniform crossover is always preferred, except for the second group of instances partitioned with respect to rho1fit, together with a relatively high crossover rate (except for the third group of instances partitioned by K). However, the best-found population size varies substantially when comparing the output of the baseline irace and the proposed methodology. We can also remark that, when

Table 2. First elite configuration found by irace for each feature cluster. The first row corresponds to the configuration found when considering the whole training set.

problem feature	cluster	feature range			# inst. (training , test)		pop. size	crossover operator	cross. rate	mut. rate
*	—				(800 ,	200)	32	uniform	0.95	5.5
N	#0: N	∈ [	501 ,	877)	(200 ,	50)	16	uniform	0.95	6.5
	#1: N	∈ [	877 ,	1 253)	(200 ,	51)	32	uniform	1.00	6.5
	#2: N	∈ [	1 253 ,	1 627)	(200 ,	49)	64	uniform	0.75	6.5
	#3: N	∈ [	1 627 ,	2 000]	(200 ,	50)	64	uniform	1.00	8.5
K	#0: K	∈ [	0 ,	3)	(218 ,	54)	256	uniform	1.00	7.5
	#1: K	∈ [	3 ,	6)	(218 ,	55)	64	uniform	0.95	7.0
	#2: K	∈ [	6 ,	9)	(219 ,	54)	32	uniform	0.30	7.0
	#3: K	∈ [	9 ,	10]	(145 ,	37)	16	uniform	1.00	6.0
type	#0: type	=	NK$_q$		(400 ,	100)	32	uniform	0.75	7.0
	#1: type	=	NK$_p$		(400 ,	100)	64	uniform	0.85	7.5
p | q	#0: param	∈ [	0.600 ,	0.765)	(200 ,	50)	64	uniform	1.00	8.0
	#1: param	∈ [	0.765 ,	0.930]	(200 ,	50)	32	uniform	0.95	7.0
	#2: param	∈ [	2.000 ,	6.000)	(222 ,	55)	32	uniform	0.80	6.5
	#3: param	∈ [	7.000 ,	10.000)	(178 ,	45)	64	uniform	0.95	7.0
avg fitness	#0: mufit	∈ [	0.031 ,	0.117)	(200 ,	49)	64	uniform	0.95	7.5
	#1: mufit	∈ [	0.117 ,	0.486)	(200 ,	51)	64	uniform	0.75	8.0
	#2: mufit	∈ [	0.486 ,	0.501)	(200 ,	59)	32	uniform	0.90	6.5
	#3: mufit	∈ [	0.501 ,	0.519]	(200 ,	41)	32	uniform	0.85	7.5
r1 fitness	#0: rho1fit	∈ [	0.955 ,	0.985)	(200 ,	50)	32	uniform	0.95	6.5
	#1: rho1fit	∈ [	0.985 ,	0.989)	(200 ,	60)	32	1−point	0.90	7.5
	#2: rho1fit	∈ [	0.989 ,	0.993)	(200 ,	46)	64	uniform	1.00	7.5
	#3: rho1fit	∈ [	0.993 ,	0.998]	(200 ,	44)	32	uniform	0.95	7.5
neutral rate	#0: rateeq	∈ [	0.000 ,	0.044)	(205 ,	55)	16	uniform	0.80	7.0
	#1: rateeq	∈ [	0.044 ,	0.085)	(197 ,	48)	64	uniform	0.90	6.5
	#2: rateeq	∈ [	0.085 ,	0.193)	(198 ,	47)	16	uniform	1.00	6.5
	#3: rateeq	∈ [	0.193 ,	0.841]	(200 ,	50)	128	uniform	0.95	7.5

adopting a feature-based tuning methodology, the mutation rate is higher compared against the baseline setting. Although it is difficult to correlate these observations with the considered NK$_{q|p}$-landscapes, we can clearly see that irace is able to seemingly find different configurations, depending on how the input training test is partitioned. We attribute this to the fact that instances belonging to the same group are expected to expose less heterogeneity for the configuration procedure.

To go further into the analysis, we evaluate, for each individual feature, how the feature-based methodology performs against the configuration obtained when mixing all the instances as in baseline irace. To do so, we examine the performance of the MA when experimented on 200 independently-generated testing instances. We execute the MA with every configuration for 30 runs on each test instance, while subtracting the cost of computing the features to the budget allocated to MA whenever necessary. In Fig. 4, we report the number of test instances where the configuration found by feature-based irace allows the MA to perform significantly better (resp. worst, and insignificantly different) than when configured using the output of baseline irace. For the pairwise comparison of configurations on the same instance, we use a Wilcoxon signed rank statistical test with a p-value of 0.05 and a Bonferroni correction. Overall, the

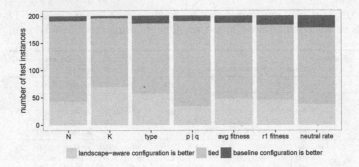

Fig. 4. Number of test instances where the landscape-aware configuration with respect to each feature is significantly better, tied or worse than the baseline configuration.

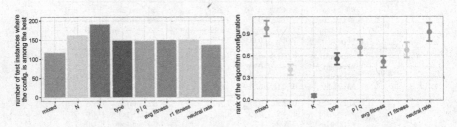

Fig. 5. Number of test instances (out of 200) where the baseline configuration (mixing all training instances) and each feature-based configuration (partitioning training instances) is not statistically outperformed by any other (left), and rank of each configuration over all test instances (right).

proposed methodology appears to effectively enhance the baseline one, since the number of instances on which the feature-based configuration provides better results is significantly higher than the baseline configuration, independently of the considered feature. This is confirmed by the basic statistics reported in Fig. 5, comparing baseline `irace` against `irace` using the feature-based partitioning. More precisely, on the left subfigure, we show the number of instances where the corresponding configuration is not statistically outperformed by any other. In the right subfigure, we report the number of times a given MA configuration is statistically outperformed by another. For a given configuration, a dot corresponds to the average rank over all test instances, where a value of 0 means that a specific configuration was actually never outperformed by any other on any test instance. Interestingly, baseline `irace` appears to identify the configuration with the largest rank. We can also see that the feature-based configuration methodology performs at its best when using K, which suggests that the non-linearity and the ruggedness of the instances is one of the most important feature one has to take into account when configuring the MA. The problem size N and the average fitness value `avg fitness` are also among the most insightful features when searching for a good configuration of the MA. Notice also that feature `rho1fit`, which is intended to approximate the ruggedness of an instance, does not allow irace to perform as well as with K, although it still has a better overall ranking compared to baseline irace. This suggests that alternative features that could approximate the ruggedness of a given instance more accurately would be worth investigating in the future.

The previous statistics aggregate the instances over the whole test set. In Fig. 6, we report a more detailed description on the relative behavior of feature-based irace. Specifically, the x-axis of each subfigure refers to the corresponding feature values from all test instances. Then, for each instance, the y-axis indicates whether configuring the MA with baseline irace provides statistically better (resp. worst, tied) performance than the proposed methodology. This allows us to investigate in more details the distribution of instances where we are able to improve or to worsen the performance of baseline irace by feature values. We clearly see that, overall, the feature-based methodology allows to enhance irace, independently of the feature values, and then independently of the characteristics of the considered instance. This is of high importance, since we can then claim that a landscape-aware automatic algorithm configuration effectively allows to improve parameter accuracy for a relatively large spectrum of heterogeneous instances.

At last, we report in Fig. 7 the results of cross-validating the performance of the different configurations that irace is able to obtain for each partition, with respect to a particular feature. Specifically, the x-axis refers to the group of test instances obtained by partitioning, i.e. four groups except for type. Then, for each group of test instances, we compare all other configurations that irace is able to find when considering either the whole set of training instances or a specific subgroup of training instances. The number of test instances where

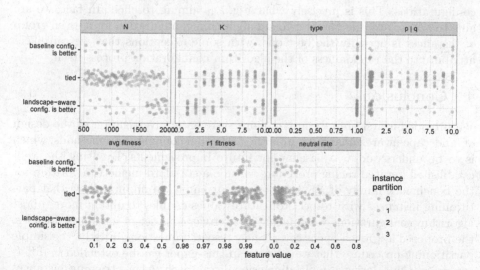

Fig. 6. Detailed distribution of test instances where the landscape-aware configuration with respect to each feature is significantly better, tied or worse than the baseline configuration, as a function of the feature value.

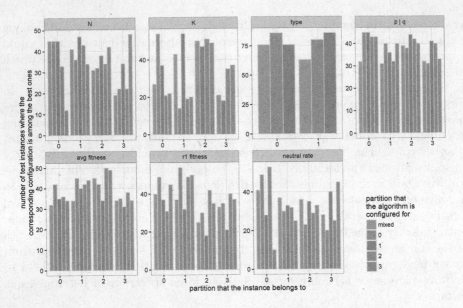

Fig. 7. Number of test instances where each landscape-aware configuration is not outperformed by any other, as a function of the feature group.

the corresponding configuration is not statistically outperformed by any other is reported in the y-axis. One should expect that, when running the algorithm configuration obtained specifically for the group of training instances to which the test instance belongs to, the performance is at its best relatively to other configurations. This is precisely what Fig. 7 is aiming to elicit. In fact, we are able to appreciate that the best-found algorithm configuration for a given group of instances is actually the best one, with some exceptions that we can likely attribute to the randomness of the algorithm configuration process itself.

4   Conclusions

We provided a first step towards a more systematic investigation of the design of landscape-aware enhanced automatic algorithm configuration methods, which is to be understood as a baseline for future improvements. By using the well-established iterated racing procedure to tune a standard memetic algorithm for the benchmark family of NK-landscapes, our empirical findings show that partitioning instances with respect to feature values enables to obtain more robust algorithm configurations when facing a heterogeneous set of instances. Besides, the proposed approach opens several new research questions. Firstly, the simple partitioning procedure that we adopted in this paper can be extended in different ways. Considering a multi-dimensional approach, where training instances are clustered by using multiple landscape features simultaneously, is of special interest in order to capture the similarities and differences of instances from different inter-dependent and orthogonal perspectives. Additionally, the number of groups was fixed empirically in our study, such as the global budget allowed for the whole tuning process. We believe that a more systematic investigation on the granularity of the partitioning procedure and its relation with the available budget will lead to new insightful results on the accuracy of landscape-aware algorithm configuration. Notice that the granularity of the partitioning actually opens nice opportunities for distributing the flow of the tuning procedure over different parallel cooperating entities, thus improving the quality and runtime of offline algorithm configuration, which is actually known to be time consuming. Secondly, the methodology adopted in this work does not change the way the tuning process is conducted, but simply considers the tuning procedure as a black-box mechanism. Nevertheless, we believe that the same idea of using landscape analysis to characterize instances can be seemingly used inside the tuning procedure itself, thus ending-up with new algorithm configuration methods. With respect to iterated racing, one particularly promising idea consists in carefully choosing the instances where some configuration should race at every iteration based on the features values of the instances experimented in previous iterations. At last, it would be interesting to benchmark and extend our work with other scenarios, such as different algorithms, different problems, different domains, or different tuners, and to compare our methodology with approaches from [10–12]. A particularly challenging issue is to highlight which general-purpose features can allow to provide the highest insights, and then the most accurate configurations.

Acknowledgments. We are grateful to M. López-Ibáñez for fruitful suggestions on the paper.

References

1. Birattari, M., Stützle, T., Paquete, L., Varrentrapp, K.: A racing algorithm for configuring metaheuristics. In: Genetic and Evolutionary Computation Conference, pp. 11–18 (2002)
2. Birattari, M.: Tuning Metaheuristics: A Machine Learning Perspective. Springer, Heidelberg (2009)
3. Bartz-Beielstein, T.: Experimental Research in Evolutionary Computation. Springer, Heidelberg (2006)
4. Adenso-Diaz, B., Laguna, M.: Fine-tuning of algorithms using fractional experimental designs and local search. Oper. Res. **54**(1), 99–114 (2006)
5. Hutter, F., Hoos, H.H., Leyton-Brown, K., Stützle, T.: ParamILS: an automatic algorithm configuration framework. J. Artif. Int. Res. **36**(1), 267–306 (2009)
6. López-Ibáñez, M., Dubois-Lacoste, J., Cáceres, L., Birattari, M., Stützle, T.: The irace package: iterated racing for automatic algorithm configuration. Oper. Res. Perspect. **3**, 43–58 (2016)
7. Merz, P.: Advanced fitness landscape analysis and the performance of memetic algorithms. Evol. Comput. **12**(3), 303–325 (2004)
8. Richter, H., Engelbrecht, A. (eds.): Recent Advances in the Theory and Application of Fitness Landscapes. Emergence Complexity and Computation. Springer, Heidelberg (2014)

9. Smith-Miles, K., Lopes, L.: Measuring instance difficulty for combinatorial optimization problems. Comput. Oper. Res. **39**(5), 875–889 (2012)
10. Xu, L., Hoos, H.H., Leyton-Brown, K.: Hydra: automatically configuring algorithms for portfolio-based selection. In: Conference on Artificial Intelligence, pp. 210–216 (2010)
11. Hutter, F., Hoos, H.H., Leyton-Brown, K.: Sequential model-based optimization for general algorithm configuration. In: Coello, C.A.C. (ed.) LION 2011. LNCS, vol. 6683, pp. 507–523. Springer, Heidelberg (2011). doi:10.1007/978-3-642-25566-3_40
12. Kadioglu, S., Malitsky, Y., Sellmann, M., Tierney, K.: ISAC - instance-specific algorithm configuration. In: European Conference on Artificial Intelligence, pp. 751–756 (2010)
13. Kauffman, S.A.: The Origins of Order. Oxford University Press, Oxford (1993)
14. Weinberger, E.D.: Correlated and uncorrelatated fitness landscapes and how to tell the difference. Biol. Cybern. **63**(5), 325–336 (1990)
15. Verel, S., Collard, P., Clergue, M.: Scuba search: when selection meets innovation. In: Congress on Evolutionary Computation, pp. 924–931 (2004)
16. Marmion, M.-E., Dhaenens, C., Jourdan, L., Liefooghe, A., Verel, S.: On the neutrality of flowshop scheduling fitness landscapes. In: Coello, C.A.C. (ed.) LION 2011. LNCS, vol. 6683, pp. 238–252. Springer, Heidelberg (2011). doi:10.1007/978-3-642-25566-3_18
17. Newman, M., Engelhardt, R.: Effect of neutral selection on the evolution of molecular species. Proc. R. Soc. London B. **256**, 1333–1338 (1998)
18. Barnett, L.: Ruggedness and neutrality - the NKp family of fitness landscapes. In: International Conference on Artificial Life, pp. 18–27 (1998)
19. Pelikan, M.: Analysis of estimation of distribution algorithms and genetic algorithms on NK landscapes. In: Genetic and Evolutionary Computation Conference, pp. 1033–1040 (2008)

Understanding Phase Transitions with Local Optima Networks: Number Partitioning as a Case Study

Gabriela Ochoa[1]([⊠]), Nadarajen Veerapen[1], Fabio Daolio[1], and Marco Tomassini[2]

[1] Computing Science and Mathematics, University of Stirling,
Stirling, Scotland, UK
gabriela.ochoa@cs.stir.ac.uk
[2] Faculty of Business and Economics, Information Systems Department,
University of Lausanne, Lausanne, Switzerland

Abstract. Phase transitions play an important role in understanding search difficulty in combinatorial optimisation. However, previous attempts have not revealed a clear link between fitness landscape properties and the phase transition. We explore whether the global landscape structure of the number partitioning problem changes with the phase transition. Using the local optima network model, we analyse a number of instances before, during, and after the phase transition. We compute relevant network and neutrality metrics; and importantly, identify and visualise the funnel structure with an approach (*monotonic sequences*) inspired by theoretical chemistry. While most metrics remain oblivious to the phase transition, our results reveal that the funnel structure clearly changes. Easy instances feature a single or a small number of dominant funnels leading to global optima; hard instances have a large number of suboptimal funnels attracting the search. Our study brings new insights and tools to the study of phase transitions in combinatorial optimisation.

1 Introduction

It has been recognised that phase transitions play an important role in analysing combinatorial optimisation problems; yet a clear link between fitness landscape structure and the phase transition phenomenon is still lacking. We use the local optima networks model to analyse and visualise the global structure of Number Partitioning fitness landscapes.

The Number Partitioning Problem (NPP) is defined as follows. Given a set of N positive numbers $L = \{r_1, r_2, \ldots, r_N\}$, find a partition $A \cup B = L$ such that the partition difference

$$D = \left| \sum_{r_i \in A} r_i - \sum_{r_i \in B} r_i \right|$$

© Springer International Publishing AG 2017
B. Hu and M. López-Ibáñez (Eds.): EvoCOP 2017, LNCS 10197, pp. 233–248, 2017.
DOI: 10.1007/978-3-319-55453-2_16

is minimised. The decision version of the NPP belongs to the class of NP-complete problems which appear to require a super-polynomial amount of computation time in the instance input size [1,2]. NP-hard optimisation problems are at least as hard as the corresponding decision problems. Many important practical optimisation problems are NP-hard and it is thus important to find efficient approximate methods to solve them. Though requiring exponential time to be solved in the worst case, several hard constraint-satisfaction problems show an instance-dependent *computational phase transition*, meaning that below some critical point, instances are typically easy to solve while they become hard to solve above such a point. Well-known examples are the Boolean Satisfaction Problem (SAT) [1,3], the Graph-Colouring problem [1,4], and the NPP [1,5]. The control parameter is a problem-dependent quantity that must be suitably defined; for example, in SAT, it is the ratio of the number of clauses to the number of variables, and the phase transition phenomenon has been observed in both random and structured instances [6,7].

For problems undergoing a computational hardness phase transition it is of interest to understand how it arises and what are the problems' features that characterise the transition. The original methodology was developed by physicists and it is based on the statistical mechanics approach to physical phase transitions such as the ferromagnetic/paramagnetic transition. An introduction to this rather technical field as applied to hardness phase transitions can be found in [8]. We consider, instead, fitness landscape analysis as a tool for revisiting the phase transition phenomenon. In particular, the Local Optima Network (LON) model [9,10]. Local optima networks compress the whole search space into a graph, where nodes are local optima and edges are transitions among them with a given search operator. Local optima are key features of fitness landscapes as they can be seen as obstacles for reaching high quality solutions. The local optima networks model emphasises the number, distribution and most importantly, the connectivity pattern of local optima in the underlying search space. They are therefore an ideal tool for modelling and visualising the global structure of fitness landscapes. Among local optima network metrics, we particularly study the presence and distribution of so-called funnels in the landscape.

The term 'funnel' was introduced in the protein folding community to describe "a region of configuration space that can be described in terms of a set of downhill pathways that converge on a single low-energy structure or a set of closely-related low-energy structures" [11]. It has been suggested that the energy landscape of proteins is characterised by a single deep funnel, a feature that underpins their ability to fold to their native state. In contrast, some shorter polymer chains (polypeptides) that misfold are expected to have other funnels that can act as traps. Energy landscapes are conceptually related to fitness landscapes, and funnel structures have also been studied in heuristic continuous optimisation [12,13], and more recently in combinatorial optimisation [14–16].

The next section overviews previous and related work. Section 3 presents relevant definitions and algorithms related to local optima networks. Section 4 presents our fitness landscape analysis and visualisation. Finally, Sect. 5 summarises our main findings and suggests directions of future work.

2   Background and Related Work

2.1   The Number Partitioning Fitness Landscape

The existence of the NPP hardness phase transition was first demonstrated numerically by Gent and Walsh [5], who introduced the control parameter k and estimated the transition point to occur around $k_c = 0.96$. The control parameter k corresponds to the number of significant bits in the encoding of the input numbers r_i divided by N (the instance size), specifically $k = log_2(M)/N$, where M is the largest number in the set $L = \{r_1, r_2, \ldots, r_N\}$. For $log_2(M)$ and N tending to infinity, the transition occurs at the critical value of $k_c = 1$, such that for $k < 1$, there are many perfect partitions with probability tending to 1, whereas for $k > 1$, the number of perfect partitions drops to zero with probability tending to 1 [5].

Further studies within the physics community, have confirmed the existence of the NPP phase transition characterising it rigorously [17,18]. However, they provide no direct answer to the question of what features of the corresponding fitness landscapes, if any, are responsible for the widely different observed behaviour. A step in this direction was taken by Fontanari et al. [19] who studied various landscape features before and after the transition, in particular considering *barrier trees* [20]. However, they were unable to find any effect of the transition on barrier tree features and other landscape metrics. Likewise, Alyahya and Rowe [21] performed an exhaustive statistical analysis of NPP landscapes for instances of size $N = 20$ and several number distributions, but did not find any significant correlation between most landscape features and easy or hard instances. They observed differences only in the number of global optima, which is high before the phase transition and low after it, and in the existence of neutral networks which are abundant in the easy phase and tend to disappear in the hard phase.

Considering these studies and given the lack of a clear picture, we decided to investigate additional landscape features based on local optima networks, and the recently proposed approach to identifying multiple funnels in combinatorial search spaces.

2.2   Multiple Funnels in Combinatorial Landscapes

The big-valley hypothesis [22] suggests that on the travelling salesman problem (TSP) and other combinatorial optimisation problems, local optima are not randomly distributed, instead they are clustered around one central global optimum. Recent studies on TSP landscapes, however, have revealed a more complex picture [14,15,23]. The big-valley seems to decompose into several sub-valleys or multiple funnels. This helps to explain why certain iterated local search heuristics can quickly find high-quality solutions, but fail to consistently find the global optimum.

The procedure for identifying funnels on the TSP has evolved in recent work, ranging from visual inspection of the fitness distance correlation plots [23], connected components in the local optima networks [14], and 3D LON visualisation [15]. We propose here to use the notion of *monotonic sequences* from theoretical chemistry [24], which describes a sequence of local minima where the energy of minima is always decreasing. We adapt this notion to the context of fitness landscapes and consider a monotonic sequence as a sequence of local optima where the fitness (costs) of solutions is non-deteriorating. The set of monotonic sequences leading to a particular optimum has been termed 'basin' [24], 'monotonic sequence basin' [25] and 'super-basins' [26], in the theoretical chemistry literature. We chose here to call them 'funnel basins' or simply 'funnels' borrowing from the protein folding literature. We can distinguish the *primary* funnel, as the one involving monotonic sequences that terminate at the global optimum (there can be more than one). The primary funnel is separated from other neighbouring secondary funnels by transition states laying on a so-called 'primary divide' [24]. Above such a divide, it is possible for a local optima to belong to more than one funnel through different monotonic sequences.

The presence of multiple funnels has also been recently observed on binary search spaces (*NK* landscapes) [16], where the authors observed a connection between groupings (communities) in local optima networks and the notion of funnels. Results confirm that landscapes consists of several clusters and the number of clusters increases with the epistasis level. A higher number of clusters leads to a higher search difficulty, measured by the empirical success rate of an iterated local search implementation. The success rate was also found to strongly correlate with the size of the cluster containing the global optimum.

There is evidence of clustering of solutions in Random Satisfiability problems [27] but a study of the funnel structure of SAT instances has not yet been conducted.

3   Definitions and Algorithms

This section overviews the definitions and algorithms constituting the local optima network model for the number partitioning problem.

3.1   Preliminaries

Fitness Landscape. A landscape [28] is a triplet (S, V, f) where S is a set of potential solutions, i.e., a search space; $V : S \longrightarrow 2^{|S|}$, a neighbourhood structure, is a function that assigns to every $s \in S$ a set of neighbours $V(s)$, and $f : S \longrightarrow \mathbb{R}$ is a fitness function that can be pictured as the *height* of the corresponding solutions.

In our study, the search space is composed of binary strings of length N, therefore its size is 2^N. The neighbourhood is defined as the 1-move or bit-flip operation, but definitions can be generalised to larger neighbourhoods.

Neutral Neighbour. A neutral neighbour of s is a neighbour configuration x with the same fitness $f(s)$.

$$V_n(s) = \{x \in V(s) \mid f(x) = f(s)\}$$

The neutral degree of a solution is the number of its neutral neighbours. A fitness landscape is neutral if there are many solutions with high neutral degree. The landscape is then composed of several sub-graphs of configurations with the same fitness value.

Plateau. A plateau, also known in the literature as a neutral network [29, 30], is a set of connected configurations with the same fitness value. Two vertices in a plateau are connected if they are neutral neighbours, that is, if they differ by one point mutation. With the bit-flip mutation operator, for all solutions x and y, if $x \in V(y)$ then $y \in V(x)$. So in this case, the plateaus are the equivalence classes of the relation $R(x, y)$ iff $(x \in V(y)$ and $f(x) = f(y))$.

Local Optimum. A local optimum, which in the NPP case is a minimum, is a solution s^* such that $\forall s \in V(s^*)$, $f(s^*) \leq f(s)$. Notice that the inequality is not strict, in order to allow the treatment of the neutral landscape case.

In the presence of neutrality, local minima are identified by a stochastic hill-climber h that, starting from any solution s, chooses the next best-improving mutant at each iteration by splitting ties at random, until convergence on a local optimum plateau.

Local Optimum Plateau. A plateau is a local optimum if all its configurations are local optima.

3.2   Local Optima Networks

In order to construct the networks, we need to define their nodes and edges. Nodes are local optima and edges represent escape probabilities. Local optima networks for neutral landscapes have been studied before by Verel et al. [10]; we borrow their notation and definitions, but name a sequence of connected solutions with the same fitness as *plateaus* rather than as *neutral networks*, to avoid confusion with the local optima network terminology.

Since we are interested in determining the landscape's funnel structure using the notion of monotonic sequences, we only consider transitions between local optima where fitness is non-deteriorating. This leads to a variant of the model which we term *Monotonic Local Optima Networks (M-LON)*. Furthermore, our experiments revealed that neutrality is also present at the level of local optima transitions, that is, there are connected components in the M-LON which share the same fitness value. This leads us to define an even coarser model of the landscape, where these M-LON plateaus are compressed into single nodes, we termed this new model *Compressed Monotonic Local Optima Networks (CM-LON)*. Relevant formal definitions are given below.

LON Nodes. The set of local optimum plateaus (formed of one or more local optima), $LOp = \{lop_1, lop_2, \ldots, lop_n\}$ corresponds to the node set of the local optima network. The *basin of attraction* of a lop_i is the set of solutions $b_i = \{s \in S \mid h(s) = lop_i$ with probability $p_i(s) > 0\}$ and its size is $|b_i| = \sum_{s \in S} p_i(s)$.

Monotonic Edges. The set of monotonic edges, ME is defined according to a distance function d (minimal number of moves between two solutions), and a positive integer $D > 0$. Edges account for the chances of jumping from a local optimum plateau lop_i into the basin of a non-deteriorating local optimum plateau lop_j after a controlled perturbation. Namely, if we perturb a solution $s \in lop_i$ by applying D random moves, we obtain a solution s' that will belong to another basin b_j with probability p_j: that is, $h(s') = lop_j$ with probability p_j. The probability to go from s to b_j is then $p(s \rightarrow b_j) = \sum_{s' \in b_j} p(s \rightarrow s') p_j(s')$, where $p(s \rightarrow s') = P(s' \in \{\dot{z} \mid d(z, s) \leq D\})$ is the probability for s' to be within D moves from s and can be evaluated in terms of relative frequency. Therefore, we can draw an edge e_{ij} between lop_i and lop_j with weight $w_{ij} = p(lop_i \rightarrow b_j) = \frac{1}{|lop_i|} \sum_{s \in lop_i} p_i(s) p(s \rightarrow b_j)$.

Monotonic Local Optima Network (M-LON). The weighted, oriented local optima network M-LON $= (LOp, ME)$ is the graph where the nodes $lop_i \in LOp$ are the local optimum plateaus, and there is an edge $e_{ij} \in ME$, with weight w_{ij}, between two nodes lop_i and lop_j if $w_{ij} > 0$.

M-LON Plateau. Is a set of connected nodes in the M-LON with the same fitness value. Two nodes are connected if there is a monotonic edge between them.

Compressed LON Nodes. The set of M-LON plateaus, $CLOp = \{clop_1, clop_2, \ldots, clop_n\}$ corresponds to the node set of the compressed local optima network.

Compressed Monotonic Local Optima Network (CM-LON). The weighted, oriented local optima network $CM\text{-}LON = (CLOp, ME)$ is the graph where the nodes $clop_i \in CLOp$ are the M-LON plateaus. Weighted edges correspond to the aggregation of the multiple edges from nodes in a plateau to single edges in the compressed network. The weights of the multiple edges are added to constitute the weight of the mapped edge.

3.3 Detecting the Funnel Structures

To detect the funnel structures we first identify the funnels' 'ends' or 'bottoms'. To do so, we take advantage of the Compressed Monotonic Local Optima Networks. CM-LONs are directed graphs without loops. In a directed graph, one can distinguish the outdegree (number of outgoing edges) from the indegree (number of incoming edges); a *source* node is a node with indegree zero, while a *sink* node is a node with outdegree zero. We consider the CM-LONs sinks as the funnel bottoms.

We thus define the *funnel sinks* as the CM-LON nodes without outgoing edges. Once the funnel sinks are detected, we can proceed to identify the funnel basins (see Algorithm 1). This is done by finding all nodes in the CM-LON graph which are reachable from each funnel sink. Breadth-First-Search is used for this purpose. The set of unique nodes in the combined paths to a given funnel sink corresponds to the funnel basin. The cardinality of this set corresponds to the funnel size. Notice that the membership of a node to a funnel might be overlapping, that is, a node may belong to more than one funnel, in that there are paths from that node to more than one funnel sink. The relative size of the primary funnel (or any other secondary funnel) is calculated as its size divided by the total number nodes in the graph.

Data: CM-LON: Compressed monotonic local optima network, S: funnel sinks
Result: *bsizes*: funnel basin sizes vector, *basins*: funnel basins vector

$i \leftarrow 0$
for $s \in S$ **do**
 $basin[i] \leftarrow$ breadthFirstSearch(CM-LON, s)
 $bsize[i] \leftarrow$ length($fbasin[i]$)
 $i \leftarrow i + 1$
end

Algorithm 1: Identifying funnel basins.

4 Results and Analysis

One advantage of modelling landscapes as complex networks is the possibility of visualising them. After describing the experimental setting, we visualise a set of selected instances before, during, and after the phase transition. We continue with a study of local optima network metrics, including the new set of funnel measurements, and explore how they relate to the phase transition.

4.1 Experimental Setting

In order to minimise the influence of the random creation of landscapes, we considered 30 different and independent landscapes for each parameter combination: N and k. Measurements consider the distribution of values across these 30 landscapes. The empirical study considers $N \in \{10, 15, 20\}$, where $N = 20$ is the largest possible value allowing practical exhaustive enumeration of the search space. The parameter k was varied from 0.4 to 1.2 in steps of 0.1. For each landscape, we extract the full local optima network using code adapted from Daolio et al. [31,32]. We then construct both the monotonic local optima networks (M-LON) and the compressed monotonic local optima networks (CM-LON). When extracting the local optima networks, we set the parameter D for the maximum escape distance to $D = 2$.

4.2 Visualisation

We visualise CM-LONs for selected instances with $N = 15$ and $k \in \{0.4, 0.6, 0.8, 1.0\}$. Due to space constraints, the instance with $k = 1.2$ is not shown, but it reflects a similar structure to that of $k = 1.0$

Network plots were produced using the R statistical language together with the igraph and rgl packages. Graph layouts consider *force-directed* methods. Networks are decorated to reflect features relevant to search dynamic. Red nodes correspond to global sinks, while blue nodes to suboptimal sinks; all other nodes are grey. An edge's width is proportional to its weight, which indicates the probability of transitions. That is, the most probable transitions are thicker in the plots.

We explored two ways of visualising nodes. First, as rectangles (Fig. 1) with lengths proportional to plateau sizes (i.e. the number of single local optima within a plateau). As the plots in Fig. 1 illustrate, for low values of k the landscape global optima form a large plateau, and there are several other large plateaus in the vicinity. With increasing k, the plateaus shrink, with nodes becoming single local optima for $k \geq 8$. Neutrality at the optima network level is, therefore, high for low values of k, gradually decreases with intermediate values of k and finally disappears for $k \geq 0.8$.

A second alternative is to visualise nodes with sizes proportional to their incoming strength (weighted incoming degree), as in Fig. 2. Incoming strength is relevant as it reflects the extent to which a node 'attracts' the search dynamics; that is, it conveys the combined probability of a stochastic search process reaching it. We present both 2D and 3D images. In the 3D visualisations, the x and y coordinates are determined by the force-directed graph layout algorithm; while fitness is visualised as the z coordinate. This provides a clearer representation of the funnel and sink concepts, bringing an almost tangible aspect to these metaphors.

As Fig. 2 illustrates, for $k = 0.4$ there is a single funnel structure easily guiding the search to the single global optimum. For $k = 0.6$, a single dominant central structure is still visible, but several different unconnected global optima now stem out from it. For $k \leq 0.6$ only optimal (red) sinks are observed, indicating that instances are easy to solve. When k increases over 0.6, suboptimal (blue) sinks start to emerge; initially only a few of them, but the number increases with increasing k. The number of optimal (red) sinks decreases and rapidly becomes only two. Search thus become harder, as can be inferred from the 2D and 3D visualisations of the landscape with $k = 1.0$; 16 blue sinks are observed and their combined incoming strength exceeds that of the 2 red sinks. Moreover, as indicated by the 3D image, some suboptimal blue sinks are deep, that is, they are close in fitness to the optimal solution.

4.3   Metrics

Due to space constraints, we can only visualise a few examples. Therefore, we turn to the statistical analysis of the complete dataset. Many features can be collected from fitness landscapes and local optima networks [9,10]. Moreover, a new set of metrics can be gathered from computing the landscape sinks and funnel structure. We selected a subset of metrics after some preliminary experiments, including some that corroborate previous findings, and new local optima network metrics that intuitively relate to search dynamics. Figure 3 summarises the results, showing metrics for local optima network cardinality (1st row), neutrality (2nd

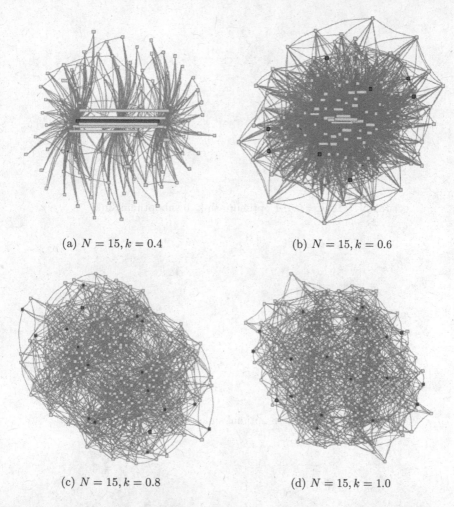

(a) $N = 15, k = 0.4$ (b) $N = 15, k = 0.6$

(c) $N = 15, k = 0.8$ (d) $N = 15, k = 1.0$

Fig. 1. Local optima networks (CM-LONs) for selected NPP instances with $N = 15$. Nodes are local optimum plateaus visualised as rectangles with length proportional to their size (i.e. number of single local optima on them). Long rectangles indicate large plateaus, while squares indicate single local optima. For $k = 0.4$ the whole network is visualised, while for $k \in \{0.6, 0.8, 1.0\}$, the fittest part of the network is shown.

row), sinks and funnels (3rd and 4th rows). The last row in Fig. 3 shows the empirical search cost of an Iterated Local Search (ILS) implementation, using a single bit-flip best-improvement hill-climber and a two bit-flip random perturbation.

Plot (a) confirms the surprising result, noted in previous studies [19, 21], that the number of local optima remains virtually invariable across different values of k. Indeed most of the landscape metrics studied before: the size of the global and local basins, the correlation between basin size and fitness [21], and several barrier-tree metrics [19], are oblivious to the hardness phase transition. An exception is

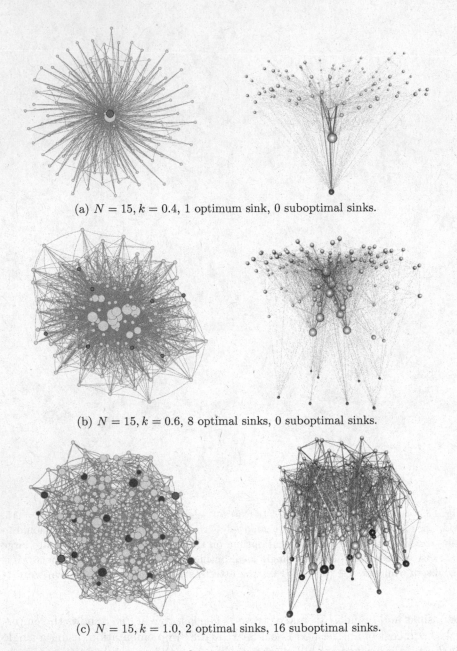

(a) $N = 15, k = 0.4$, 1 optimum sink, 0 suboptimal sinks.

(b) $N = 15, k = 0.6$, 8 optimal sinks, 0 suboptimal sinks.

(c) $N = 15, k = 1.0$, 2 optimal sinks, 16 suboptimal sinks.

Fig. 2. Local optima networks (CM-LONs) for selected NPP instances with $N = 15$. Images are shown in 2D and a 3D projection (where the vertical dimension corresponds to fitness). Node sizes are proportional to their incoming strength, and edge thickness to their weight. Red nodes correspond to globally optimal sinks, while blue nodes to suboptimal sinks. For $k = 0.4$ the whole network is visualised, while for $k \in \{0.6, 1.0\}$, the fittest part of the network is shown. (Color figure online)

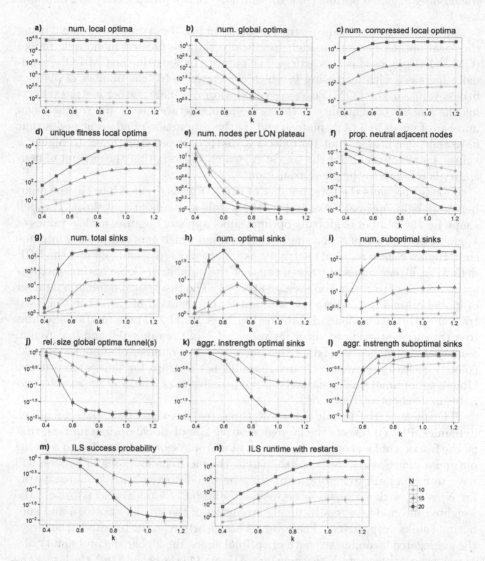

Fig. 3. Local optima network features averaged over 30 instances per value of k (x-axis) and N (legend). The first row of plots illustrate cardinality metrics, the second describes neutrality, and the third statistics on the number of sinks. The fourth row shows metrics that relate to search dynamics: (j) the fraction of local optima that lie on monotonic sequences to a global optimum sink, (k) the aggregated incoming strength of optimal sinks, and (l) the equivalent measure for suboptimal sinks. The last row shows the empirical search cost of an ILS algorithm on the same problem instances in terms of: (m) success probability, and (n) number of function evaluations with restarts. All plots are in semi-log scale.

the number of global optima (plot b), which decreases progressively to 2 at about $k \approx 1$.

The rest of the landscape metrics reported in Fig. 3 can only be gathered using the local optima network model, specifically the compressed monotonic model (CM-LON) proposed in this article. The number of nodes in the CM-LON, gradually increases with increasing k, which correlates to (d) the number of distinct fitness values in the LON (Pearson's correlation $r \approx 0.82$). Other metrics reflecting the amount of neutrality at the local optima network level are: (e) the mean number of nodes in a LON plateau and (f) the proportion of adjacent nodes that have the same fitness. Plot (e) reflects a sharp decrease from lower to higher k values, more noticeable for the largest N; which suggests that the amount of neutrality is relevant to the phase transition.

Plot (g) presents the total number of sinks. This is divided into (h) the number of globally optimal sinks and (i) the number of suboptimal sinks. The bell shape for the number of globally optimal sinks appears because, for low values of k, global optima are part of a single LON plateau which gets compressed into one sink. Higher k show reduced neutrality, as seen in plots (d, e and f). From the NPP definition, higher values of k also mean that the number of global optima progressively decreases to reach 2 on average at $k \geq 1$. The theoretical minimum number of global optima is 2, where one solution is the negation of the other. An illustration of this phenomenon is provided in Figs. 1 and 2. These metrics also hint to a transition starting to occur for values of k in the range 0.6 and 0.8. Again, a sharper change is observed for the largest N. The number of suboptimal sinks (plot i) is clearly relevant to search, as sinks act as traps for the search process. Once a suboptimal sink is reached it is not possible to escape, and the search stagnates in a suboptimal solution.

The plots in the fourth row show three metrics that also relate to search dynamics. Plot (j) reflects the average relative size of global optima funnels, that is, the fraction of local optima that lie on monotonic sequences leading to a global optimum sink. Clearly, the larger this value, the more chances a search process will have to find a path to an optimum. This metric decreases with k, more sharply for $N = 20$, with a transition between 0.6 and 0.8. Plots (k) and (l) report the weighted incoming degree (incoming strength) of the globally optimal and suboptimal sinks, respectively. These values are clearly relevant to search, the larger the aggregated incoming strength of optimal sinks, the higher the probability of a search process successfully reaching one of them. On the other hand, the larger the incoming strength of suboptimal sinks, the higher the changes of getting trapped. Again, a transition gradually occurs for values of k between 0.6 and 0.1.

Finally, the last row summarises the empirical cost of 1000 ILS runs in terms of (m) probability of success with a stopping condition consisting of 2^{15} function evaluations, and (n) number of function evaluations when the ILS is combined with random restarts until a global optimum is found [33]. We can notice a clear relationship between the aggregate incoming strength of globally optimal sinks and the probability of success. Figure 4a highlights this relationship with a scatter plot fitted with a univariate linear regression model. This confirms the strong

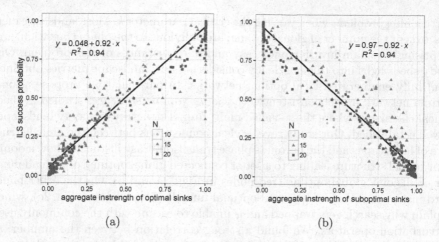

Fig. 4. ILS success probability against (a) the aggregate incoming strength of the optimal sinks and (b) the aggregate incoming strength of the suboptimal sinks. The lines are univariate linear regression models. The equation and coefficient of determination of each line are given in the plots.

correlation, with coefficient of determination $R^2 \approx 0.94$ (which corresponds to Pearson's r correlation). Figure 4b is the scatter plot for the probability of success against the aggregate incoming strength of suboptimal sinks. Here, as we might expect, the relation is reversed. The number of suboptimal sinks is also inversely correlated to the probability of success ($r \approx -0.68$). These are useful relationships for future work on performance prediction since an approximation of the number of sinks and their incoming strength might be estimated using some sampling method.

Several of the metrics studied seem to reflect and explain the known hardness NPP phase transition. The transition, however, seems to appear earlier than the theoretical expected value of $k = 1.0$ [5,17,18], and is not exactly bracketed. However, these trends can be explained in the following way. In theory, the NPP computational phase transition becomes sharp only in the limit of infinite system size N and for $log_2 M$, the number of bits in the input numbers, tending to infinity as well keeping k finite. In practice, we can only simulate finite systems because of computational limitations. A semi-rigorous argument [34] shows that in this case the transition point, k_c, becomes $k_c \approx 1 - \frac{log_2 N}{2N}$ where the second term accounts for finite-size effects. For example, with $N = 15$, k_c is around 0.87 instead of one. This means that the phase transition is observed earlier and that it is not sharp, rather the system changes more gradually approaching it, and this is what we qualitatively observe in our numerical study.

5   Conclusions

Previous studies have failed to reveal clear links between the structure of fitness landscapes and the hardness phase transition known to exist on number partitioning problems when varying the critical parameter k. Most landscape metrics, except the number of global optima, are oblivious to the phase transition, and surprisingly remain invariable for easy and hard instances of this problem. Our study sheds light into this puzzle, by considering new landscape metrics obtained from fully enumerated local optima networks. In particular we propose a local optima network model consistent with the *monotonic sequences* studied in theoretical chemistry, where the so-called multi-funnel structure of energy landscapes is well established. Our study reveals clear connections between the global structure of landscapes and the hardness phase transition. Easy instances show a dominant funnel structure leading to a set or connected global optima, or a small number or disjoint global optima (red nodes in Figs. 2a and b). On the other hand, hard instances reveal multiple suboptimal funnels (blue nodes in Fig. 2c), which explain why search gets trapped and is unable to escape with the commonly used perturbation operators. We found a strong correlation between the number, as well as the combined attracting strength, of suboptimal (blue) sinks and empirical search difficulty on the studied instances. Another important contribution of this work is to bring a more accessible visual approach to understanding search difficulty in combinatorial optimisation.

Future work will consider larger NPP instances using sampling, probe other number distributions, and most importantly, study whether other constraint satisfaction problems such as MAX-SAT reveal a similar global funnel structure explaining the hardness phase transition.

Acknowledgements. This work was supported by the Leverhulme Trust [award number RPG-2015-395] and by the UK's Engineering and Physical Sciences Research Council [grant number EP/J017515/1].

Data Access. All data generated during this research are openly available from the Stirling Online Repository for Research Data (http://hdl.handle.net/11667/85).

References

1. Garey, M.R., Johnson, D.S.: Computers and Intractability. Freeman, San Francisco (1979)
2. Papadimitriou, C.H., Steiglitz, K.: Combinatorial Optimization: Algorithms and Complexity. Prentice-Hall, Englewood Cliffs (1982)
3. Gent, I.P., Walsh, T.: The SAT phase transition. In: Proceedings of ECAI 1996, vol. 94, pp. 105–109. PITMAN (1994)
4. Culberson, J., Gent, I.P.: Frozen development in graph coloring. Theor. Comput. Sci. **265**(1), 227–264 (2001)
5. Gent, I.P., Walsh, T.: Phase transitions and annealed theories: number partitioning as a case study. In: Proceedings of ECAI 1996, pp. 170–174. PITMAN (1996)
6. Gomes, C., Walsh, T.: Randomness and structure. In: Rossi, F., van Beek, P., Walsh, T. (eds.) Handbook of Constraint Programming, vol. 2, pp. 639–664. Elsevier, New York (2006)
7. Kambhampati, S.C., Liu, T.: Phase transition and network structure in realistic SAT problems. In: Proceedings of the Twenty-Seventh AAAI Conference on Artificial Intelligence, AAAI 2013, pp. 1619–1620. AAAI Press (2013)

8. Martin, O.C., Monasson, R., Zecchina, R.: Statistical mechanics methods and phase transitions in optimization problems. Theor. Comput. Sci. **265**(1), 3–67 (2001)
9. Tomassini, M., Vérel, S., Ochoa, G.: Complex-network analysis of combinatorial spaces: the NK landscape case. Phys. Rev. E **78**(6), 066114 (2008)
10. Verel, S., Ochoa, G., Tomassini, M.: Local optima networks of NK landscapes with neutrality. IEEE Trans. Evol. Comput. **15**(6), 783–797 (2011)
11. Doye, J.P.K., Miller, M.A., Wales, D.J.: The double-funnel energy landscape of the 38-atom Lennard-Jones cluster. J. Chem. Phys. **110**(14), 6896–6906 (1999)
12. Lunacek, M., Whitley, D., Sutton, A.: The impact of global structure on search. In: Rudolph, G., Jansen, T., Beume, N., Lucas, S., Poloni, C. (eds.) PPSN 2008. LNCS, vol. 5199, pp. 498–507. Springer, Heidelberg (2008). doi:10.1007/978-3-540-87700-4_50
13. Kerschke, P., Preuss, M., Wessing, S., Trautmann, H.: Detecting funnel structures by means of exploratory landscape analysis. In: Proceedings of the 2015 Annual Conference on Genetic and Evolutionary Computation, GECCO 2015, pp. 265–272. ACM, New York (2015)
14. Ochoa, G., Veerapen, N.: Deconstructing the big valley search space hypothesis. In: Chicano, F., Hu, B., García-Sánchez, P. (eds.) EvoCOP 2016. LNCS, vol. 9595, pp. 58–73. Springer, Heidelberg (2016). doi:10.1007/978-3-319-30698-8_5
15. Ochoa, G., Veerapen, N.: Additional dimensions to the study of funnels in combinatorial landscapes. In: Proceedings of the Genetic and Evolutionary Computation Conference 2016, GECCO 2016, pp. 373–380. ACM, New York (2016)
16. Herrmann, S., Ochoa, G., Rothlauf, F.: Communities of local optima as funnels in fitness landscapes. In: Proceedings of the Genetic and Evolutionary Computation Conference 2016, GECCO 2016, pp. 325–331. ACM, New York (2016)
17. Ferreira, F.F., Fontanari, J.F.: Probabilistic analysis of the number partitioning problem. J. Phys. A: Math. Gen. **31**(15), 3417 (1998)
18. Mertens, S.: Phase transition in the number partitioning problem. Phys. Rev. Lett. **81**(20), 4281–4284 (1998)
19. Stadler, P.F., Hordijk, W., Fontanari, J.F.: Phase transition and landscape statistics of the number partitioning problem. Phys. Rev. E **67**(5), 056701 (2003)
20. Flamm, C., Hofacker, I.L., Stadler, P.F., Wolfinger, M.T.: Barrier trees of degenerate landscapes. Z. Phys. Chem. (Int. J. Res. Phys. Chem. Chem. Phy.) **216**(2/2002), 155–173 (2002)
21. Alyahya, K., Rowe, J.E.: Phase transition and landscape properties of the number partitioning problem. In: Blum, C., Ochoa, G. (eds.) EvoCOP 2014. LNCS, vol. 8600, pp. 206–217. Springer, Heidelberg (2014). doi:10.1007/978-3-662-44320-0_18
22. Boese, K.D., Kahng, A.B., Muddu, S.: A new adaptive multi-start technique for combinatorial global optimizations. Oper. Res. Lett. **16**(2), 101–113 (1994)
23. Hains, D.R., Whitley, L.D., Howe, A.E.: Revisiting the big valley search space structure in the TSP. J. Oper. Res. Soc. **62**(2), 305–312 (2011)
24. Berry, R.S., Kunz, R.E.: Topography and dynamics of multidimensional interatomic potential surfaces. Phys. Rev. Lett. **74**, 3951–3954 (1995)
25. Wales, D.J.: Energy landscapes and properties of biomolecules. Phys. Biol. **2**(4), S86–S93 (2005)
26. Becker, O.M., Karplus, M.: The topology of multidimensional potential energy surfaces: theory and application to peptide structure and kinetics. J. Chem. Phys. **106**(4), 1495 (1997)
27. Mézard, M., Mora, T., Zecchina, R.: Clustering of solutions in the random satisfiability problem. Phys. Rev. Lett. **94**, 197205 (2005)

28. Stadler, P.F.: Fitness landscapes. Appl. Math. Comput. **117**, 187–207 (2002)
29. Huynen, M.A., Stadler, P.F., Fontana, W.: Smoothness within ruggedness: the role of neutrality in adaptation. Proc. Nat. Acad. Sci. U.S.A. **93**(1), 397–401 (1996)
30. Barnett, L.: Ruggedness and neutrality - the NKp family of fitness landscapes. In: Adami, C., Belew, R.K., Kitano, H., Taylor, C. (eds.) Proceedings of the Sixth International Conference on Artificial Life, ALIFE VI, pp. 18–27. The MIT Press, Cambridge (1998)
31. Daolio, F., Verel, S., Ochoa, G., Tomassini, M.: Local optima networks of the quadratic assignment problem. In: 2010 IEEE Congress on Evolutionary Computation (CEC), pp. 1–8 (2010)
32. Daolio, F., Tomassini, M., Vérel, S., Ochoa, G.: Communities of minima in local optima networks of combinatorial spaces. Phys. A: Stat. Mech. Appl. **390**(9), 1684–1694 (2011)
33. Auger, A., Hansen, N.: Performance evaluation of an advanced local search evolutionary algorithm. In: The 2005 IEEE Congress on Evolutionary Computation, vol. 2, pp. 1777–1784. IEEE (2005)
34. Mertens, S.: The easiest hard problem: number partitioning. In: Percus, A., Istrate, G., Moore, C. (eds.) Computational Complexity and Statistical Physics. The Santa Fe Institute Studies in the Sciences of Complexity, vol. 125, pp. 125–139. Oxford University Press, New York (2006)

Author Index